FOOD BIOACTIVES

Functionality and Applications in Human Health

FOOD BIOACTIVES

Functionality and Applications in Human Health

Edited by

Sankar Chandra Deka

Dibyakanta Seth

Nishant Rachayya Swami Hulle

AAP APPLE ACADEMIC PRESS

Apple Academic Press Inc.
3333 Mistwell Crescent
Oakville, ON L6L 0A2, Canada

Apple Academic Press Inc.
1265 Goldenrod Circle NE
Palm Bay, Florida 32905, USA

Exclusive worldwide distribution by CRC Press, a member of Taylor & Francis Group

International Standard Book Number-13: 978-1-77188-799-1 (Hardcover)
International Standard Book Number-13: 978-0-42924-279-3 (eBook)

Library and Archives Canada Cataloguing in Publication

Title: Food bioactives : functionality and applications in human health / edited by Sankar Chandra Deka, Dibyakanta Seth, Nishant Rachayya Swami Hulle.

Other titles: Food bioactives (Oakville, Ont.)

Names: Deka, Sankar Chandra, 1965- editor. | Seth, Dibyakanta, 1980- editor. | Hulle, Nishant Rachayya Swami, 1985- editor.

Description: Includes bibliographical references and index.

Identifiers: Canadiana (print) 20190149663 | Canadiana (ebook) 20190149728 | ISBN 9781771887991 (hardcover) | ISBN 9780429242793 (ebook)

Subjects: LCSH: Functional foods. | LCSH: Plant bioactive compounds.

Classification: LCC QP144.F85 F66 2019 | DDC 613.2—dc23

Library of Congress Cataloging-in-Publication Data

Names: Deka, Sankar Chandra, 1965- editor. | Seth, Dibyakanta, 1980- editor. | Hulle, Nishant Rachayya Swami, 1985- editor.

Title: Food bioactives : functionality and applications in human health / edited by Sankar Chandra Deka, Dibyakanta Seth, Nishant Rachayya Swami Hulle.

Description: Oakville, ON ; Palm Bay, Florida : Apple academic Press, [2020] | Includes bibliographical references and index. | Summary: "Food Bioactives: Functionality and Applications in Human Health helps to address the growing consumer demand for novel functional food products and for high value and nutritionally rich products by focusing on the sources and applications of bioactives from food. The chapters in the book describe functional properties and discuss applications of the selected food ingredients obtained from various sources, including culinary banana, phalsa, pseudocereals, roselle calyces, asparagus, and more. Several chapters address the resurgence of interest in pseudocereals due to their excellent nutritional and biological values, gluten-free composition, and the presence of some health-promoting compounds. The book also looks at utilizing industrial byproducts for making functional and nutraceutical ingredients. The chapters on prebiotics and probiotics highlight different functional properties, and a chapter on food allergens discusses advancements in detection and management in the food manufacturing industries. With contributions from experts in their respective fields, this book provides topics covering selected important areas of food science, technology and processing and their application for developing novel food products"-- Provided by publisher.

Identifiers: LCCN 2019030394 (print) | LCCN 2019030395 (ebook) | ISBN 9781771887991 (hardcover) | ISBN 9780429242793 (ebook)

Subjects: LCSH: Functional foods. | Bioactive compounds. | Food industry and trade.

Classification: LCC QP144.F85 F6635 2020 (print) | LCC QP144.F85 (ebook) | DDC 613.2--dc23

LC record available at https://lccn.loc.gov/2019030394

LC ebook record available at https://lccn.loc.gov/2019030395

Apple Academic Press also publishes its books in a variety of electronic formats. Some content that appears in print may not be available in electronic format. For information about Apple Academic Press products, visit our website at **www.appleacademicpress.com** and the CRC Press website at **www.crcpress.com**

About the Editors

Sankar Chandra Deka, PhD

Sankar Chandra Deka, PhD, is presently working as a Senior Professor in the Department of Food Engineering and Technology, Tezpur University Assam, India. He has more than 29 years of teaching and research experience. He has guided the theses work of more than 35 students for their BTech, MSc, and MTech degrees, and has worked with seven PhD students to date. Dr. Deka has successfully handled more than 15 research projects funded by various government funding agencies. He has published more than 96 research papers in journals of national and international repute and about 20 book chapters. His area of interest is food quality, food chemistry, and fermented foods and processing.

Dibyakanta Seth, PhD

Dibyakanta Seth, PhD, is presently an Assistant Professor in the Department of Food Engineering and Technology, Tezpur University, Tezpur, Assam, India. He has published several research papers in national and international journals. He has participated in national/international conferences and attended training programs in the area of food processing and has been bestowed with Young Scientist and Young Educator Awards at the International Conference on Food Properties held at Bangkok, Thailand, in 2016, and Sharjah, United Arab Emirates, in 2017, respectively. He has been associated with an e-course development program of the University Grants Commission (UGC), ePG-PATHSHALA in the capacity of Paper Coordinator as well as Content Writer on the subject of Unit Operations in Food Processing. He has handled two funded research projects and is an active member of scientific bodies such as the Association of Food Scientists and Technologists (India) Mysore, Indian Society of Agricultural Engineers New Delhi, and Swedish South Asian Studies Network. His research interests include dairy engineering and technology, spray drying, optimization, and transfer processes. He teaches subjects such as fluid mechanics, mechanical operations in food processing, and dairy products technology at Tezpur University.

Nishant Rachayya Swami Hulle, PhD

Nishant R. Swami Hulle, PhD, is an Assistant Professor in the Department of Food Engineering and Technology, Tezpur University, Assam, India. He has published several research articles in international peer-reviewed journals, co-authored book chapters published by CRC Press, and has presented at various conferences. His areas of research interest are nonthermal processing, fruit, and vegetable processing, and extraction of bioactives from plant sources. He has received a Young Scientist Award at the Third International Conference on Food Properties (iCFP, 2018), Sharjah, United Arab Emirates. He teaches subjects such as fruit and vegetable process technology; processing technology of meat, poultry, and fish; and thermal operations in food processing at Tezpur University.

Contents

Contributors

Bhanja Amrita
Food Microbiology and Bioprocess Laboratory, Department of Life Science,
National Institute of Technology, Rourkela, Odisha–769008, India

Pitambar Baishya
Cancer Genetics and Chemoprevention Research Group, Department of Molecular Biology and Biotechnology, Tezpur University, Napaam, Tezpur 784028, Assam, India

Kiran Bala
Department of Food Technology, Guru Jambheshwar University of Science and Technology,
Hisar (Haryana)–125001, India

Vasudha Bansal
Department of Food Engineering and Nutrition, Center of Innovative and Applied Bioprocessing (CIAB), Mohali, Punjab, India, E-mail: vasu22bansal@gmail.com

Utpal Bora
Department of Biosciences and Bioengineering, Center for the Environment,
Indian Institute of Technology Guwahati–781039, Assam, India, E-mail: ubora@iitg.ernet.in

Nidhi Budhalakoti
Department of Food Engineering and Nutrition, Center of Innovative and Applied Bioprocessing (CIAB), Mohali, Punjab, India

Papori Buragohain
Department of Biosciences and Bioengineering, Indian Institute of Technology Guwahati–781039, Assam, India

Monoj Kumar Das
Cancer Genetics and Chemoprevention Research Group, Department of Molecular Biology and Biotechnology, Tezpur University, Napaam, Tezpur–784028, Assam, India

Sankar Chandra Deka
Department of Food Engineering Technology, Tezpur University, Napaam, Tezpur–784028, Assam, India

K. Dihingia
Department of Food Engineering and Technology, Tezpur University, Napaam, Assam–784028, India

Francis Dutta
Department of Horticulture, Assam Agricultural University, Jorhat–785013, Assam, India

Khalid Gul
Food Process Engineering Laboratory, School of Applied Life Sciences,
Gyeongsang National University, 900 Gajwa-Dong, Jinju, 660701, Republic of Korea

R. Jayabalan
Food Microbiology and Bioprocess Laboratory, Department of Life Science,
National Institute of Technology, Rourkela, Odisha–769008, India, E-mail: jayabalanr@nitrkl.ac.in

Piyush K. Jha
ONIRIS-GEPEA (UMR CNRS 6144), Site de la Géraudière CS 82225, 44322 Nantes Cedex 3, France

Jon Jyoti Kalita
Department of Biosciences and Bioengineering, Indian Institute of Technology Guwahati–781039, Assam, India

B. S. Khatkar
Department of Food Technology, Guru Jambheshwar University of Science and Technology, Hisar–125001 (Haryana), India, E-mail: bskhatkar@yahoo.co.in

Alain Le-Bail
ONIRIS-GEPEA (UMR CNRS 6144), Site de la Géraudière CS 82225, 44322 Nantes Cedex 3, France

Nisar Ahmad Mir
Department of Food Engineering and Technology, Sant Longowal Institute of Engineering and Technology, Longowal, Punjab–148106, India

Chandrasekar Chandra Mohan
Center for Food Technology, Anna University, Chennai, Tamil Nadu, India

Pradeep K. Das Mohapatra
Department of Microbiology, Vidyasagar University, Midnapore–721102, West Bengal, India

Keshab C. Mondal
Department of Microbiology, Vidyasagar University, Midnapore–721102, West Bengal, India

Ponnala Vimal Mosahari
Center for the Environment, Indian Institute of Technology Guwahati–781039, Assam, India

S. Muchahary
Department of Food Engineering and Technology, Tezpur University, Napaam, Assam–784028, India

Prakash Kumar Nayak
Department of Food Engineering and Technology, Central Institute of Technology, Kokrajhar, BTAD, Assam–783370, India

Saswati Parua
Department of Physiology, Bajkul Milani Mahavidyalaya, Purba Medinipur, West Bengal, India

Pramod K. Prabhakar
Department of Food Science and Technology, National Institute of Food Technology Entrepreneurship and Management, Kunldi, Sonepat, Haryana, India, E-mail: pramodkp@niftem.ac.in, pkprabhakariitkgp@gmail.com

Kesavan Radhakrishnnan
Assistant Professor, Department of Food Engineering and Technology, Central Institute of Technology, Kokrajhar, BTAD, Assam–783370, India, Tel.: +91-84738-21333, Fax: +91-03661-277143, E-mail: k.radhakrishnan@cit.ac.in

Paulraj Rajamani
School of Environmental Sciences, Jawaharlal Nehru University, New Delhi, 10067, India

C. T. Ramachandra
Department of Agricultural Engineering, University of Agricultural Sciences, Bengaluru, GKVK, Bengaluru–560065, Karnataka, India, E-mail: ramachandract@gmail.com

Anand Ramteke
Cancer Genetics and Chemoprevention Research Group, Department of Molecular Biology and Biotechnology, Tezpur University, Napaam, Tezpur–784028, Assam, India

Anamika Ranjan
2413 Via Palermo APT 1623, Fort Worth, Texas, 76109, USA

Ashish Rawson
IIFPT, MOFPI, Pudukkottai Road, Thanjavur – 613005, Tamil Nadu, India

Aradhita Barman Ray
Department of Food Technology, Guru Jambheshwar University of Science and Technology, Hisar (Haryana)–125001, India, E-mail: dhitaray@gmail.com

Charanjit Singh Riar
Department of Food Engineering & Technology, Sant Longowal Institute of Engineering & Technology, Longowal, Punjab–148106, India

Winny Routray
Marine Bioprocessing Facility, Center for Aquaculture and Seafood Development, Fisheries and Marine Institute, Memorial University of Newfoundland, P.O. Box 4920, St. John's, NL, A1C 5R3, Canada, E-mail: routrayw@yahoo.com

Ananta Saikia
Department of Horticulture, Assam Agricultural University, Jorhat–785013, Assam, India

Pankaj Preet Sandhu
Department of Food Engineering and Nutrition, Center of Innovative and Applied Bioprocessing (CIAB), Mohali, Punjab, India

V. H. Shruthi
Department of Agricultural Engineering, University of Agricultural Sciences, Bengaluru, GKVK, Bengaluru–560 065, Karnataka, India, E-mail: shruthihgowda@gmail.com

Ritu Sindhu
Center of Food Science and Technology, Chaudhary Charan Singh Haryana Agricultural University, Hisar–125001, Haryana, India

Neelu Singh
School of Environmental Sciences, Jawaharlal Nehru University, New Delhi, 10067, India

Sukhcharn Singh
Department of Food Engineering and Technology, Sant Longowal Institute of Engineering and Technology, Longowal, Punjab–148106, India

Basharat Yousuf
Department of Post Harvest Engineering and Technology, Faculty of Agricultural Sciences, Aligarh Muslim University, Aligarh–202002, India

Abbreviations

2D-PAGE	Two-dimensional polyacrylamide gel electrophoresis
ACE	angiotensin I-converting enzyme
ALP	alkaline phosphatase
ALT	alanine aminotransferase
AP	apple pomace
AuNP	gold nanoparticles
BOD	biological oxygen demand
BPS	bathophenanthrolinedisulfonic acid
BSHs	bile salt hydrolases
BV	breakdown viscosity
CDKs	cyclin-dependent kinases
CE	capillary electrophoresis
CH	chalcone
CHD	coronary heart disease
CLA	conjugated linoleic acid
CMC	carboxymethyl cellulose
COD	chemical oxygen demand
COX	cyclooxygenase
CS	chitosan
CVD	cardiovascular diseases
DC	Dietitians of Canada
DEN	diethylnitrosamine
DHA	docosahexaenoic acid
DM	dry matter
DPPH	2,2-diphenyl-1-picrylhydrazyl
DS	degree of substitution
DSC	differential scanning calorimetry
EAST	enzyme allergo sorbent test
EDTA	ethylene diamine tetraacetic acid
EFSA	European Food Safety Authority
ELISA	enzyme-linked immunosorbent assay
ELOSA	enzyme-linked oligosorbent assay
EMB	eosin methylene blue
EPA	eicosapentaenoic acid

ER	estrogen receptor
ESI	electrospray ionization
FALCPA	Food Allergy Labeling and Consumer Protection Act
FAO	Food and Agricultural Organization
FC	flavylium cation
FFA	free fatty acid
FFF	field-flow fractionation
FLISA	fluorescent-linked immunosorbent assay
FNAB	fermented non-alcoholic beverages
FOSHU	foods for specified health uses
FPHs	fish protein hydrolysates
FRAP	ferric reducing antioxidant power
FSMA	food safety modernization act
FTICR	Fourier-transform ion cyclotron resonance
FTIR	Fourier transform infrared
FV	final viscosity
GBF	green banana flour
GBPF	green banana peel flour
GCE	glassy carbon electrode
GI	geographical indication
GI	glycemic index
GIT	gastrointestinal tract
GMO	genetically modified organism
GPx	glutathione peroxidase
GR	glutathione reductase
H2SO4	sulphuric acid
HAs	hibiscus anthocyanins
HCl	hydrochloric acid
HCMV	HSV-2, and cytomegalovirus in humans
HDL	high-density lipoprotein
HIV	AIDS virus
HPLC	high-performance liquid chromatography
HSV-1	herpes type 1 virus
HV	hold viscosity
iSPR	imaging SPR
IT	ion-trap
LAB	lactic acid bacteria
LC	liquid chromatography
LDL	low-density lipoprotein
LFD	lateral flow devices

LFIA	lateral flow immunoassay
LOAELs	lowest observed adverse effect levels
LSPR	localized surface plasmon resonance
MALDI	matrix-assisted laser desorption ionization
MAO-A	monoamine oxidase
MAPKs	mitogenic-activated protein kinases
MT	million ton
MUB	mucus binding protein
NF-κB	nuclear factor-kappa B
NOAELs	no observed adverse effect levels
NPs	nanoparticles
NSAIDs	non-steroidal anti-inflammatory drugs
PAL	precautionary advisory labeling
PB	pseudobase
PFF	peptide-fragment fingerprint
PMF	peptide mass fingerprinting
PNA	peptide nucleic acids
POV	peroxide value
PPARs	peroxisome proliferator-activated receptors
PT	pasting temperature
PUFA	polyunsaturated fatty acid
PV	peak viscosity
Q	quadrupole
QB	quinonoidal base
QCM	quartz crystal microbalance
QD	quantum dots
qPCR	quantitative PCR
RAST	radio-allergosorbent test
REA	resonance enhanced absorption
RIE	rocket immuno-electrophoresis
ROS	reactive oxygen species
RTS	ready-to-serve
RVA	rapid visco-analyzer
SA	scavenging activity
SCFAs	short-chain fatty acids
SDAP	structural database of allergen proteins
SDF	soluble dietary fiber
SEF	surface-enhanced fluorescence
SEM	scanning electron microscope
SERS	surface-enhanced Raman scattering

SFCs	supercritical fluids
STAT3	signaling transducers and activator of transcription 3
SME	small and medium scale industries
SOD	superoxide dismutase
SPE	solid-phase extraction
SPR	surface plasmon resonance
SV	setback viscosity
TC	total cholesterol
TCBPF	tender core of banana pseudo-stem flour
TGA	Thermal Gravimetric Analysis
TOF	time-of-flight
TPC	total polyphenols content
UAE	ultrasound-assisted extraction
UGC	University Grants Commission
UHT	ultra-high temperature
UNU	United Nations University
UPLC	ultra performance liquid chromatography
USFDA	United States Food and Drug Administration
VEGF-A	vascular endothelial growth factor
VITAL	voluntary incidental trace allergen labeling
VLDL	very high-density lipoprotein
WHO	World Health Organization
WVTR	water vapor transmission rate
XRD	x-ray diffraction

Preface

The Studies related to food processing technologies are essential components of undergraduate and graduate studies. Food science and technology is a multidisciplinary research area that covers a wide area of study.

This present book covers the functional properties of food products from various sources. The functional properties covered from various plant-based products are banana starch, phalsa, pseudocereals, roselle calyces, and asparagus. Phalsa is an underutilized indigenous minor berry that possesses very high nutritional as well as medicinal components. Rosella is a very popular vegetable in tropical countries and has attractive red color due to the presence of anthocyanins, which could be used in various food products as functional ingredients. Recently, there is a resurgence of interest in pseudocereals due to their excellent nutritional and biological values, gluten-free composition, and presence of some health-promoting compounds. The chapter on industrial byproducts discusses the use of different industrial byproducts for making functional and nutraceutical ingredients. The chapters on prebiotics and probiotics highlight different functional properties. The chapter on food allergens discusses advancements in analytical methods for the detection of food allergens and its management in the food manufacturing industries.

This present book is aimed to act as a reference source for students, researchers, and teachers.

We hope the contributions made by various learned authors will help students, teachers, and researchers working in the area of functional foods.

—Sankar Chandra Deka, PhD
Dibyakanta Seth, PhD
Nishant R. Swami Hulle, PhD

CHAPTER 1

Chemical Modification and Characterization of Culinary Banana (*Musa* ABB) Starch and Its Application

K. DIHINGIA, S. MUCHAHARY, and S. C. DEKA

Department of Food Engineering and Technology, Tezpur University, Napaam, Assam–784028, India

ABSTRACT

The chemical modification involved incorporation of functional groups to the main starch molecule that affected the physicochemical properties of the starch. The modification did not change the starch type, which was found to be similar, i.e., C-type to the native starch. The culinary banana (*Musa* ABB) peel showed impressive antimicrobial property against *E. coli* and thus effective against bacterial growth in films. The addition of sodium alginate showed improved mechanical property since it has unique colloidal property. Oxidative stability of butter stored in the prepared films showed better results in room temperature than the regular butter paper.

1.1 INTRODUCTION

Banana plants are considered as the world's biggest herbs, grown in many countries. Culinary bananas, also called as plantains, are mostly evolved from the edible varieties of two species *Musa acuminate* (genome "A") and *Musa balbisiana* (genome "B") (Stover and Simmonds, 1987). The culinary banana (*Musa*ABB) of Assam and Northeast India (locally known as kachkal) is an important and cheap source of vegetable in the diet of local people. It is considered as one of the potential sources of carbohydrates, starch, polyphenols, micronutrients, and functionally important bioactive compounds (Khawas and Deka, 2016a).

Starch is the major polysaccharide in plants and is in the form of granules that exists naturally within the plant. Starch is considered as the principal food reserve polysaccharides in the plant kingdom. It is the main storage carbohydrate in plants. In human nutrition, starch plays an important role. It is the major carbohydrate in the human diet and is the basic source of energy. The popularity of starch is because of its use as a major food ingredient in our diet system and its diverse functionalities, year-round availability, and low cost (Thomas and Atwell, 1999).

Starch granules are primarily composed of two kinds of polysaccharides, amylose (20–30%) and amylopectin (70–80%). Amylopectin is a semi-crystalline, highly branched polysaccharide, while amylose is amorphous. The substantial difference in the properties of starch obtained from different sources depends on the differences in the ratio of amylose to amylopectin. The characteristic properties of each fraction in terms of molecular weight, branch length, degree of branching and the physical manner in which these constituents are arranged within the starch granules, and the presence of naturally occurring non-starch components such as lipids, proteins, and phosphate groups in the starch granules (Leach et al., 1959).

Native starches have limited use in food and other industrial applications due to some weaknesses such as narrow peak viscosity (PV) range, poor process tolerance, low shear stress resistance, thermal decomposition, high retrogradation and syneresis, where cooked starches will form a weak, cohesive, and rubbery paste. These facts motivated the employment of modified starch as important functional ingredients in processed foods in recent years because of their improved functional properties over unmodified starch (Fleche, 1985). Starch modification can be physical, chemical, or enzymatic (Miyazaki et al., 2006). Starch modification, which involves the alteration of the physical and chemical characteristics of the native starch to improve its functional characteristics, can be used to tailor starch to specific food applications (Hermansson and Svegmark, 1996). Numerous chemical modifications may be applied to starch to impart properties that are useful for particular applications. The chemical modification involves the introduction of functional groups into the starch molecule, resulting in markedly altered their physicochemical and functional properties such as gelatinization, pasting, and retrogradation behavior (Liu et al., 1999).

The chemical modification is generally achieved through acid hydrolysis, esterification, oxidation, etherification, cross-linking, dual modification, etc. There are four basic types of starch modifications, namely; chemical, physical, enzymatic, and genetic, all of which target the three available hydroxyl groups of the starch copolymer (Masina et al., 2017). Chemical modification of starch is an alternative approach to produce highly functional starch (Hu et al., 2016). The chemical modification involves the introduction of functional groups into the starch molecule, resulting in markedly altered physicochemical properties. Such chemical modification of native granular starch profoundly alters their physicochemical and functional properties such as gelatinization, pasting, and retrogradation behavior, etc. (Liu et al., 1999; Hazarika and Sit, 2016).

Edible films can be used for versatile food products to reduce the loss of moisture, to restrict absorption of oxygen, to lessen migration of lipids, to improve mechanical handling properties, to provide physical protection, or to offer an alternative to commercial packaging materials. The films can enhance the organoleptic properties of packaged foods provided that various components (such as flavorings, colorings, and sweeteners) are used. The films can be used for individual packaging of small portions of food, particularly products that are currently not individually packaged for practical reasons (Bourtoom, 2009). Edible films prepared from hydrocolloids like alginate form strong films and exhibit poor water resistance, because of their hydrophilic nature (Guilbert, 1995; Kester and Fennema, 1986). A mixture of starch and alginate to form edible film has been studied (Wu et al., 2001). Alginate has a potential to form biopolymer film or coating component because of its unique colloidal properties, which include thickening, stabilizing, suspending, film forming, gel-producing, and emulsion stabilizing (King, 1982). Alginate is able to form a strong film (Guilbert, 1986; Kester and Fennema, 1989); therefore, the mixture of these two biopolymers is desirable to improve the mechanical properties of the film (Maizura et al., 2008).

1.2 MATERIALS AND METHODS

1.2.1 SAMPLE COLLECTION AND CHEMICALS

Culinary banana (*Musa*ABB) was collected from Tezpur University Campus, Sonitpur, Assam (India). All the chemicals used were of high purity analytical grade.

1.2.2 STARCH ISOLATION

The starch was isolated by the following method described by Ovando-Martinez et al. (2009). The chemical analyses viz., moisture content, ash, crude fiber, fat, and protein content were determined (AOAC, 2010). The pH of starch dispersion (8% w/v) was measured by using a pH meter.

1.2.3 AMYLOSE CONTENT

Total amylose content was determined as per the method described by McGrance et al. (1998). The amylose content was calculated from a standard curve prepared by using pure potato amylose type III.

1.2.4 DETERMINATION OF RESISTANT STARCH

The resistant starch (as total dietary fiber) of the native and modified starch samples was determined by the standard protocol (AOAC, 1997). The resistant starch was determined as the residue remaining after drying and then expressed as the percentage of starch (on dry basis).

$$\text{RS} = \frac{\text{Weight of residue}}{\text{weight of sample}} \times 100$$

1.2.5 MODIFICATION OF KACHKAL STARCH

Modification of kachkal starch by acid-alcohol treatment was done following the method of Lin et al. (2003) with slight modifications. Modification by acetylation (esterification) of starch was performed according to a slightly modified version of the method described by Saartrat et al. (2005) and Sahnoun et al. (2015).

The acetyl content (%) and degree of substitution (DS) were calculated using the following equations (Dihingia and Deka, 2019):

$$\%\ \text{Acetyl content} = \frac{(\text{b}-\text{s}) \times \text{N} \times 0.043 \times 100}{\text{W}}$$

where, '*b*' refers to the volume of 0.2 N HCl utilized to titrate blank (ml); '*s*' to the volume of 0.2 N HCl utilized to titrate sample (ml); '*N*' to the normality of 0.2 N HCl; and '*W*' to the mass of the sample (g).

$$\text{DS} = \frac{162 \times \%\text{AC}}{4300 - (42 \times \%\text{AC})}$$

1.2.6 FUNCTIONAL PROPERTIES OF KACHKAL STARCH

Swelling power and solubility were determined (Dihingia and Deka, 2019) using the method of Leach et al. (1959).

$$\text{Swelling power}\left(g/g\right)=\frac{\text{Weight of sediment}}{\text{Weight of sample}}$$

$$\text{Solubility}\%=\frac{\text{Weight of Petri dishes}\left(\text{after drying}-\text{before drying}\right)}{\text{Weight of the sample}}\times 100$$

Oil binding capacities of native and modified starches were determined according to the method described by Adeleke and Odedeji (2010). Gel consistency of both the starch samples was determined as described by Balasubaramanian et al. (2010). The paste clarity of starch suspensions was measured by a modified method of Perera and Hoover (1998) for measuring turbidity. The freeze-thaw stability of native and modified starches was evaluated by Hoover and Ratnayake (2002) method.

1.2.7 PASTING PROPERTIES

Pasting properties of starches were evaluated in rapid visco-analyzer (RVA; -4, Newport Scientific, Sydney, Australia) that is previously done by Dihingia and Deka (2019). An 8% slurry was given a programmed heating and cooling cycle set for 23 min, where the sample was held at 30°C for 1 min, heated to 95°C for 7.5 min, further held at 95°C for 5 min before cooling to 50°C within 7.5 min, and holding at 50°C for 2 min. The speed was 960 rpm for the first 10 s and 160 rpm for the remaining period of the experiment. PV, hold viscosity (HV), final viscosity (FV), breakdown viscosity (BV), setback viscosity (SV), and pasting temperature (PT) of starches were measured, and all measurements were repeated three times as described by Khawas and Deka (2016b).

1.2.8 COLOR

Hunter lab color flex colorimeter (Hunter Associates Laboratory Inc., rest on the USA) was used to measure the degree of change in color of both the starches (native and acidified methanol modified) (Bhandari et al., 2016). The color coordinates of this meter are L^* = whiteness; a^* = redness to greenness; and b^* = yellowness to blueness.

1.2.9 SEM ANALYSIS

The shape and size of the starch granules were evaluated by using a scanning electron microscope (SEM) that is previously done by Dihingia and Deka (2019). A thin layer of starch was mounted on the aluminum specimen holder by double-sided tape. The samples were coated with platinum and examined under the microscope at an accelerating voltage of 15kV with a magnification of 500X and 1000X. Size of the starch granules was determined by measuring the diameters of 3–5 randomly selected granules from the micrographs (Sit et al., 2014).

1.2.10 X-RAY DIFFRACTION (XRD) ANALYSIS

The x-ray diffraction (XRD) analysis of the starch samples was carried out by using X-ray diffractometer. The samples were exposed to an X-ray beam at 15mA and 30kV. Data was recorded over a diffraction angle (2θ) range of 0–50°C with a steep angle of 0.05. Percent crystallinity was determined by calculating the percentage ratio of the diffraction peak area to the total diffraction area (Sit et al., 2013).

1.2.11 FOURIER TRANSFORM INFRARED (FTIR) ANALYSIS

FTIR spectra were used to determine the structure of the starch. The spectra were recorded in absorbance mode from 4000 to 400 cm^{-1} (mid-infra-red region) at a resolution of 4 cm^{-1}. Samples were thoroughly ground with exhaustively dried pure KBr (1:100 w/w), and pellets were prepared by compression and analyzed. Background value from pure KBr was acquired before each sample was scanned (Sit et al., 2014).

1.2.12 THERMOGRAVIMETRIC ANALYSIS BY DIFFERENTIAL SCANNING CALORIMETRY (DSC)

Thermal gravimetric analysis (TGA) was carried by a thermogravimetric analyzer (TG50) in the temperature range 30–600°C/ min. The percentage of weight loss using the heating cycle was estimated by the associated software (Abugoch et al., 2011).

1.2.13 PREPARATION OF THE KACHKAL PEEL EXTRACT

The method of extract preparation was followed by the method described by Lopez et al. (2014).10g of culinary banana (kachkal) pulp and peel flour were mixed with 100 mL of distilled water and placed in a sonication bath (BandelinSonorex, Germany) at 50°C for 30 min at 35 kHz. The extracts were then cooled, filtered, and stored in dark glass bottles at 4°C.

1.2.14 ESTIMATION OF TOTAL POLYPHENOLS IN PEEL EXTRACT

Total polyphenols content (TPC) was determined with the Folin–Ciocalteu colorimetric method (Mallick and Singh, 1980).

1.2.15 ASSAY OF FREE RADICAL (DPPH) SCAVENGING ACTIVITY (SA)

The 2,2-diphenyl-1-picrylhydrazyl (DPPH) radical scavenging activity (SA) was measured with the method of Brand-Williams et al. (1995) and the assay is based on the ability of an antioxidant to scavenge the DPPH cation radical. SA was calculated as percent inhibition relative to control using the following equation,

$$\%\ \text{Radical scavenging activity} = \frac{A0 - A1}{A0} \times 100$$

1.2.16 ANTIMICROBIAL ACTIVITIES

The antimicrobial property of the edible film was studied by disk diffusion method (Bora and Mishra, 2016). In this method, Eosin Methylene Blue (EMB) agar was used as culture media. The pathogenic culture *E. coli* was grown in nutrient broth and incubated for 24 h. EMB agar plates were prepared and nutrient broth solution containing microorganism (100 μl) culture was poured over the plates and spread. A small portion of the edible films were cut and placed over the media. The plates were incubated at 27°C for 24 h and zone of inhibition were measured.

1.2.17 PREPARATION OF EDIBLE FILM

The edible film was prepared by following the method described by Maizura et al. (2008). Film forming solutions were prepared from a mixture of starch and sodium alginate (4:1) based on a total weight basis (5g) including 20% glycerol in 200 ml distilled water. Kachkal peel extract was incorporated into the edible film solution at various concentrations of 3%, 5% and 7% of edible film forming solution. The mixture was heated to 85°C with continuous stirring for 45 min before it was cooled to room temperature. The solutions were cast onto Petri plates followed by oven drying at 40°C for 24 h. The dry films obtained were peeled off and stored for further analysis (Table 1.1).

TABLE 1.1 Different Film Formulations with Their Compositions

Sl No.	Particulars	Formulations
1	Film 1 (NSF)	Native starch: sodium alginate:glycerol
2	Film 2 (AAS 3%)	AAS: sodium alginate:glycerol: peel extract (6 ml)
3	Film 3 (AAS 5%)	AAS: sodium alginate:glycerol: peel extract (10 ml)
4	Film 4 (AAS 7%)	AAS: sodium alginate:glycerol: peel extract (14ml)
5	Film 5 (ACS 3%)	ACS: sodium alginate:glycerol: peel extract (6ml)
6	Film 6 (ACS 5%)	ACS: sodium alginate:glycerol: peel extract (10 ml)
7	Film 7 (ACS 7%)	ACS: sodium alginate:glycerol: peel extract (14ml)

(*20 ml of film forming solution was poured in each plate).

1.2.18 PHYSICAL, CHEMICAL, MECHANICAL, AND BARRIER PROPERTIES OF KACHKAL STARCH-ALGINATE FILM

1.2.18.1 FILM THICKNESS

The thickness of the films was measured using a micrometer. Three readings were taken for each film, two from the sides and one from the center of the film.

1.2.18.2 MOISTURE CONTENT

Film samples were weighed in an aluminum moisture dish and dried at 105°C in a hot air oven for 24 h. Moisture content was determined as a percentage of the initial film weight loss during and was reported on a dry basis (Norajit et al., 2010).

1.2.18.3 WATER SOLUBILITY

The water solubility of the films was determined according to Romero-Bestida et al. (2005). Pieces of the film (2×3 cm) were cut from each film and were stored in desiccators with silica gel (0% RH) for 7 days. Samples were weighed to the nearest 0.0001 g and placed into beakers with 80 ml distilled water. The samples were maintained under constant agitation for 1h at room temperature (approximately 25°C). The remaining pieces of the film after soaking were filtered through filter paper (Whatman No.1), followed by oven drying at 60°C to constant weight.

$$\% \text{ Solubility} = \frac{\text{Initial dry weight} - \text{final dry weight}}{\text{Initial dry weight}} \times 100$$

1.2.19 WATER VAPOR PERMEABILITY AND TRANSMISSION RATE (WVTR)

The WVTR was determined according to ASTM E96–80 (ASTM, 1989), modified by Gontard et al. (1992) and calculated by using the formula:

$$\text{WVT} = \frac{\text{W} \times \text{x}}{\text{A}}$$

where, *WVT* is water vapor transmission (g H_2O mm cm^2), *x* is the average thickness of the film, and *A* is the permeation area

The water vapor transmission rate (WVTR) was calculated according to the formula:

$$\text{WVTR} = \frac{\text{W} \times \text{x}}{\text{A} \times \text{t}}$$

where, *t* is time.

1.2.20 TENSILE STRENGTH

Tensile strength and percentage elongation were measured according to ASTM-D D412–98a (ASTM, 1998) using a Mechanical Universal Testing Machine Instron 4400R, with a 50 N load cell and at a velocity of 1.0 mm/s. Test samples were cut according to the ASTM standard method.

1.2.21 THERMAL GRAVIMETRIC ANALYSIS (TGA)

TGA was carried out by a thermogravimetric analyzer (TG50). Nonisothermal experiments were performed in the temperature range of 30–600°C/min. The percentage of weight loss using the heating cycle was estimated using the associated software (Bora and Mishra, 2016).

1.2.22 PEROXIDE VALUE (POV) OF STORED BUTTER

About 1 g of sample was weighed and 20–30 ml of acetic acid chloroform mixture was added and allowed to dissolve completely. 0.5ml of the saturated KI solution was added and allowed to stand for 1 min, and then 30 ml of distilled water was added. Also, three drops of starch indicator (1%) was added and mixed well. Then it was titrated against 0.01N $Na_2S_2O_3$ with vigorous shaking to liberate from the chloroform layer until the blue Color just disappeared. The blank was titrated similarly in the absence of the oil. The peroxide values (POV) were calculated by using the formula (Manral et al., 2008):

$$\text{Peroxide value} = \frac{S \times N \times 1000}{\text{wt of sample}}$$

where, PV = milli equivalent peroxide/kg sample; S = ml $Na_2S_2O_3$ (test blank); N = normality of $Na_2S_2O_3$.

1.2.23 DETERMINATION OF FREE FATTY ACID (FFA) OF STORED BUTTER

Free fatty acid (FFA) was determined by Ranganna (2010), and the value was calculated using the formula:

$$\text{FFA} = \frac{\text{Titer value} \times \text{ N of alkali} \times 56.1}{\text{wt of sample}\,(g)}$$

1.3 RESULTS AND DISCUSSIONS

The chemical modification of culinary banana (*Musa* ABB) starch was carried out by acid alcohol treatment and acetylation.

1.3.1 CHEMICAL COMPOSITION OF NATIVE AND MODIFIED STARCH

The chemical composition of native starch has been shown in Table 1.2. The moisture content of native starch was found to be 11.36%, which is in accordance with the results reported by Khawas and Deka (2016a). According to the reports of Mweta et al. (2008) the acceptable range for storage and marketing without deterioration in the quality of starch, the moisture content should be ranged from 8.96 to 11.93%. The protein, fat, ash, and crude fiber were found to be very low, i.e., 0.29%, 0.47%, 0.37%, and 0.23%, respectively. The high ash content in kachkal starch may be indicative of the presence of more minerals like potassium and magnesium than in plantains (0.02%) as reported by Perez-Sira et al. (1997). Result of protein content was comparable with results of Khawas and Deka, (2016a). The low-fat % in kachkal starch resists to amylolysis and cause the formation of the amylose-lipid complex. Most of the regular starches contain 20–30% amylose, and in the present study of culinary banana starch, amylose content was found to be 33.24% (Table 1.2). The high amylose starch is much more resistant to digestive enzymes than the low amylose starch. The RS content in the starch sample studied was 17.79% which was on the higher side as compared to the content of 17.5% in banana flour reported by Ovando-Martinez et al. (2009) (Table 1.3).

TABLE 1.2 Chemical Composition of Native Kachkal Starch (g/100g Dry Basis)

Moisture%	Ash%	Fat%	Protein%	Crude fiber%	Resistant starch%	Amylose%
11.36	0.37	0.47	0.29	0.23	17.79	33.24

Each value in the table represents mean ± SD. Means within each column bearing different superscripts are significantly ($p < 0.05$) different.

TABLE 1.3 Moisture, Amylose Content, Resistant Starch and Yield of Chemically Modified Culinary Banana (*Musa* ABB) Starch (g/100g Dry Basis)

Sample	Moisture%	Amylose%	RS%	Yield%
AAS	10.45	35.76	19.38	82.34
ACS	10.72	24.16	13.72	81.72

Each value in the table represents mean ± SD. Means within each column bearing different superscripts are significantly ($p < 0.05$) different.

The functional properties of starch granules are greatly influenced by the amount of amylose in the starch granules. The results exhibited significant difference among native and modified starches. The amount of amylose in starch granules was reduced by chemical treatments. It could be attributed to the variance in extraction, processing, and modification procedures include acetylation, acid-thinning, and oxidation (Lawal et al., 2005; Atichokudomchai and Varavinit, 2002).

The % acetyl content in acetylated starch was found 2.52%, with the DS 0.094%. The resistant starch content was noted as 19.38% and 13.72% for AAS and ACS (Dihingia and Deka, 2019), respectively which is the desired range for starch (10–12%) used in an industrial scale (Than et al., 2007).

1.3.2 FUNCTIONAL PROPERTIES OF STARCH

1.3.2.1 SWELLING POWER AND SOLUBILITY

The swelling power and solubility of the native and modified kachkal starch are shown in Table 1.4 (Dihingia and Deka, 2019). Swelling power is obtained by measuring hydration capacity and the extent of the interaction between starch granules within the amorphous and crystalline areas (Reddy et al., 2015). The swelling power of starches is directly proportional to the increasing temperature from 70°C to 90°C. Increase in swelling power of acetylated starches was observed as compared to the native and acid alcohol treated starch. Acetylated starches showed high swelling power occurred due to the addition of bulky acetyl groups and resulted in improved water holding due to the formation of hydrogen bond within the starch granules (Tester et al., 2004). The decreased swelling power of acid-thinned and oxidized starches might be because of structural de-polymerization within the granules of the starch during the modification process (Lawal et al., 2005) (Figure 1.1).

When the starch was subjected to heating in boiling water, the leaching out of swollen starch granules contributed to the solubility of starch. Increased solubility was observed for acid-alcohol treated and acetylated (2.24%) starch compared to what is observed in the solubility of native starch (2.58%). These results are comparable with poovan banana (*Musa* AAB), and kachkal (*Musa* ABB) studied by Reddy et al. (2015) and Khawas and Deka (2016b), respectively.

1.3.2.2 OIL BINDING CAPACITY AND GEL CONSISTENCY

Oil binding capacity refers to the amount of oil held by starch in a food matrix. The acid–alcohol treatment reduced oil-binding capacity of starch, and results are in agreement with the findings of Bhandari et al. (2016) in

TABLE 1.4 Swelling Power and Solubility (%) of Native and Modified Kachkal Starch

Sample	70°C		80°C		90°C	
	Swelling power	**Solubility (%)**	**Swelling power**	**Solubility (%)**	**Swelling power**	**Solubility (%)**
NS	4.01 ± 0.005^{b}	3.45 ± 0.035^{b}	7.75 ± 0.010^{b}	7.06 ± 0.020^{b}	10.97 ± 0.010^{b}	8.75 ± 0.010^{b}
AAS	3.67 ± 0.010^{c}	4.26 ± 0.017^{a}	6.71 ± 0.011^{c}	7.96 ± 0.012^{a}	8.13 ± 0.010^{c}	9.37 ± 0.450^{a}
ACS	4.87 ± 0.010^{a}	2.56 ± 0.015^{c}	8.11 ± 0.010^{a}	6.75 ± 0.100^{c}	11.35 ± 0.010^{a}	6.88 ± 0.100^{c}

Each value in the table represents mean ± SD. Means within each column bearing different superscripts are significantly ($p < 0.05$) different.

black-eyed pea starch. This decrease might be due to the increase in the crystalline region and a decrease in the amorphous region in starch. The acetylation increased the oil binding capacity of starch. Sahnoun et al. (2015) reported that the improvement observed in oil absorption could be due to the modification of the structural arrangement of starches following the fixing of acetyl groups (Dihingia and Deka, 2019) (Table 1.5 and Figure 1.2).

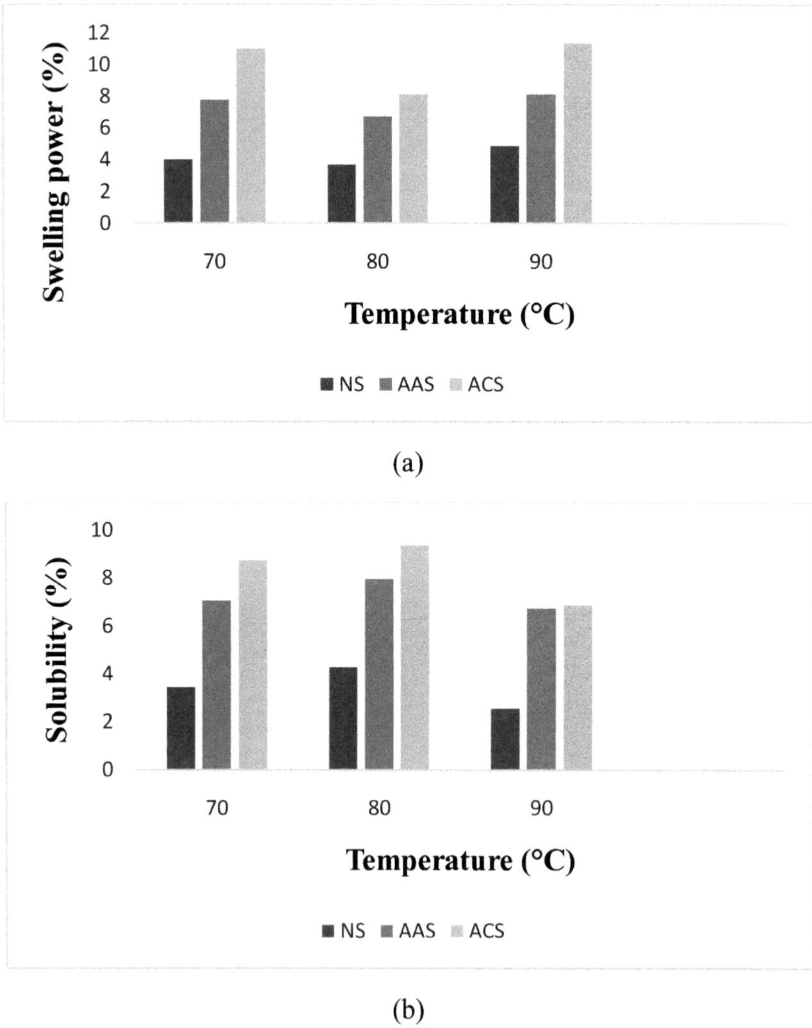

FIGURE 1.1 Swelling power and solubility (%) of kachkal starch: (a) Swelling power of native and modified kachkal starch (b) Solubility (%) of native and modified kachkal starch.

TABLE 1.5 Oil Binding and Gel Consistency Native and Modified Kachkal Starch

Particulars	NS	AAS	ACS
Oil binding capacity (g oil/g sample)	1.18 ± 0.01[b]	0.97 ± 0.02[c]	1.20 ± 0.01[a]
Distance traveled (cm)	10.53 ± 0.23[b]	11.40 ± 0.10[a]	9.63 ± 0.15[c]

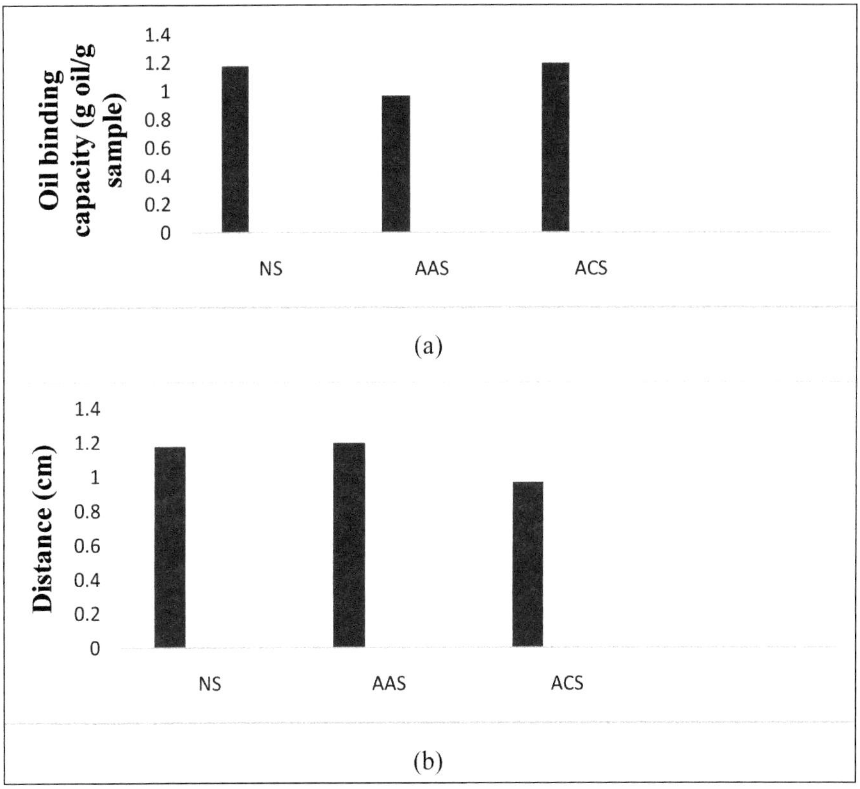

FIGURE 1.2 Oil binding and gel consistency of kachkal starch: (a) Oil binding capacity of native and modified kachkal starch, (b) Gel consistency of native and modified kachkal starch.

The gel consistency of the native and modified culinary banana (*Musa* ABB) starch is illustrated in Figure 1.2 (b). The results showed that the modified starch traveled a longer distance as compared to the native starch that indicated lower gel consistency of the modified starch than the native starch. As reported by Balasubramaniam et al. (2010) the acid-catalyzed hydrolysis of starch as opposed to enzymatic (α-amylase) hydrolysis can take place at branch points, thus reducing the degree of branching and increasing the percentage of linear segments and gel consistency.

1.3.2.3 PASTE CLARITY

The starch paste clarities are presented in Figure 1.3. Starch gel clarity is a much desirable functionality of starches for its utilization in food industries since it directly influences brightness and opacity in foods. The starch sample experienced low paste clarity (4.46–1.17% light transmittance) during storage time and the results are in line with previous reports of Khawas and Deka (2016b) and Reddy et al. (2015) that banana starch forms an opaque gel, and with increase in storage period the transmittance value decreases. This reduction in transmittance is due to retrogradation tendency of starch pastes, which means that under refrigerated condition banana starch have a tendency to retrograde. The opaqueness of starch paste has been attributed to various factors such as granule swelling, granule remnants, leached amylose, and amylopectin, amylose-amylopectin chain-length, intra- or intermolecular bonding and presence of lipids (Jacobson et al., 1997) (Table 1.6 and Figure 1.3).

TABLE 1.6 Paste Clarity of Native and Modified Kachkal Starch

Sample	Day 1	Day 2	Day 3	Day 4
NS	3.24 ± 0.01^{c}	1.91 ± 0.01^{c}	1.38 ± 0.01^{c}	1.17 ± 0.01^{c}
AAS	4.02 ± 0.01^{b}	2.78 ± 0.01^{b}	2.21 ± 0.01^{b}	1.41 ± 0.01^{b}
ACS	4.46 ± 0.01^{a}	2.83 ± 0.01^{a}	2.32 ± 0.01^{a}	1.76 ± 0.01^{a}

Each value in the table represents mean ± SD. Means within each column bearing different superscripts are significantly ($p < 0.05$) different.

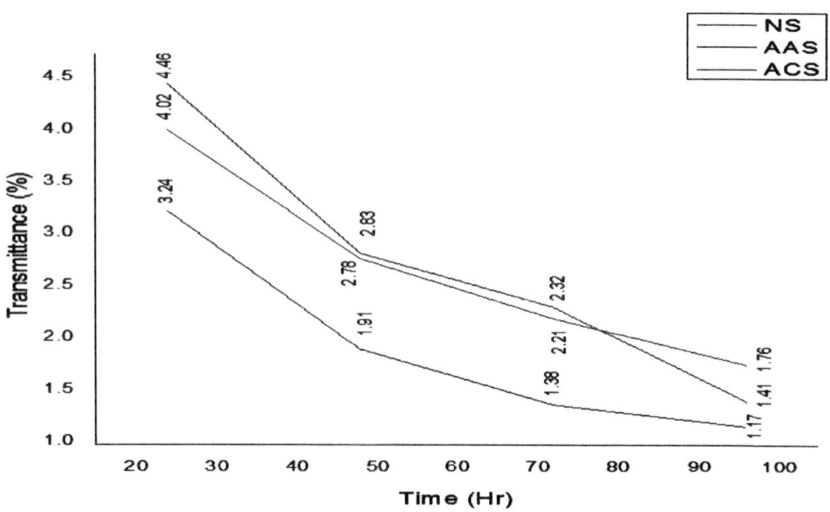

FIGURE 1.3 Pasting properties of native and modified kachkal starch.

1.3.2.4 FREEZE-THAW STABILITY

Freeze-thaw stability measures the amount of water released from the gels during storage by the degree of syneresis and is an important factor to be considered when formulating refrigerated and frozen foods (Baker et al., 1998). The freeze-thaw stability of the native and modified starch has been shown in Figure 1.4. The results showed that the acid alcohol treated starch released a higher amount of water (20.87–39.63%) (Table 1.7 and Figure 1.4).

TABLE 1.7 Syneresis (%) of Native and Modified Kachkal Starch

Sample	Cycle 1	Cycle 2	Cycle 3	Cycle 4
NS	24.53 ± 0.02[b]	33.56 ± 0.02[b]	38.32 ± 0.01[b]	44.14 ± 0.02[b]
AAS	27.20 ± 0.10[c]	36.52 ± 0.03[c]	38.64 ± 0.05[c]	45.80 ± 0.10[a]
ACS	20.13 ± 0.05[a]	19.85 ± 0.01[a]	32.24 ± 0.02[a]	36.73 ± 0.01[c]

Each value in the table represents mean ± SD. Means within each column bearing different superscripts are significantly ($p < 0.05$) different.

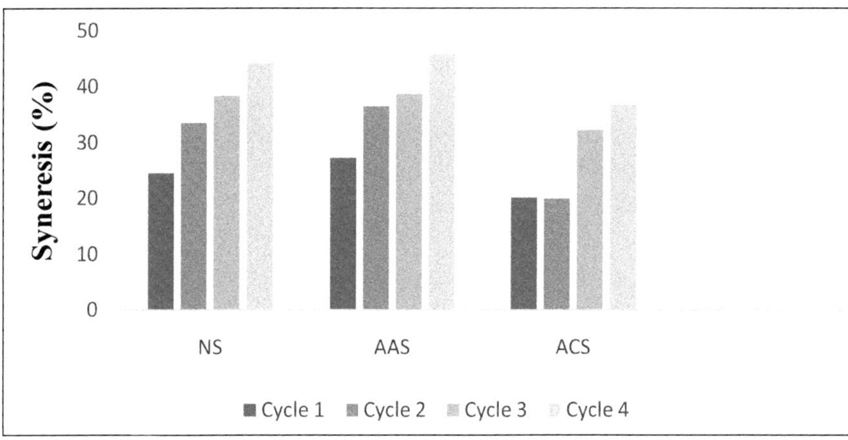

FIGURE 1.4 Syneresis (%) of native and modified kachkal starch.

1.3.2.5 PASTING PROPERTIES

The pasting properties of the native and modified kachkal banana starch are shown in Table 1.8. The results showed reduced pasting properties in acetylated and acid-alcohol treated starch. According to Wang and Wang (2002), the reduced pasting parameters in acetylated and acid thinned starches could be attributed to the addition of new functional groups (acetyl), degree of hydrolysis and increased amylose re-crystallization and the results found were in line with Reddy et al. (2015). The PT of the modified starch was higher than native starch and ranged from 82.6 to 83.2°C.

TABLE 1.8 Pasting Properties of Native and Modified Culinary Banana (Musa ABB) Starch

Particulars	NS	AAS	ACS
Pasting temperature (°C)	81.8	82.6	83.2
Peak viscosity (cP)	2080	452	1167
Final viscosity (cP)	1741	498	1162
Hold viscosity (cP)	1286	449	1018
Breakdown viscosity (cP)	29	17	242
Setback (cP)	398	71	146

1.3.2.6 COLOR

The color of the native and modified kachkal starch was studied by Hunter Color Lab, which showed that modification had changed the starch Color parameters as compared to the native starch. The acetylated starch showed the highest L* value that indicated an increase in starch whiteness and can be attributed to depolymerization of the starch chain into smaller molecules with the high refractive index (Table 1.9).

TABLE 1.9 Color of Native and Modified Kachkal Starch

Color parameters	NS	AAS	ACS
L*	81.66	81.73	86.31
a	1.73	0.41	1.46
b	5.76	2.85	3.75

The ‘a’ and ‘b’ values are the chemically modified starch decreased from 1.73 to 1.46 and 5.76 to 2.85, respectively than the native starch.

1.3.3 SCANNING ELECTRON MICROSCOPY ANALYSIS

The structural properties obtained by SEM (Dihingia and Deka, 2019) of native and modified kachkal starch has been illustrated in Figure 1.5(a), (b), (c) and (d) for NS, AAS, and ACS at 500X and 1000X magnification, respectively. The micrographs showed that the starch granules are a mixture of spherical and elliptical shaped and are in the range of 7.59–30 μm, which is considered as under range as reported by Eggleston et al. (1992). The surface of the starch sample appeared to be smooth, which indicated that the isolation and the modification process were efficient and did not cause damage to starch granules. Sizes and shapes of granules might have influenced the physicochemical and functional properties such as gelatinization, swelling, and solubility (Figure 1.5).

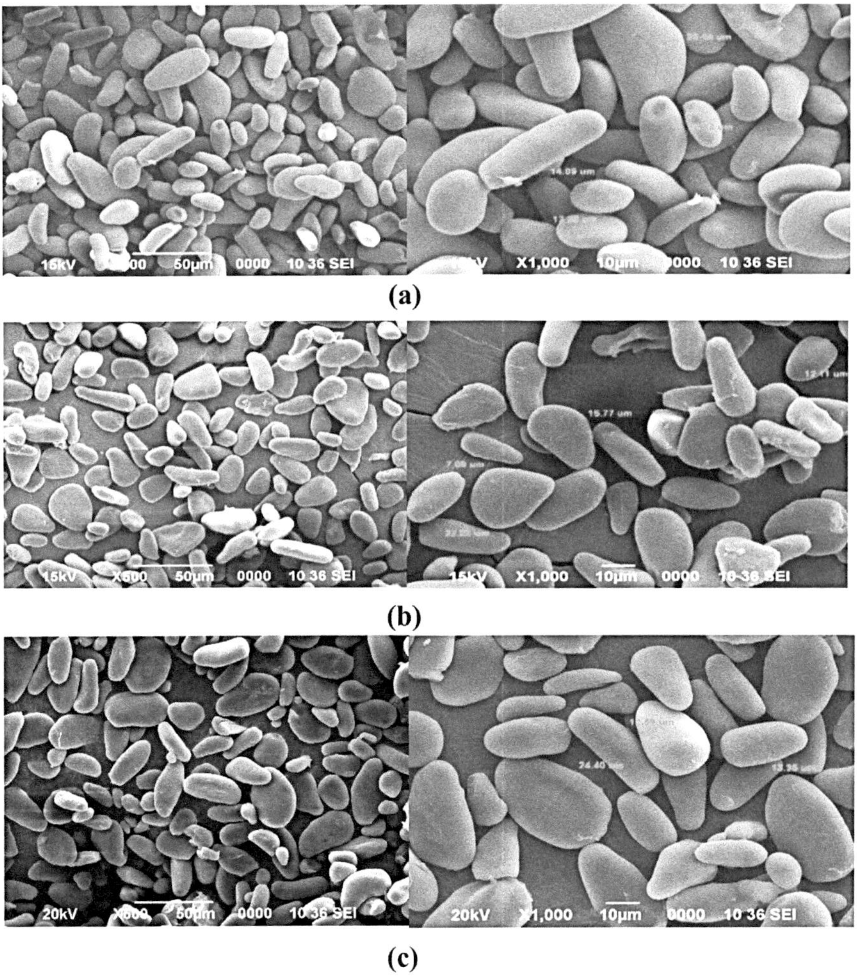

FIGURE 1.5 Scanning electron micrographs of kachkal starch (a) Native starch, (b) Acid alcohol starch and (c) Acetylated starch.

1.3.4 X-RAY DIFFRACTION (XRD) ANALYSIS OF CULINARY BANANA (*MUSA* ABB) STARCH

The position of strong and weak peaks of the native and modified kachkal starch has been presented in Figure 1.6(a), (b) and (c) for NS, AAS, and ACS, respectively (Dihingia and Deka, 2019). XRD provides an elucidation of the long-range molecular order, typically termed as crystalline, which is due to the ordered arrays of double helices formed by the amylopectin side chains (Perez and Bertoft, 2010). The crystallinity of the native kachkal starch was found to be 27.38%, and the crystallinity of the acid alcohol treated starch and acetylated starch was found to be 24.48% and 23.37%, respectively. The XRD patterns showed that the native and modified starches are a mixture of A and B-type. According to the report of Yu et al. (2013) C-type starch pattern has been considered a mixture of both A and B-types because its XRD pattern showed a combination of both.

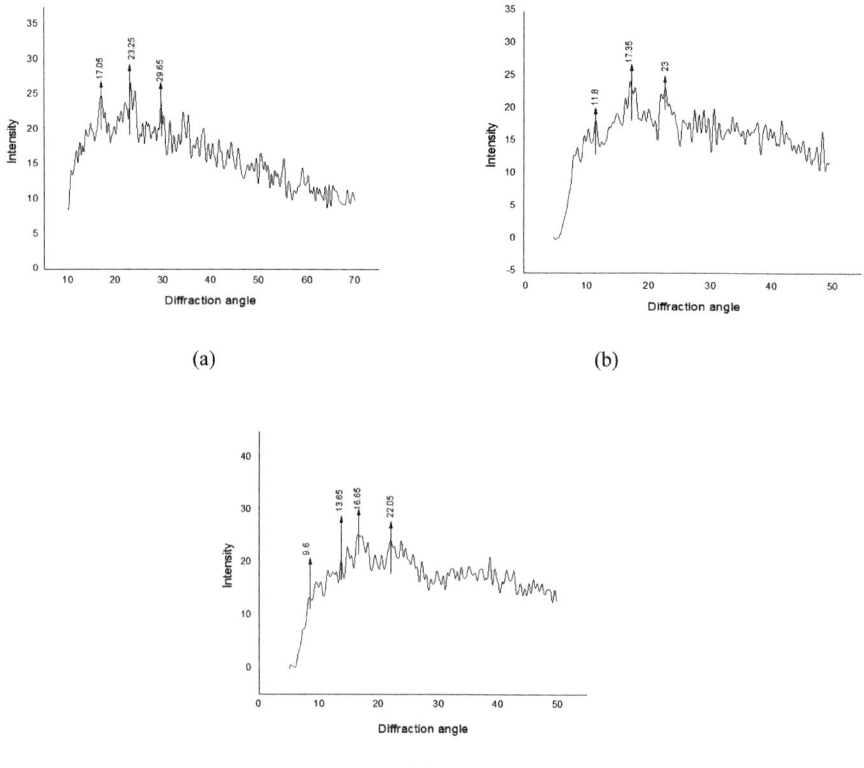

FIGURE 1.6 XRD graphs: (a) XRD graph of native kachkal, (b) XRD graph of acid alcohol treated kachkal starch, (c) XRD graph of acetylated kachkal starch.

1.3.5 FT-IR SPECTRAL PROFILE OF NATIVE AND MODIFIED STARCH

FTIR spectra of the starch samples were scanned in the 4000–400 cm^{-1} region and are shown in Figure 1.7(a), (b) and (c) for NS, AAS, and ACS, respectively. The spectra obtained for the native and modified were similar in the form and intensity of the major peaks. The spectra showed high absorption near the wave numbers 1156, 1369, 1417, 1646, 2925, and 3429 cm^{-1} confirming the carbohydrate nature of starch samples and the peaks at 2926, 2929 and 2931 could be attributed to O-H and H-C-H bond stretching (Perez and Bertoft, 2010; Gallant et al., 1997). Peaks observed at 1397 and 1396 cm^{-1} were attributable to the bending modes of H–C–H and C–H symmetric bending of CH_3. The peaks at 1069 were attributed to C-O-H (Zeng et al., 2011 and Fan et al., 2012). The band at 1648 in acetylated starch was due to COO- vibration in the carbohydrate group. The absorption peak at 1735 cm^{-1} showed the presence of CH_3COO-group as reported by Singh et al. (2008). The FT-IR spectra observed in the present study resemble the spectra obtained for plantain and banana starches (Khawas and Deka, 2016b; Reddy et al., 2015) (Figure 1.7).

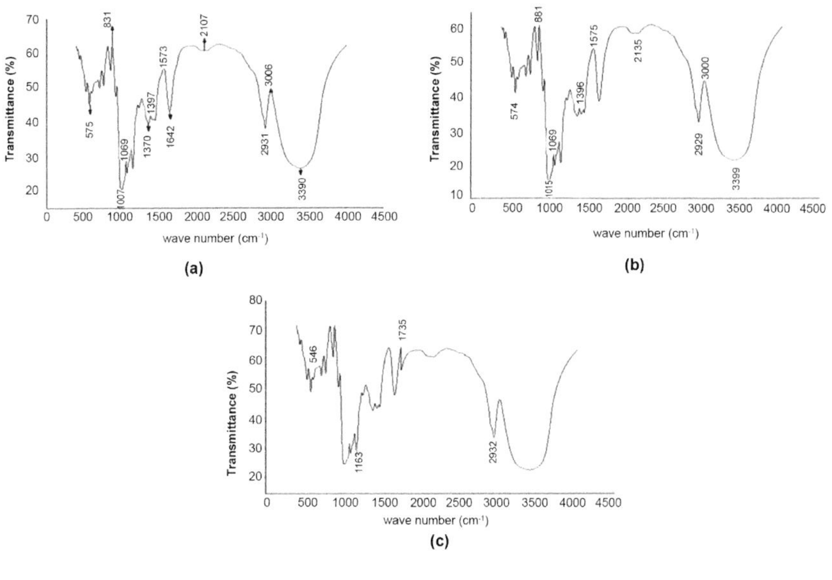

FIGURE 1.7 FTIR graphs: (a) FTIR graph of native kachkal starch, (b) FTIR graph of acid alcohol treated kachkal starch c) FTIR graph of acetylated kachkal starch.

1.3.6 DIFFERENTIAL SCANNING CALORIMETRY (DSC) ANALYSIS

Thermal properties of the modified starch were analyzed by DSC and as presented in Table 1.10. The results showed a distinctive difference in peak temperature between native and modified starch. It was probably due to the introduction of new groups resulting in destabilization of granular structure, which leads to the early rupture of the amylopectin double helices and a decrease in peak temperature. After hydrolysis of starch by methanol-HCl, onset temperature (To) significantly changed, and this was greatly due to hydrolysis of amorphous regions of starch and formation of the amylose-methanol complex (Gope et al., 2016) (Table 1.10).

TABLE 1.10 Thermal Properties of Native and Modified Starch

Gelatinization temperature (°C)	NS	AAS	ACS
Peak	84.6	87.3	61.7
Onset	54.7	54.2	43.7

1.3.7 TOTAL POLYPHENOLS AND DPPH SCAVENGING ACTIVITY (SA) AND ANTIMICROBIAL PROPERTY OF CULINARY BANANA PEEL

Antioxidant activities of plant extracts are related to their phenolic, ascorbic acid and carotenoids. Hydrogen donating characteristic of the plant-derived phenolic compounds is responsible for quenching the reactive oxygen species (ROS) from the body system and boost the homeostasis mechanism in cells for neutralizing the free radicals (Fuente and Victor, 2000).

The total polyphenol content of the peel extract was calculated using gallic acid standard curve. The total polyphenol content and DPPH SA was found to be 831.32 mg GAE/100g of dry matter (DM) and 67.21%, respectively which is under the range of total polyphenol content and DPPH SA of kachkal peel at the maturity stage 4 as reported by Khawas and Deka (2016a).

1.3.7.1 ANTIMICROBIAL PROPERTY

The antimicrobial property of the peel extract was studied by the disk diffusion method. The peel extract revealed a distinctive zone of inhibition against *E. coli*. The zone of inhibition was found to be 5.76 mm, which was higher than the zone of inhibition showed by the films. The kachkal peel extract is a rich source of polyphenols such as hydroquinone, catechol, vanillic acid, p-coumaric acid, ferulic acid, quinic acid, apigenin, salicylic acid, chlorogenic acid, quercetin, caffeic acid as reported by Khawas and Deka (2016a). The vanillic acid and p-coumaric acid showed antimicrobial properties against *E. coli* (Alves et al., 2013). According to Lou et al. (2012) p-coumaric acid changes the permeability of the bacterial cell membrane and has the capacity to bind DNA, inhibiting cell function. The presence of vanillic acid and p-coumaric acid in kachkal peel extract strongly showed antimicrobial property against *E. coli* (Figure 1.8).

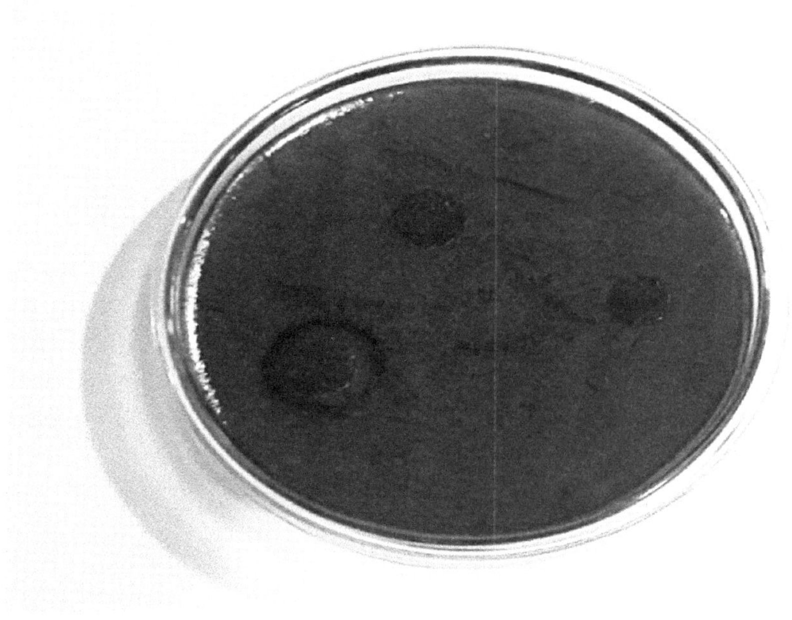

FIGURE 1.8 Zone of inhibition of kachkal peel extract against *E. coli.*

1.3.7.2 CHARACTERIZATION OF EDIBLE FILM

Starch alginate edible film incorporated with culinary banana (*Musa* ABB) peel extract were prepared and characterized. The physical properties like thickness, moisture content, water solubility, swelling ratio, were analyzed. The mechanical, thermal, and barrier properties like tensile strength and water vapor permeability were observed.

1.3.7.3 FILM THICKNESS

The film thickness of all the films was measured in micrometer (Table 1.11). The values were used in finding the water vapor permeability of the films. The thickness of the films prepared from native starch and modified starch had not much difference with one another.

1.3.7.4 MOISTURE CONTENT

The moisture content of the films was determined (Table 1.12), and the film thickness affected the moisture retention of the films. In the present study, the moisture content of the native starch film was highest compared to the other films. The films of acid alcohol treated starch (F2, F3, F4) had the lowest moisture retention than the films prepared from native (F1) and acetylated starch (F5, F6, F7).

1.3.7.5 WATER SOLUBILITY

Solubility test of the edible film was carried out at 25°C. The different films showed a variation in the water solubility, and the solubility of the films depends upon various factors like starch type, culinary banana (*Musa* ABB) peel extract. The acid alcohol treated starch–alginate films showed higher water solubility compared to the other films due to the higher starch solubility.

1.3.7.6 SWELLING RATIO

The swelling of the prepared films was studied for 2 min (Table 1.11). The acetylated starch-alginate films (F5, F6, and F7) showed higher swelling power compared to the other films. The acetylation increased the solubility of starch, and the alginate is highly soluble in water and increased the films' water solubility. The swelling ratio is accompanied by a decrease in mechanical properties.

1.3.7.7 WATER VAPOR PERMEABILITY AND TRANSMISSION OF EDIBLE FILMS

The water vapors permeability and transmission of the edible films are listed in Table 1.11. The results revealed that the native starch film had low water vapors permeability than the other films containing starch, alginate, and culinary banana peel extract. Considering the hydrophilic nature of these films, it was expected that the water vapor barrier (Tapia et al., 2012) weakened in the systems made from modified starches (Gutierrez et al., 2014; Kester and Fennema, 1986).

TABLE 1.11 Thickness, Moisture Content, Water Solubility and Swelling Ratio of Culinary Banana Starch –Alginate Edible Films

Films	Thickness (mm)	Moisture content (%)	Water solubility (%)	Swelling ratio (%)
Film 1	0.182 ± 0.001^{d}	53.02 ± 0.020^{b}	43.83 ± 0.020^{f}	31.20 ± 0.100^{g}
Film 2	0.095 ± 0.002^{e}	47.47 ± 0.015^{e}	57.63 ± 0.025^{a}	38.22 ± 0.010^{f}
Film 3	0.093 ± 0.001^{e}	47.56 ± 0.005^{e}	54.36 ± 0.025^{c}	39.76 ± 0.010^{d}
Film 4	0.095 ± 0.001^{e}	48.30 ± 0.200^{d}	54.83 ± 0.030^{b}	39.61 ± 0.015^{e}
Film 5	0.187 ± 0.001^{c}	52.38 ± 0.015^{c}	51.26 ± 0.152^{d}	47.75 ± 0.025^{b}
Film 6	0.192 ± 0.002^{b}	53.15 ± 0.015^{b}	51.33 ± 0.015^{d}	49.31 ± 0.010^{a}
Film 7	0.195 ± 0.001^{a}	53.92 ± 0.010^{a}	50.86 ± 0.025^{e}	47.43 ± 0.026^{c}

Each value in the table represents mean ± SD. Means within each column bearing different superscripts are significantly ($p < 0.05$) different.

1.3.7.8 MECHANICAL PROPERTIES OF THE EDIBLE FILMS

The mechanical properties of the films were presented in terms of tensile strength and elongation at break (Table 1.12). The tensile strength accounts for film's mechanical resistance due to cohesion forces between chains, while elongation at break measures its plasticity, which is a capacity of film to extend before breaking. The thickness of the films also affected the tensile strength (Figure 1.9).

TABLE 1.12 Mechanical and Barrier Properties of Edible Films

Particulars	WVP (g H_2O mm/cm^2)	WVTR (g H_2O mm/h. cm^2)	Tensile strength (MPa)
Film 1	1.3×10^{-4}	5.42×10^{-5}	9.41 ± 0.020^{a}
Film 2	1.4×10^{-5}	5.91×10^{-6}	6.23 ± 0.002^{d}
Film 3	1.5×10^{-5}	6.08×10^{-5}	6.03 ± 0.152^{e}
Film 4	1.5×10^{-5}	6.13×10^{-5}	5.85 ± 0.020^{f}
Film 5	1.5×10^{-4}	4.29×10^{-5}	6.72 ± 0.015^{c}
Film 6	1.7×10^{-4}	4.37×10^{-4}	6.73 ± 0.025^{c}
Film 7	1.8×10^{-4}	4.25×10^{-5}	6.89 ± 0.020^{b}

Each value in the table represents mean ± SD. Means within each column bearing different superscripts are significantly ($p < 0.05$) different.

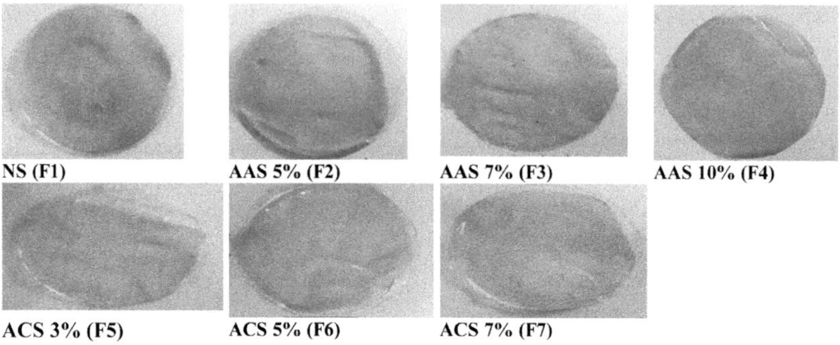

FIGURE 1.9 Films prepared from native, acetylated, and acid alcohol treated kachkal starch.

1.3.7.9 THERMOGRAVIMETRIC ANALYSIS OF THE EDIBLE FILM

The thermogravimetric curves of the films are shown in Figure 1.10. All the films showed a gradual loss in weight at around 200° C due to loss of moisture from the films. Mass loss in all the films was similar, and the weight loss might be due to water loss, the release of volatiles or decomposition of the material as the temperature increased (Figure 1.10).

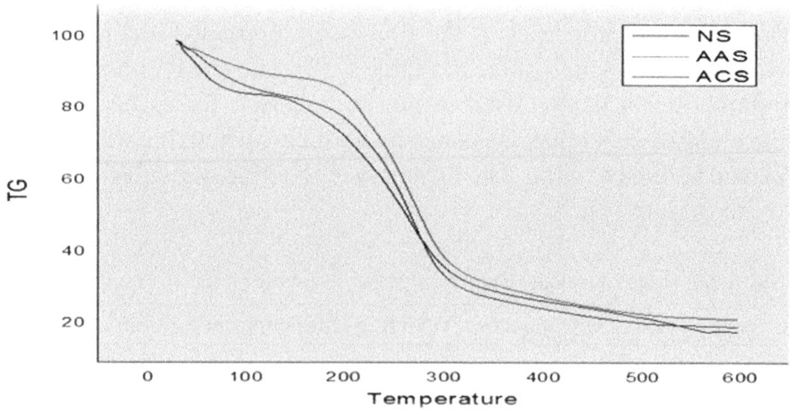

FIGURE 1.10 TGA graph of the edible films.

1.3.7.10 ANTIMICROBIAL PROPERTY OF THE FILMS

The antimicrobial property of the films was experimented by the disk diffusion method against *E. coli* pathogen. The native starch alginate film did not show any inhibition zone that indicated that film1 had no antimicrobial properties. Addition of culinary banana peel extract in the films showed a clear zone of inhibition against *E. coli* (Figure 1.11 and Table 1.13).

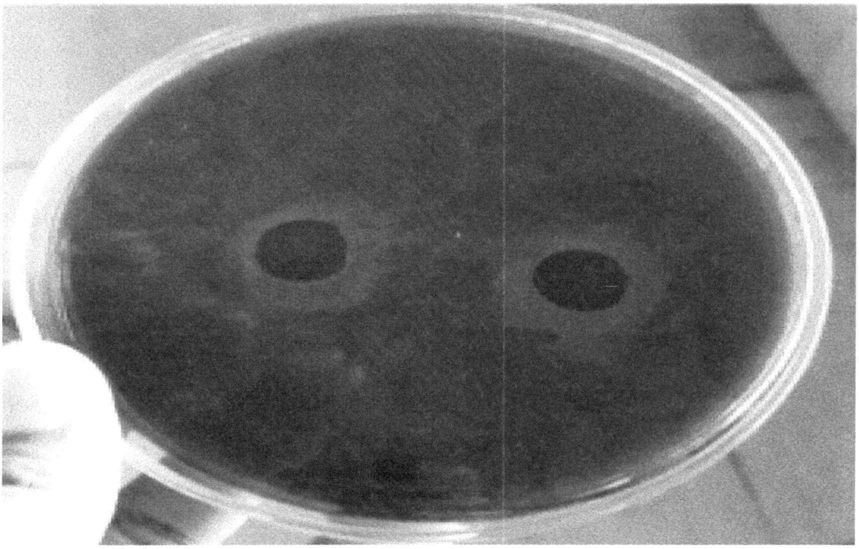

FIGURE 1.11 Zone of inhibition observed in films incorporated with kachkal peel extract against *E. coli*.

TABLE 1.13 Zone of Inhibition Against *E. coli*

Sl no.	Particulars	Zone of inhibition (mm)
1	Film 1	0
2	Film 2	1.7
3	Film 3	2.3
4	Film 4	4.2
5	Film 5	2.0
6	Film 6	2.8
7	Film 7	4.2

1.3.7.11 OXIDATIVE STABILITY OF BUTTER PACKED IN EDIBLE FILMS

The storage study and quality test of butter in edible films were done in both 4°C and 25°C. To check the quality of butter POV, FFA value tests were done for each sample at an interval of 7 days.

1.3.7.12 EFFECT OF DIFFERENT STORAGE ENVIRONMENT ON STORAGE STABILITY OF FILMS

The POV and FFA value of the stored butter is presented in Table 1.14. The POV and FFA were increased with the time for all the films. The native starch-alginate film (F1) showed the highest POV and FFA than the other films. And the storage temperature showed an important effect on POV and FFA values. The butter stored at low temperature (4°C) showed low POV and FFA values as compared to the butter stored at 25°C (Tables 1.15, Figures 1.12 and 1.13).

TABLE 1.14 Peroxide Value of Butter Stored in Edible Films

Particulars	No. of days	POV at 4°C (meq O_2/kg fat)	POV at 27°C (meq O_2/kg fat)
	0	5.76 ± 0.00	5.76 ± 0.000
Control	7	5.81 ± 0.010	6.80 ± 0.010
	14	5.86 ± 0.015	8.96 ±0.015
	0	5.76 ± 0.000	5.76 ± 0.000
Film 1	7	5.91 ± 0.015	6.82 ± 0.015
	14	6.02 ± 0.010	8.90 ± 0.110
	0	5.76 ± 0.000	5.76 ± 0.000
Film 2	7	5.77 ± 0.010	7.10 ± 0.010
	14	5.83 ±0.010	8.77 ± 0.010
	0	5.76 ± 0.000	5.76 ± 0.000
Film 3	7	5.82 ± 0.010	6.17 ± 0.010
	14	5.90 ± 0.011	8.45 ±0.560
	0	5.76 ± 0.000	5.76 ± 0.000
Film 4	7	5.75 ± 0.010	6.21 ± 0.015
	14	5.91 ± 0.020	8.17 ± 0.620
	0	5.76 ± 0.000	5.76 ± 0.000
Film 5	7	5.82 ± 0.010	6.64 ± 0.010
	14	5.97 ± 0.005	8.86 ± 0.089
	0	5.76 ± 0.000	5.76 ± 0.000
Film 6	7	5.77 ± 0.015	6.54 ± 0.020
	14	5.80 ± 0.005	8.89 ± 0.508
	0	5.76 ± 0.000	5.76 + 0.000
Film 7	7	5.82 ± 0.005	5.84 ± 0.035
	14	5.91 ±0.010	8.92 ± 0.007

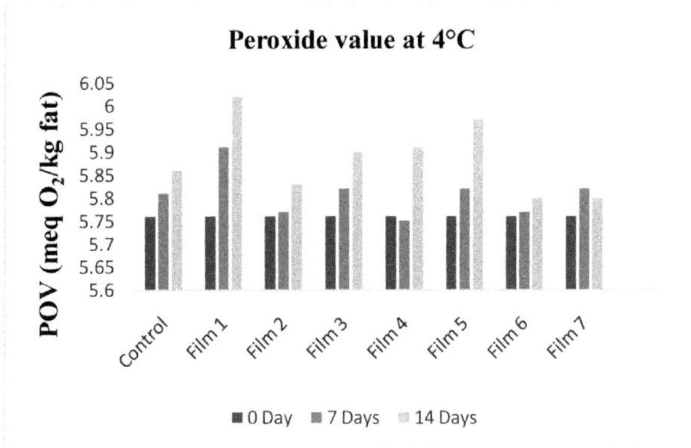

(a)

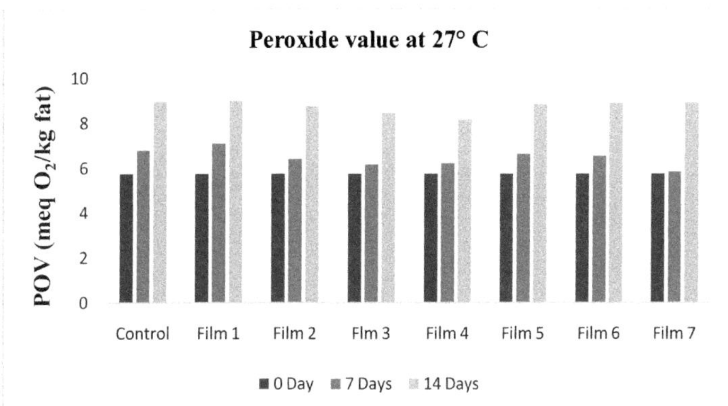

(b)

FIGURE 1.12 Peroxide values: (a) Peroxide value of stored butter at 4°C and (b) Peroxide value of stored butter at 27°C.

TABLE 1.15 Free Fatty Acid of Butter Stored in Edible Films

Particulars	Days	FFA at 4°C (%)	FFA at 27°C (%)
	0	1.12 ± 0.000	1.12 ± 0.000
Control	7	3.36 ± 0.010	5.22 ± 0.010
	14	3.61 ± 0.010	7.86 ± 0.010
	0	1.12 ± 0.000	1.12 ± 0.000
Film 1	7	3.46 ± 0.005	5.75 ± 0.020
	14	3.71 ± 0.011	7.22 ± 0.010
	0	1.12 ± 0.000	1.12 ± 0.000
Film 2	7	3.13 ± 0.005	4.36 ± 0.010
	14	3.22 ± 0.010	6.27 ± 0.010

TABLE 1.15 *(Continued)*

Particulars	Days	FFA at 4°C (%)	FFA at 27°C (%)
		1.12 ± 0.000	1.12 ± 0.000
Film 3	7	3.07 ± 0.010	4.12 ± 0.020
	14	3.20 ± 0.015	5.96 ± 0.010
	0	1.12 ± 0.000	1.12 ± 0.000
Film 4	7	4.01 ± 0.010	5.26 ± 0.010
	14	4.12 ± 0.010	5.82 ± 0.010
	0	1.12 ± 0.000	1.12 ± 0.000
Film 5	7	3.18 ± 0.005	4.80 ± 0.010
	14	3.23 ± 0.015	6.12 ± 0.020
	0	1.12 ± 0.000	1.12 ± 0.000
Film 6	7	3.23 ± 0.001	5.14 ± 0.010
	14	3.51 ±0.010	6.25 ± 0.010
	0	1.12 ± 0.000	1.12 ± 0.000
Film 7	7	3.15 ± 0.002	4.46 ± 0.010
	14	3.54 ± 0.015	5.36 ± 0.020

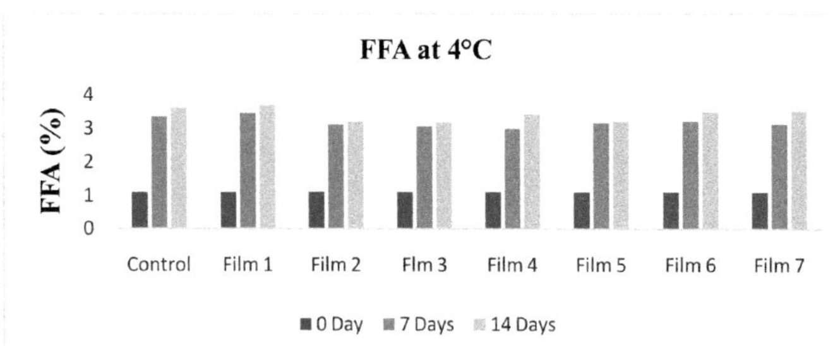

(a)

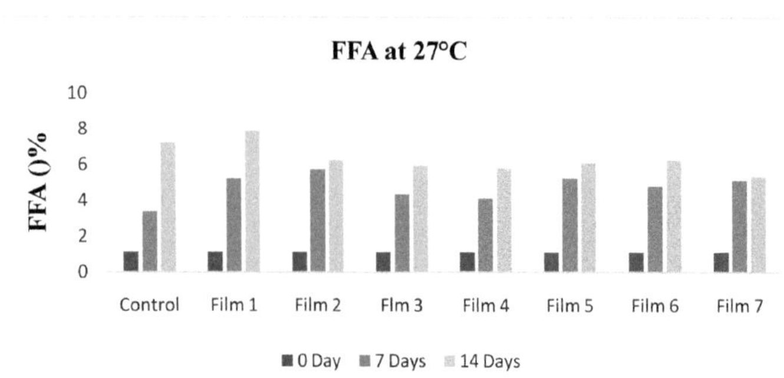

(b)

FIGURE 1.13 Free fatty acid values: (a) Free fatty acid value of stored butter at 4°C (b) Free fatty acid value of stored butter at 27°C.

ACKNOWLEDGMENT

The financial help received from DRDO, Ministry of Defense, Govt. of India, is duly acknowledged.

KEYWORDS

- **free fatty acid**
- **peroxide value**
- **tensile strength**
- **thermal gravimetric analysis**
- **water solubility**
- **water vapor permeability and transmission rate**

REFERENCES

Abugoch, L. E., Tapia, C., Villamán, M. C., Yazdani-Pedram, M., & Díaz-Dosque, M., (2011). Characterization of quinoa protein–chitosan blend edible films. *Food Hydrocolloids, 25*(5), 879–886.

Adeleke, R. O., & Odedeji, J. O., (2010). Functional properties of wheat and sweet potato flour blends. *Pakistan Journal of Nutrition, 9*(6), 535–538.

Alves, A. S., Adão, H., Ferreroc, T. J., Marques, J. C., Costa, M. J., & Patrícioa, J., (2013). Benthic meiofauna as indicator of ecological changes in estuarine ecosystems: The use of nematodes in ecological quality assessment. *Ecological Indicators, 24*, 462–475.

AOAC, (1997). *Association of Official Analytical Chemists International Official Methods of Analysis* (16th edn.). AOAC, Arlington.

AOAC, (2010). *Association of Official Analytical Chemists Official Method of Analysis of the Association of Official Analytical Chemists* (18th edn.). AOAC International, Washington DC.

Atichokudomchai, N., & Varavinit, S., (2002). Characterization and utilization of acid-modified cross-linked tapioca starch in pharmaceutical tablets. *Carbohydrate Polymers, 53*(3), 263–270.

Baker, L. A., & Rayas-Duarte, P., (1998). Freeze-thaw stability of amaranth starch and the effects of salts and sugars. *Cereal Chem., 75*(3), 301–307.

Balasubramanian, N., Bai, P., Buchek, G., Korza, G., & Weller S. K., (2010). Physical interaction between the herpes simplex virus type 1 exonuclease. *UL12, and the DNA Double-Strand Break-Sensing MRN Complex, 84*(24), 12504–12514.

Bhandari, L., Sodhi, N. S., & Chawla, P., (2016). Effect of acidified methanol modification on physicochemical properties of black-eyed pea (vignaunguiculata) starch. *International Journal of Food Properties, 19*(12), 2635–2648.

Bora, A., & Mishra, P., (2016). Characterization of casein and casein-silver conjugated nanoparticle containing multifunctional (pectin–sodium alginate/casein) bilayer film. *Journal of Food Science and Technology, 53*(10), 3704–3714.

Bourtoom, T., (2009). Edible protein films: Properties enhancement. *International Food Research Journal, 16*(1), 1–9.

Brand-Williams, W., Cuvelier, M. E., & Berset, C. L. W. T. (1995). Use of a free radical method to evaluate antioxidant activity. *LWT-Food Science and Technology, 28*(1), 25–30.

Dihingia, K., & Deka, S. C., (2019). Effect of acetylation on the physicochemical properties of culinary banana (*Musa* ABB) starch. In: *Innovations in Food Processing Technologies, NIPA*. ISBN: 9789386546517.

Eggleston, G., Swennen, R., & Akoni, S., (1992). Physicochemical studies on starches isolated from plantain cultivars, plantain hybrids and cooking bananas. *Starch-Stärke, 44*(4), 121–128.

Fan, D., Ma, W., Wang, L., Huang, J., Zhao, J., Zhang, H., & Chen, W., (2012). Determination of structural changes in microwaved rice starch using Fourier transform infrared and Raman spectroscopy. *Starch-Stärke, 64*(8), 598–606.

Fleche, G., (1985). Chemical modification & degradation of starch. *Food Technol., 14*, 73–99.

Fuente, M., & Victor, V. M., (2000). Anti-oxidants as modulators of immune function. *Immunol. Cell Biology, 78*(1), 49–54.

Gontard, N., Guilbert, S., & Cuq, J. L., (1992). Edible wheat gluten film: Influence of the main process variable on film properties using response surface methodology. *Journal of Food Science, 57*, 190–195.

Gope, S., Samyor, D., Paul, A. K., & Das, A. B., (2016). Effect of alcohol-acid modification on physicochemical, rheological and morphological properties of glutinous rice starch. *International Journal of Biological Macromolecules, 93*, 860–867.

Guilbert, S., (1986). Technology and application of edible protective films. In: Mthlouthi, M., (ed.), *Food Packaging and Preservation* (pp. 371–394). New York: Elsevier.

Guilbert, S., Gontard, N., & Cuq, B., (1995). Technology and applications of edible protective films. *Packaging Technology and Science, 8*(6), 339–346.

Gutierrez, T. J., Tapia, M. S., Perez, E., & Fam, L., (2015). Structural and mechanical properties of edible films made from native and modified cush-cush yam and cassava starch. *Food Hydrocolloids, 45*, 211–217. ISSN–0268-005X.

Hazarika, B. J., & Sit, N., (2016). Effect of dual modification with hydroxypropylation and cross-linking on physicochemical properties of taro starch. *Carbohydrate Polymers, 140*, 269–278.

Hermansson, A. M., & Svegmark, K., (1996). Developments in the understanding of starch functionality. *Trends in Food Science & Technology, 7*(11), 345–353.

Hoover, R., & Ratnayake, W. S., (2002). Starch characteristics of black bean, chickpea, lentil, navy bean and pinto bean cultivars grown in Canada. *Food Chemistry, 78*(4), 489–498.

Hoover, R., Sinnott, A. W., & Perera, C. (1998). Physicochemical characterization of starches from *Amaranthus cruentus* grains. *Starch–Stärke, 50*(11–12), 456–463.

Hu, J., Jiao, D., Xu, Q., Ying, X., Liu, W., Chi, Q., Ye, Y., Li, X., & Cheng, L., (2016). Identification of proteasome subunit beta type 2 associated with deltamethrin detoxification in Drosophila Kc cells by cDNA microarray analysis and bioassay analyses. *Gene, 582*(1), 85–93. (Export to RIS)

Jacobson, M. R., Obanni, M., & Bemiller, J. N., (1997). Retrogradation of starches from different botanical sources. *Cereal Chemistry, 66*, 173–182.

Kester, J. J., & Fennema, O. R., (1986). Edible Films and Coatings: A Review. *Food Technology (USA), 40*(12), 47–59.
Kester, J. J., & Fennema, O., (1989). Resistance of lipid films to water vapor transmission. *Journal of the American Oil Chemists' Society, 66*(8), 1139–1146.
Khawas, P., & Deka, S. C., (2016a). Comparative nutritional, functional, morphological, and diffractogram study on culinary banana (*Musa* ABB) peel at various stages of development. *International Journal of Food Properties, 19*(12), 2832–2853.
Khawas, P., & Deka, S. C., (2016b). Isolation and characterization of cellulose nanofibers from culinary banana peel using high-intensity ultrasonication combined with chemical treatment. *Carbohydrate Polymers, 137*, 608–616.
King, P., (1982). Biotechnology: An industrial view. *Journal of Chemical Technology and Biotechnology, 32*, 2–8.
Lawal, O. S., Adebowale, K. O., Ogunsanwo, B. M., Barba, L. L., & Ilo, N. S., (2005). Oxidized and acid thinned starch derivatives of hybrid maize: Functional characteristics, wide-angle x-ray diffractometry and thermal properties. *International Journal of Biological Macromolecules, 35*(1), 71–79.
Leach, H. W., McCowen, L. D., & Schoch, T. J., (1959). Structure of the starch granule. I. Swelling and solubility patterns of various starches. *Cereal Chemistry, 36*, 534–544.
Liu, H., Ramsden, L., & Corke, H., (1999). Physical properties and enzymatic digestibility of hydroxypropylated ae, wx, and normal maize starch. *Carbohydrate Polymers, 40*(3), 175–182.
López-Martínez, G., Borrull, A., Poblet, M., Roy, N. R., & Cordero-Otero, R., (2014). Metabolomic characterization of yeast cells after dehydration stress. *Int. Microbiol., 17*(3), 131–139.
Lou, E., Fujisawa S., Barlas A., Romin Y., Manova-Todorova K., & Moore M. A., (2012). Tunneling nanotubes: A new paradigm for studying intercellular communication and therapeutics in cancer. *Commun. Integr. Biology, 5*(4), 399–403.
Maizura, M., Fazilah, A., Norziah, M. H., & Karim, A. A., (2008). City of modified sago starch-alginate based edible film incorporated with lemongrass 233 (2008) short communication antibacterial activity of modified sago starch-alginate based edible film incorporated with lemongrass (Cymbopogon citrates) Oil. *International Food Research Journal, 15*(2), 233–236.
Mallick, C. P., & Singh, M. B., (1980). *Plant Enzymology & Histoenzymology (end)* (p. 286). Kalyani Publishers, New Delhi.
Manral, M., Pandey, M. C., Jayathilakan, K., Radhakrishna, K., & Bawa, A. S., (2008). Effect of fish (Catlacatla) frying on the quality characteristics of sunflower oil. *Food Chemistry, 106*(2), 634–639.
Masina, N., Choonara, Y. E., Kumar, P., Du Toit, L. C., Govender, M., Indermun, S., & Pillay, V., (2017). A review of the chemical modification techniques of starch. *Carbohydrate Polymers, 157*, 1226–1236.
McGrance, S. J., & Cornell, H. J., & Colin, R., (1998). A simple and rapid colorimetric method for the determination of amylose in starch products. *Starch –Stärke, 50*, 158–163. 10.1002/(SICI)1521–379X(199804)50:43.0.CO, 2–7.
Miyazaki, M., Van Hung, P., Maeda, T., & Morita, N., (2006). Recent advances in application of modified starches for bread making. *Trends in Food Science & Technology, 17*(11), 591–599.

Mweta, D. E., Labuschagne, M. T., Koen, E., Benesi, I. R. M., & Saka, J. D. K., (2008). Some properties of starches from cocoyam (Colocasiaesculenta) and cassava (Manihoteesculentra Crantz.) grown in Malawi. *African Journal of Food Science, 2*, 102–111.

Norajit, K., Kim, K. M., & Ryu, G. H., (2010). Comparative studies on the characterization and antioxidant properties of biodegradable alginate films containing ginseng extract. *Journal of Food Engineering, 98*(3), 377–384.

Ovando-Martinez, M., Sáyago, A. S., Agama-Acevedo, E., Goñi, I., & Bello-Pérez, L. A., (2009). Unripe banana flour as an ingredient to increase the undigestible carbohydrates of pasta. *Journal of Food Chemistry, 113*, 121–126.

Perez, S., & Bertfort, E., (2010). The molecular structures of starch components and their contribution to the architecture of starch granules: A comprehensive review. *Starch-Starke, 62*(8), 389–420.

Perez-Sira, E. E., Lic, L. M., & Gonzalez, Z. M., (1997). Characterization of starch isolated from white and dark sorghum. *Starch/Starke, 49*, 103–106.

Ranganna, S., (2004.). *Handbook of Analysis and Quality Control for Fruit and Vegetable Products* (2nd edn.). ISBN 10: 0074518518 / ISBN 13: 9780074518519.

Reddy, C. K., Vidya, P. V., Vijina, K., & Haripriya, S., (2015). Modification of poovan banana (*Musa* AAB) starch by c-irradiation: effect on *in vitro* digestibility, molecular structure and physicochemical properties. *International Journal of Food Science Technology, 50*(8), 1778–1784.

Romero-Bastida, C. A., Bello-Pérez, L. A., García, M. A., Martino, M. N., Solorza-Feria, J., & Zaritzky, N. E., (2005). Physicochemical and microstructural characterization of films prepared by thermal and cold gelatinization from non-conventional sources of starches. *Carbohydrate Polymers, 60*(2), 235–244.

Saartrat, S., Puttanlek, C., Rungsardthong, V., & Uttapap, D., (2005). Paste and gel properties of low-substituted acetylated canna starches. *Carbohydrate Polymers, 61*(2), 211–221.

Sahnoun, M., Ismail, N., & Kammoun, R., (2016). Enzymatically hydrolyzed, acetylated and dually modified corn starch: Physicochemical, rheological and nutritional properties and effects on cake quality. *Journal of Food Science and Technology, 53*(1), 481–490.

Singh, A. L., Hariprassanal, K., & Solanki, R. M., (2008). Screening and selection of groundnut genotypes for tolerance of soil salinity. *Australian Journal of Crop Science Southern Cross Journals, 1*(3), 69–77.

Sit, N., Misra, S., & Deka, S. C., (2013). Physicochemical, functional, textural and color characteristics of starches isolated from four taro cultivars of North-East India. *Starch-Stärke, 65*(11/12), 1011–1021.

Sit, N., Misra, S., & Deka, S. C., (2014). Characterization of physicochemical, functional, textural and color properties of starches from two different varieties of taro and their comparison to potato and rice starches. *Food Science and Technology Research, 20*(2), 357–365.

Stover, R. H., & Simmonds, N. W., (1987). *Bananas (No. Ed. 3)*. Longman Scientific & Technical.

Tapia, J. C., Kasthuri, N., Hayworth, K. J., Schalek, R., Lichtman, J. W., Smith, S. J., & Buchanan, J., (2012). High-contrast en bloc staining of neuronal tissue for field emission scanning electron microscopy. *Nature Protocols, 7*(2), 193.

Tester, R. F., Karkalas, J., & Qi, X., (2004). Starch composition, fine structure, and architecture. *Journal of Cereal Science, 39*(2), 151–165.

Than, P. P., Jeewon, R., Hyde, K. D., Pongsupasamit, S., Mongkolporn, O., & Taylor, P. W. J. (2008). Characterization and pathogenicity of Colletotrichum species associated with anthracnose on chilli (*Capsicum* spp.) in Thailand. *Plant Pathology, 57*(3), 562–572.

Thomas, D. J., & Atwell, W., (1999). *Practical Guides the Food Industry in Starch*. St. Paul: Eagan.

Wang, Y. J., & Wang, L., (2002). Characterization of acetylated waxy maize starches prepared under catalysis by different alkali and alkaline-earth hydroxides. *Starch-Stärke, 54*(1), 25–30.

Yu, H., Cheng, L., Yin, J., Yan, S., Liu, K., Zhang, F., Xu, B., & Li, L., (2013). Structure and physicochemical properties of starches in lotus (*Nelumbonuceifera*Gaertn). *Rhizome. Food Science and Nutrition, 1*(4), 273–283.

Zeng, J., Li, G., Gao, H., & Ru, Z., (2011). Comparison of A and B starch granules from three wheat varieties. *Molecules, 16*(12), 10570–10591.

CHAPTER 2

Functional Foods from Different Sources

PRAKASH KUMAR NAYAK,[1] CHANDRASEKAR CHANDRA MOHAN,[2] and KESAVAN RADHAKRISHNNAN[1]

[1]*Department of Food Engineering and Technology, Central Institute of Technology, Kokrajhar, BTAD, Assam-783370, India, Tel.: +91-84738-21333, Fax: +91-03661-277143, E-mail: k.radhakrishnan@cit.ac.in*

[2]*Center for Food Technology, Anna University, Chennai, Tamil Nadu, India*

ABSTRACT

Over the last few decades, the consumer's preference to choose healthier foods has been increased enormously in several industrialized countries. As a result, food industries encountering various challenges in order to meet the consumer's expectations. It stimulated the development of new food products that provide beneficial health effects (decreasing health risks/improving health or well-being) beyond basic nutrition and commonly known as functional foods. It's becoming the fastest growing sector among the different categories in food industries. The effectiveness of functional foods depends on the active compounds known as functional ingredients, and it includes, carotenoids, flavonoids, betalains, phenols, phenolic acids, phytosterols, alkaloids, phytoestrogens, and dietary fiber, etc., These functional ingredients (also known as bioactives) occur naturally or added to functional foods. In this chapter, functional foods from different sources are discussed, along with their potential health benefits.

2.1 INTRODUCTION

The awareness of consumers on the intake of healthier foods, comprising bioactive compounds has been increased tremendously. Those foods are termed as functional foods, designer foods, or nutraceuticals (Hasler, 1998). In 1984, the word "functional food" was used first in Japan. It was emerged out of a research work where the correlations of nutrition, sensory acceptance, the fortification was investigated in a food matrix enriched with special constituents (Alzamora et al., 2005; Bigliardi and Galati, 2013). The term "functional food" was not defined universally though it has been used broadly (Health Canada, 1998). In many countries, functional foods are not defined statutorily as the differentiation of conventional and functional foods was not easier (Niva, 2007). Few of the definitions given by various organizations are listed in Table 2.1.

2.2 FUNCTIONAL FOODS FROM DIFFERENT PLANT SOURCES

The data from various research studies shown that the diet rich in plant products may lower the likelihoods of various health disorders, especially cancer. Block et al. (1992) discovered that people are consuming higher amounts of plant foods having a lesser chance of developing cancer in comparison with the people taking lower quantities of fruits and vegetables. Plant foods consist of numerous bioactive compounds in addition to nutrient constituents. Steinmetz and Potter (1991) identified these health-promoting beneficial constituents from plants and labeled as "phytochemicals." Unfortunately, they were considered as irrelevant compounds, which may cause harmful effects to human health (Rodriguez et al., 2006).

2.2.1 SOY

The interest in the research activities with soy has started from the 1990s. It has been known as a very good source of protein. It also has its role in the decreasing the risk of cardiovascular diseases (CVD) and cancer. Among its effects, the cholesterol-reducing capacity was a well-known physiological role. The results of meta-analysis shown that intake of soy protein accomplished the lowering of total cholesterol (TC), low-density lipoprotein (LDL) cholesterol, triglycerides as 9.3%, 12.9%, and 10.5%, respectively. It also showed an increase of high-density lipoprotein (HDL) cholesterol as 2.4% (Anderson et al., 1995). Various classes of anticancer compounds have been acknowledged in soybeans, including protease inhibitors, phytosterols, saponins, phenolic acids, phytic acid, and isoflavones (Messina and Barnes, 1991). Recently, the focus has been shifted to one of the important

TABLE 2.1 Definition of Functional Foods by Different Organizations

Organization	Definition	Reference
Japanese Ministry of Health, Labor, and Welfare–Foods for Specified Health Uses (FOSHU)	Foods for Specified Health Uses (FOSHU) refer to foods containing an ingredient with functions for health and officially approved to claim its physiological effects on the human body. FOSHU is intended to be consumed for the maintenance/promotion of health, or special health uses by people who wish to control health conditions	FOSHU, 1991
Food and Nutrition Board	Food that encompasses potentially helpful products, including any modified food or food ingredient that may provide a health benefit beyond that of the traditional nutrient it contains	National Academy of Sciences, 1994
Health Canada	A functional food is similar in appearance to, or maybe, a conventional food consumed as part of a usual diet, and is demonstrated to have physiological benefits and/or reduce risk of chronic disease beyond basic nutritional functions	Health Canada, 1998
National Institute of Nutrition	Foods or food components that may have health benefits that reduce the risk of specific diseases or other health concerns	National Institute of Nutrition, 2000
CSIRO	Foods that may be eaten regularly as part of a normal diet that have been designed specifically to provide a physiological or medical benefit by regulating body functions to protect against or retard the progression of diseases such as coronary heart disease, cancer, hypertension, diabetes, and osteoporosis.	CSIRO Human Nutrition, 2004
Food and Agricultural Organization (FAO) of the United Nations	Functional foods should be a food similar in appearance to a conventional food (beverage, food matrix), consumed as part of the usual diet which contains biologically active components with demonstrated physiological benefits and offers the potential of reducing the risk of chronic disease beyond basic nutritional functions.	FAO, 2007
Academy of Nutrition and Dietetics (formerly the American Dietetic Association)	All foods are functional at some physiological level, but it is the position of the Academy of Nutrition and Dietetics that functional foods that include whole foods and fortified, enriched, or enhanced foods have a potentially beneficial effect on health when consumed as part of a varied diet on a regular basis, at effective levels.	American Dietetic Association, 2009
Dietitians of Canada (DC)	Functional foods are foods that offer unique health benefits that go beyond simply meeting basic nutrient needs. Many also help to reduce chronic disease risk. Functional foods contain." bioactive compounds," or naturally occurring chemicals that act on our bodies. It is these bioactive compounds that offer the health and wellness benefits that have been linked to functional foods.	DC, 2010

components in soy, known as isoflavones (Potter, 1998). As isoflavones were weak estrogens, they may serve as antiestrogens by competing with the strong endogenous estrogens. It may show the effect of constituents present in soy on reducing the risk factors of estrogen-dependent cancer.

Another significant role of soy proteins is strengthening the bones (Anderson and Garner, 1997). The effect of soy on bone health was studied by taking 40 g of isolated soy protein per day, and it was found that the bone strength was increased significantly due to the rise in bone mineral content and density on the lumbar spine after 6 months. In another study, it was shown that the daily consumption of ISP (60 g/day) for 3 months has decreased hot flashes by 45% in 104 postmenopausal women (Albertazzi et al., 1998).

2.2.2 FLAXSEED

Flaxseed oil contains higher quantities of omega-3 fatty acid, a-linolenic acid. But, the research studies are concentrated on lignan, a fiber component. Plant lignans are acting as a resource for the synthesis of two key mammalian lignans, namely, enterodiol, and enterolactone by bacteria present in GI tract (Setchell et al., 1981; Thompson et al., 1991). As naturally occurring enterodiol and enterolactone are identical to synthetic estrogens, they found to have lesser estrogenic and antiestrogenic activities. It may act as an important factor in reducing the possibilities of estrogen-dependent cancers. However, no studies were reported to confirm the hypothesis till now. Some studies conducted in rodent systems showed to reduce the likelihoods of the colon and mammary gland cancers (Thompson, 1995) and lung cancer (Yan et al., 1998).

The research studies also shown the effect of flaxseed on CVD risk by reducing total and LDL cholesterol (Bierenbaum et al., 1993) along with the reduction of platelet aggregation (Allman et al., 1995). Further, the effects of flaxseed on reducing the cancer risk was studied. In which, the consumption of 10 g of flaxseed per day caused some hormonal changes associated with the reduction of breast cancer was found (Phipps et al., 1993). Adlercreutz et al. (1982) reported that the levels of urinary lignan were less in postmenopausal breast cancer patients with respect to the controls having a normal diet.

2.2.3 TOMATOES

Tomatoes are rich in lycopene, an important carotenoid present in the fruits (Gerster, 1997) which plays a vital role in the prevention of cancer

(Weisburger, 1998). The research work of Giovannucci et al. (1995) conveyed that the regular intake of tomatoes reduced the possibilities of prostate cancer development. The higher levels of lycopene in serum or tissue also resulted in the reduced the risk of cancers like breast, digestive tract, cervix, bladder, and skin (Clinton, 1998) and lung (Li et al., 1997). The antioxidant activity of lycopene is the major reason for its cancer preventive action. Lycopenes are very much effective against singlet O_2 in various biological systems (Di Mascio et al., 1989). The radical scavenging ability of lycopene is also associated with the reduction of myocardial infarction (Kohlmeier et al., 1997).

2.2.4 APPLES

Apple has been consisting of various compounds such as hydroxycinnamic acids, dihydrochalcones, flavonols, catechins, triterpenoids, and oligomeric procyanidins and anthocyanins (red apples). The studies conducted on the bioactive compounds in apple and apple products are responsible for decreasing the risk factors of CVD, asthma, and pulmonary dysfunction, diabetes, obesity, and cancer. Gerhauser (2008), found that the compound–oligomeric procyanidins present in apple juice, apple peel, apple extracts were responsible for its cancer preventive effect. It has been processed through various mechanisms including, antimutagenic activity, modulation of carcinogen metabolism, antioxidant activity, anti-inflammatory mechanisms, and modulation of signal transduction pathways, antiproliferative, and apoptosis-inducing activity.

2.2.5 OATS

Whole grains of oatmeal may have a potential effect on decreasing the likelihoods of coronary heart disease (CHD) and extensively studied as a source of β-glucan. Kelly et al. (2007) studied the effect of consuming whole grain foods on CHD, and the trails were conducted on the patients previously diagnosed with CHD or with existing risk factors for CHD. From the results, it was found that the intake of oatmeal could lower the levels of total and LDL cholesterol, which may reduce the possibilities of CHD. It was also seen that the high content of β-glucan in oatmeal was responsible for lessening postprandial glycemic response of glucose in patients with type 2 diabetes.

2.2.6 GARLIC

Garlic (*Allium sativum*) is commonly considered as a significant herb for its therapeutic roles. (Nagourney, 1998). The health effects of garlic include cancer chemopreventive, antibiotic, anti-hypertensive, and cholesterol-lowering properties (Srivastava et al., 1995). One of the main constituents of garlic is allicin, and it is responsible for its chemopreventive role, and it has been studied by various researchers (Reuter et al., 1996).

Garlic contains inulin, a non-digestible polysaccharide, in the levels of 26–30% fresh garlic cloves and 77% in dried garlic. It is the compound responsible for its prebiotic activity. The study by Kannar et al. (1998) recommended that the intake of garlic and inulin supplement may influence the growth of microflora, specifically in foods which have higher amounts of fats, carbohydrate, and dietary fiber.

2.2.7 ALMOND

In a study, it was found that the consumption of almond may increase the antioxidant defense in the body and reduce the risk of CVD and cancer. These beneficial effects were associated with the polyphenols present in the almond skins. It has been found that the quinone reductase activity was increased by these polyphenols, and it should be noted that the effect was based on the levels of polyphenols, extraction method, and the presence of vitamins.

2.2.8 BROCCOLI AND GREEN LEAFY VEGETABLES

Broccoli belongs to the family, Brassicaceae. It contains compounds such as vitamin C, soluble fiber, diindolylmethane, and selenium. From the epidemiological studies, it was found that the regular intake of cruciferous vegetables was associated with the reduction in cancer risk. The anti-cancerous properties of cruciferous vegetables were related to the high levels of glucosinolates (Verhoeven et al., 1997), and they can be found in the cell vacuoles of leafy vegetables.

The anti-cancerous effect was shown by a group of natural and synthetic isothiocyanates for preventing cancers in animals (Hecht, 1995). In particular, a specific type of isothiocyanate, identified as sulforaphane in broccoli, was responsible for decreasing the possibilities of cancers. It was observed that sulforaphane was acting as a key source of enzyme, known as quinone reductase. In addition, diindolylmethane found in broccoli is acting as an important controller of the innate immune response system with anti-viral, anti-bacterial, and anti-cancer activity.

2.2.9 CITRUS FRUITS

Citrus fruits are rich in various nutrients such as vitamin C, folate, and fiber. Especially, the antioxidant properties depend on flavonoids, in addition to vitamin C. It was also found that the citrus fruits are playing a key role in reducing the probabilities of various cancers (Gould, 1997). This anti-cancerous property of citrus fruits was related to a group of phytochemicals labeled as limonoids (Hasegawa and Miyake, 1996). The cancer preventive effect was studied by Crowell (1997), and he found that it was effective against both spontaneous and chemically-induced rodent tumors.

2.2.10 BERRIES

Berries are rich in anthocyanin, antioxidants, and have a variety of preventive activities against CVD, oxidative stress, inflammatory responses, etc. Berries also associated with improvement of neuronal and cognitive brain functions, ocular health as well as protecting genomic DNA integrity. Zafra-Stone et al. (2007) studied the effect of combined berry extracts and it was found that the extracts displayed a variety of functions including, high oxygen radical quenching capacity, antiangiogenic, and antiatherosclerotic activity.

2.2.11 TEA

Tea constituents have been studied widely for its various activities, and the main focus has been pointed at the polyphenolic compounds from green tea (Harbowy and Balentine, 1997). The polyphenol content of fresh tealeaves (dry weight basis) was varied up to 30%. The most important and predominant polyphenol of tea is catechins. It comprises of epigallocatechin-3-gallate, epigallocatechin, epicatechin-3-gallate, and epicatechin.

Reports from several works suggested that green tea may decrease the possibilities of CVD and cancer, as well as it may have an influence on bone health, cognitive function, tooth problems, and kidney stones. The findings from a study of Dreosti et al. (1997) supported the anti-cancerous effect of tea compounds. Tannins present in green tea may possess significant antioxidant activity, and it has been confirmed in a study by Nakagawa and Yokozawa (2002).

The research data from Hertog et al. (1993) supported the effectiveness of tea consumption on CVD. In this work, the effect of flavonoids (quercetin, kaempferol, myricetin, apigenin, and luteolin) on a group of elderly men was tested, and the results were found to be satisfactory for the reduction CVD health risk.

2.2.12 GRAPES

The potent bioactive compounds in the grape extracts are polyphenols such as resveratrol, phenolic acids, anthocyanins, and flavonoids. Grapes are found to decrease the risk factor associated with the inflammatory diseases like CVD. The positive effect on CVD risk factors is due to the ability of grapes or grape juices in reducing the LDL-cholesterol levels (Frankel et al., 1993) and decreasing platelet aggregation. Another compound present in red wine is trans-resveratrol, a phytoalexin found in grape skins (Creasy and Coffee, 1988). The estrogenic properties of resveratrol have been associated with decreasing the risk factors of cancers (Jang et al., 1997).

2.3 FUNCTIONAL FOODS FROM ANIMAL SOURCES

The functionally active compounds of plant origin are numerous. Still, animal foods also contain numerous physiologically-active components that have a significant effect on health promotion (Prates and Mateus, 2002).

2.3.1 MEAT AND MEAT PRODUCTS

Meat and meat products are commonly considered as good resources of protein. It also considered as a rich source of nutrients such as fatty acids, iron, zinc, vitamin B12, and folic acid (USDA/HHS Dietary guidelines Americans, 2010). The study report of Mann (2000), stated that the diets rich in lean meat may influence the decrease in plasma cholesterol levels.

Currently, the research has been focused on conjugated linoleic acid (CLA), which was reported for its anti-carcinogenic properties in 1987 (Ha et al., 1987). CLA indicates a mixture of linoleic acid (C18:2, *n*–6) isomers where the double bonds are conjugated than existing as a typical methylene interrupted arrangement (Prates and Mateus, 2002). The rich sources of CLA are ruminant animals, including beef and lamb. CLA can be found in beef fat in the range of 3.1 to 8.5 mg per gram of fat (Krumhout et al., 1985). From the research studies, it was found that CLA was effective in preventing for stomach tumors in mice, aberrant colonic crypt foci in rats, and mammary carcinogenesis in rats (Ip and Scimeca, 1997). The effectiveness of CLA was also studied for changing body composition, and the results suggested that it has been acting as an agent in decreasing the fat levels and elevating lipolysis in adipocytes (Park et al., 1997). Some studies reported that the occurrence of colorectal cancer might be lowered by the proper consumption of CLA (Larsson et al., 2005). CLA also found to be exhibit anti-oxidative and CLA has also been immune modulative properties.

The anti-oxidative effect of meat depends on two histidyl dipeptides known as carnosine (β-alanyl-L-histidine) and anserine (N-β-alanyl-1-methyl-L-histidine). Chicken meat contains carnosine in the level of 500 mg/g and a higher level of anserine in comparison with carnosine. These peptides having the potential to chelate metals like copper (Brown, 1981). Research findings reported that these compounds acting as a resistance to various disorders and health issue associated with oxidative stress (Hipkiss and Brownson, 2000).

2.3.2 FISH

Fish is an important source of omega-3 (n-3) fatty acids, an important class of polyunsaturated fatty acids (PUFA). It includes eicosapentaenoic acid (EPA; C20:5, *n*–3) and docosahexaenoic acid (DHA; C22:6, *n*–3). The research works on seafood has been focused mainly on these PUFAs due to their various health beneficial effects (Bahri et al., 2002). Research works also carried out to find the relationship of PUFAs in the prevention of various diseases including, CVD (Ness et al., 2002), high blood pressure (Morris et al., 1993); blood clotting (Murphy et al., 1999); cancer (Wolk et al., 2006). The health benefits of *n*–3-PUFAs are due to their ability to change the lipid fractions in the membrane and causing changes in metabolic and signal-transduction pathways (Huang et al., 2009).

DHA is a vital component for the proper functioning and growth of the brain and central nervous system. It also plays an important role in the growth of these systems in newborns or in expectant mothers (Nys and Debruyne, 2011). Results from studies also shown the beneficial role of *n*–3 PUFAs against dementia. A research work conducted by Nehru Science Center, Mumbai showed that DHA rich diet has reduced the progress of neurodegenerative disorders in geriatric patients. The study of Tan et al. (2012) has confirmed the effect of EPA/DHA enriched diet on defending against brain aging. Another work by Virtanen et al. (2013) recommended that circulating *n*–3 PUFAs could be associated in decreasing the possibilities of certain subclinical brain abnormalities.

Fish also comprises of anti-hypertensive peptides called angiotensin I-converting enzyme (ACE) inhibitors (Gormley, 2006). ACE compounds have been found in hydrolysates from fish waste, and Kitts and Weiler (2003) have suggested the suitable bioprocess methods for the separation of these peptides. ACE peptides are involved in the lowering of blood pressure by restricting the vasoconstrictor effects of Angiotensin II and endorsing the vasodilatory effects of bradykinin (De-Leo et al., 2009). The review of Vercruysse et al. (2005) has confirmed the antihypertensive effect of ACE inhibitors that have been isolated from fish muscle. Enzyme-treated fish protein hydrolysates (FPHs) have the potential to be used as cardio-protective (anti-atherogenic) components as a part of a nutraceutical or pharmaceutical (Berge, 2005).

2.3.3 EGG

Generally, eggs are not considered as functional foods due to their negative impact on the blood serum cholesterol levels. But, the regular intake of eggs may not influence the cholesterol levels according to the study of Hasler, (2000). Eggs are serving as a good source of protein, sphingolipids, choline, *n*–3 PUFA, and lutein/zeaxanthin. So, the egg may be seen as one of the vital components in changing the face of functional foods (Hasler, 2000).

Egg is one of the important source of sphingolipids apart from other sources like dairy products and soybean (Vesper et al., 1999). Sphingolipids are known to be associated with the cellular functions including, growth regulation, differentiation, and apoptosis. Form the animal studies, it was identified that supplementation of sphingolipids inhibits colon carcinogenesis, reduces serum LDL-cholesterol levels, and elevates HDL cholesterol levels (Vesper et al., 1999).

Choline is present in egg at the level of 2000 mg, which is sufficient to fulfill the daily intake level as suggested by the National Academy of Sciences in 1998 (Food and Nutrition Board, 1998). The research studies are focused in choline for its health benefits associated with the cognitive function (Blusztajn, 1998), especially in the early brain development. Eggs are also considered as an excellent source of lutein and zeaxanthin, which has been associated with reducing the risk factors of age-related macular degeneration, the foremost cause of irreversible blindness in the US (Klein et al., 1997).

Through diet modification, enriched eggs with *n*–3 PUFA can be produced. The modified eggs have the *n*–3 PUFA level of 350 mg in comparison with the normal egg containing 60 mg. The individuals consuming *n*–3 enriched eggs for four weeks were shown insignificant changes in plasma TC levels and LDL cholesterol levels (Simopoulos, 1999).

2.3.4 MILK AND MILK PRODUCTS

The effects of milk and milk products on human health are well documented, and it may be due to the bioactive compounds in milk and to the probiotic bacteria that present in the fermented milk products. Apart from nutrients like protein, carbohydrates, and lipids, milk contain numerous bioactive constituents including, immunoglobulins, enzymes, anti-microbial peptides, oligosaccharides, hormones, cytokines, and growth factors (Donovan, 2006).

Lactose, the sugar component of milk, has been converted to lactic acid, which decreases the pH and modifies the physical properties of casein. Thereby it improves digestibility, increases the utilization of calcium, and prevents the growth of bacteria. Lactulose was produced during the thermal treatment of milk, and it improves the health status of an individual by allowing the growth of probiotic bacteria (bifidobacteria and lactobacilli) alone (Gibson, 2004). The sour dairy products have also contains some other polysaccharide and their hydrolyzed products like kefiran, one of the compounds in kefir (Farnworth, 2005). Kefir has many health benefits against various problems like metabolic disorders, atherosclerosis, allergic diseases, tuberculosis, cancer, and gastrointestinal disorders (Otles and Ozlem, 2003).

The protein components of milk are caseins, β-lactoglobulin, α-lactalbumin, immunoglobulins, lactoferrin, and serum albumin. The health benefits of protein also depend on their different peptides (protein degradation components). These peptides include casomorphins, cytokinins, immunopeptides, lactoferrin, lactoferricin, and phosphopeptides. The functions of the peptides include immunomodulation, anti-microbial activity, anti-thrombotic activity, and blood pressure regulation (Meisel, 1998). Whey proteins such as α-lactalbumin, β-lactoglobulin, lactoferrin, lactoperoxidase, immunoglobulins occur in larger amounts in fermented dairy products. It also have a range of health-promoting functions, such as digestive functions and anti-carcinogenic activity (McIntosh et al., 1998).

One of the protein component, β-lactoglobulin has been involved in the actions like emulator, and immunomodulator (Beaulieu et al., 2006). Apart from these functions, it can also produce some other activities including, antihypertensive, antithrombotic, opioid, antimicrobial, immunomodulant, and hypocholesterolemic properties, all β-lactoglobulin-derived peptides also display radical-scavenging activity (SA) (Hernandez-Ledesma et al., 2007). The other protein component, α-lactalbumin, has been used as a nutrient in infant foods due to its low allergy-inducing potential. In addition, it also have other biological functions such as anticancer activity, immunomodulatory effects, anti-microbial activity (Pellegrini, 2003) and anti-ulcerative properties (Mezzaroba et al., 2006).

Milk contains immunoglobulins, including IgG1, IgM, IgA, and IgG2. These immunoglobulins exert immunological functions like defense against pathogenic organisms, activation of complement, stimulation of phagocytosis, preventing adhesion of microbes and neutralization of viruses and toxins. They also improve the levels of glutathione, which acts as a potent anti-oxidant. They also involved in the prevention of various microbial infections (Mehra et al., 2006).

Lactoferrin has been occurring in lower amounts in milks and have various physiological functions such as regulation of iron homeostasis, protection against various microbial infections, anti-inflammatory, and anti-cancer activity. It can also perform either as immunosuppressive, anti-inflammatory, or immunostimulatory agent. One of the peptides formed after the degradation of lactoferrin is lactoferricin, which is the primary factor for its activities. Lactoferrin in combination with lactoferricin may display anti-viral action against hepatitis C (Isawa et al., 2002), human papillomavirus (Mistry et al., 2007), herpes simplex virus (Jenssen, 2005), chronic hepatitis C along with interferon and ribavirin (Kaito et al., 2007).

The bioactive peptides from milk proteins are efficient in lowering the likelihoods of obesity and type 2 diabetes (Haque and Chand, 2008). Through fermentation also bioactive peptides can be produced in products such as various cheese types and fermented milk (Fitzgerald and Murray, 2006). From cheese ripening, biologically active peptides also can be produced, and these peptides have recognized in the products like comet, and cheddar cheese (Singh et al., 1997). In addition, the number of bioactive peptides were produced through secondary hydrolysis as the cheese ripening proceeds further. So, it can be understood that the functioning of peptides were based on the ripening period. Other research works reported the presence of bioactive peptides in fermented milk products (yogurt, sour milk, and Dahi). The presence of ACE-inhibitory, immunomodulatory, and opioid peptides can be seen in yogurt and in milk added with the cultures of *Lactobacillus casei* ssp. Rhamnosus strain (Rokka et al., 1997). They can also be found in yogurt prepared from bovine milk (Chobert et al., 2005) and in kefir produced from caprine milk (Quiros et al., 2007).

Among the several milk fat components, including CLA, butyric acid, ether lipids, β-carotene, and vitamins A and D, have the ability to reduce the various cancers (Khanal and Olson, 2004). The cancer preventive action of CLA has been explained in many studies, along with the ability to prevent atherosclerosis and in controlling some parts of the immune system (MacDonald, 2000). Another compound, sphingolipids, and their metabolite were involved in various health promoting effects such as prevention of cancer, antimicrobial, and immunomodulatory activities, inhibition of cholesterol adsorption (Akalin et al., 2006). Butyric acid is assumed as an anticancer agent and along with etheric lipids, vitamins (A, D, E) and linoleic acid, make a resistive layer that fights against various non-communicable disorders (Parodi, 2004). Caprylic and capric acid may show antiviral activities (Thormar et al., 1994). Lauric acid (C12:0) may act as an antiviral and antibacterial agent (Thormar and Hilmarsson, 2007) along with it may show anticaries and antiplaque activity (Schuster et al., 1980).

2.4 FUNCTIONAL FOODS FROM ALGAE

The marine ecosystem is not reconnoitered well as a prospective source of bioactive compounds, which may be utilized in various fields such as pharmaceuticals, cosmetics, and food. Among the marine sources, macroalgae (or seaweeds) and microalgae are the two mostly studied sources of bioactives. Suitable extraction methods should be employed to obtain value added products from macro and microalgae (Herrero et al., 2015).

Generally, algae have been considered as a rich source of nutrients such as proteins, carbohydrates, fiber, minerals, and vitamins, with low content of fats. Algae has been characterized as a good source of fiber, where the soluble portion mainly of sulfated galactan as agar or carragenates (in red algae) and alginates, fucans, and laminarin (in brown algae) (Plaza et al., 2008). Intake of dietetic fiber may show protective effects such as decreasing the possibilities of colon cancer, constipation, hypercholesterolemia, obesity, and diabetes. Apart from the antioxidant activity of dietetic constituents, it shows immunological activity also. In this connection, *Undaria. pinnatifida* (wakame) found to have a beneficial effect on some of the CVD (Ikeda et al., 2003). This alga mainly comprises dietetic fiber (as a primary compound) and alginic acid, which was found to be decreasing the risk factors of hypertension (Ikeda et al., 2003).

Another group of compounds present in algae of potential impact was polysaccharides. Many algae contain polysaccharides which are very good anti-viral agents. Among them, *Sargassum vulgare* comprises alginic acid, xylofucans, and two species of fucans (Dietrich et al., 1995), whereas *Undariapinnatifida (*brown alga) has good amounts of sulfated polysaccharides, particularly, sulfated fucans (fucoidans) (Hemmingson et al., 2006) and sulfate of galactofucan (Thompson and Dragar, 2004). These components have been shown potential anti-viral response against herpes type 1 virus (HSV-1), HSV-2, and cytomegalovirus in humans (HCMV). The fucoidans, may have the potential to act as an anticoagulant and antithrombotics agents (Lee et al., 2004). Porphyran, a sulfated polysaccharide was present in *Porphyra* sp. a red alga. It has effective apoptotic activity (evaluated using AGS cells from a human gastric cancer) inducing the death of the carcinogenic cells (Kwon and Nam, 2006).

One of the significant bioactive compound present in marine sources is carotenoid. Apart from serving as a coloring agent, it possesses biological functions such as antioxidant, antiproliferative, anti-inflammatory, provitamin A activity, and even protection of macular degeneration (Fernandez-García et al., 2012). Some macro and microalgae have been considered as a vital source (*Dunaliella salina,* a green microalga) of carotenoid (Tafreshi and Shariati, 2009). β-carotene produced from *Dunaliella* is a combination of all-trans, 9-cis, 15-cis, and other minor isomers. Another type of microalgae studied for the source of carotenoid (xanthophyll) is *Haematococcus pluvialis* (Jaime et al., 2010).

Algae are also an excellent source of PUFA, as the EPA described in *Himanthalia elongata*, *Undaria pinnatifida*, and *Porphyra sp*. PUFA have the ability to reduce the occurrence of diseases such as coronary diseases, thrombosis, and arteriosclerosis (Simopoulos, 2004). Sterols are another category of compounds, which are present in many algae species. The experiments described the effectiveness of sterols in decreasing the cholesterol levels in the blood. Further, sterols also possess multiple bioactive functions such as anti-inflammatory, antibacterial, antifungicidal, antiulcerative, and antitumoral activity (Sanchez-Machado et al., 2004).

2.5 CONCLUSION

The results of various studies associated with the functioning of bioactive compounds from different sources suggested that these compounds might improve the health status of an individual. The bioactive constituents include: saponins, phytic acid, isoflavones, omega-3 fatty acid, lignan precursors, lycopene, flavonols (quercetin glycosides), catechins, triterpenoids, oligomeric procyanidins, anthocyanins, β-glucan, allicin, vitamin C, soluble fiber, diindolylmethane, and selenium sulforaphane, folate, catechins, resveratrol from plants; CLA and carnosine from meat and meat products; *n*–3 fatty acids, angiotensin, FPHs from fish; sphingolipids, choline, *n*–3 PUFA and lutein/zeaxanthin from eggs;Lactulose, β-lactoglobulin, α-lactalbumin, immunoglobulins, lactoferrin, serum albumin, CLA, butyric acid, ether lipids, β-carotene and vitamins A and D, calcium, probiotics, whey proteins and whey peptides, from dairy products; fiber, polysaccharides, carotenoid, and PUFA from algae. These bioactive compounds possess different biological functions, which were effective in reducing the risk of various diseases. Additionally, the research works on the functional foods, and functional ingredients were required in order to study the stability and association of bioactive compounds with other food constituents during processing and storage.

KEYWORDS

- **animal sources**
- **bioactive compounds**
- **functional foods**
- **nutraceuticals**
- **plants sources**

REFERENCES

Adlercreutz, H., Heikkinen, R., Woods, M., Fotsis, T., Dwyer, J. T., Goldin, B. R., & Gorbach, S. L., (1982). Excretion of the lignans enterolactone and enterodiol and of equol in omnivorous and vegetarian postmenopausal women and in women with breast cancer. *The Lancet, 320*(8311), 1295–1299.

Akahn, S., Gönc, S., & Ünal, G., (2006). Functional properties of bioactive components of milk fats and metabolism. *Pakistan Journal of Nutrition*, *5*(3), 194–197.
Albertazzi, P., Pansini, F., Bonaccorsi, G., Zanotti, L., Forini, E., & De Aloysio, D., (1998). The effect of dietary soy supplementation on hot flushes. *Obstetrics & Gynecology*, *91*, 6–11.
Allman, M. A., Pena, M. M., & Pang, D., (1995). Supplementation with flaxseed oil versus sunflower seed oil in healthy young men consuming a low-fat diet: Effects on platelet composition and function. *European Journal of Clinical Nutrition*, *49*(3), 169–178.
Alzamora, S. M., Salvatori, D., Tapia, S., Lopez-Malo, M. A., Welti-Chanes, J., & Fito, P., (2005). Novel functional foods from vegetable matrices impregnated with biologically active compounds. *Journal of Food Engineering*, *67*, 205–214.
American Dietetic Association, (2009). Position of the American Dietetic Association: Functional foods. *Journal of the American Dietetic Association*, *109*, 735–746.
Anderson, J. J. B., & Garner, S. C., (1997). The effects of phytoestrogens on bone. *Nutrition Research*, *17*, 1617–1632.
Anderson, J. W., Johnstone, B. M., & Cook-Newell, M. E., (1995). Meta-analysis of the effects of soy protein intake on serum lipids. *The New England Journal of Medicine*, *333*, 276–282.
Bahri, D., Gusko, A., Hamm, M., Kasper, H., Klor, H. U., Neuberger, D., & Singer, P., (2002). Significance and recommended dietary intake of long-chain omega-3 fatty acids-A consensus statement of the omega-3 working group. *Ernahrungs-Umschau*, *49*(3), 94–98.
Beaulieu, J., Dupont, C., & Lemieux, P., (2006). Whey proteins and peptides: Beneficial effects on immune health. *Therapy*, *3*(1), 69–78.
Berge, R., (2005). *Fish Protein Hydrolysates*. Patent number WO 2005002605 A1.
Bierenbaum, M. L., Reichstein, R., & Watkins, T. R., (1993). Reducing atherogenic risk in hyperlipemic humans with flaxseed supplementation: A preliminary report. *Journal of the American College of Nutrition*, *12*(5), 501–504.
Bigliardi, B., & Galati, F., (2013). Innovation trends in the food industry: The case of functional foods. *Trends in Food Science & Technology*, *31*(2), 118–129.
Block, G., Patterson, B., & Subar, A., (1992). Fruit, vegetables, and cancer prevention: A review of the epidemiological evidence. *Nutrition and Cancer*, *18*, 1–29.
Blusztajn, J. K., (1998). Choline, a vital amine. *Science*, *281*, 794–795.
Brown, C. E., (1981). Interactions among carnosine, anserine, ophidine and copper in biochemical adaptation. *Journal of Theoretical Biology*, *88*(2), 245–256.
Chobert, J. M., El-Zahar, K., Sitohy, M., Dalgalarrondo, M., Métro, F., Choiset, Y., & Haertlé, T., (2005). Angiotensin I-converting enzyme (ACE)-inhibitory activity of tryptic peptides of ovine beta-lactoglobulin and of milk yogurts obtained by using different starters. *Le Lait.*, *85*(3), 141–152.
Clinton, S. K., (1998). Lycopene: Chemistry, biology, and implications for human health and disease. *Nutrition Reviews*, *56*(2), 35–51.
Creasy, L. L., & Coffee, M., (1988). Phytoalexin production potential of grape berries. *Journal of the American Society for Horticultural Science (USA)*, *113*, 230–234.
Crowell, P. L., (1997). Monoterpenes in breast cancer chemoprevention. *Breast Cancer Research and Treatment*, *46*(2/3), 191–197.
CSIRO human nutrition, (2004). *Functional Foods*. Australia: CSIRO Human nutrition.
De-Leo, F., Panarese, S., Gallerani, R., & Ceci, L. R., (2009). Angiotensin converting enzyme (ACE) inhibitory peptides. *Current Pharmaceutical Design*, *15*(31), 3622–3643.
Di Mascio, P., Kaiser, S., & Sies, H., (1989). Lycopene as the most efficient biological carotenoid singlet oxygen quencher. *Archives of Biochemistry and Biophysics*, *274*(2), 532–538.
Dietitians of Canada (DC), (2010). *What are Functional Foods and Nutraceuticals*.

Dietrich, C. P., Farias, G. G. M., Deabreu, L. R. D., Leite, E. L., Da Silva, L. F., & Nader, H. B., (1995). A new approach for the characterization of polysaccharides from algae: Presence of four main acidic polysaccharides in three species of the class Phaeophyceae. *Plant Science, 108,* 143–153.

Donovan, S. M., (2006). Role of human milk components in gastrointestinal development: Current knowledge and future needs. *The Journal of Pediatrics, 149*(5), S49–S61.

Dreosti, I. E., Wargovich, M. J., & Yang, C. S., (1997). Inhibition of carcinogenesis by tea: The evidence from experimental studies. *Critical Reviews in Food Science & Nutrition, 37*(8), 761–770.

Farnworth, E. R., (2006). Kefir–a complex probiotic. *Food Science and Technology Bulletin: Fu, 2*(1), 1–17.

Fernández-García, E., Carvajal-Lérida, I., Jarén-Galán, M., Garrido-Fernández, J., Pérez-Gálvez, A., & Hornero-Méndez, D., (2012). Carotenoids bioavailability from foods: From plant pigments to efficient biological activities. *Food Research International, 46*(2), 438–450.

Fitzgerald, R. J., & Murray, B. A., (2006). Bioactive peptides and lactic fermentation. *International Journal of Dairy Technology, 59*(2), 118–125.

Food and Agricultural Organization of the United Nations, (2007). *Report on Functional Foods.*

Food and Nutrition Board, (1998). *Institute of Medicine.* Dietary Reference Intakes for thiamin, riboflavin, niacin, vitamin B6, folate, vitamin B12, pantothenic acid, biotin, and choline. Washington, DC: National Academy Press.

Food and Nutrition Board, Institute of Medicine, National Academy of Sciences, (1994). In: Thomas, P. R., & Earl, R., (eds.), *Opportunities in the Nutrition and Food Sciences.* Washington, DC: National Academy Press.

Frankel, E. N., Kanner, J., German, J. B., Parks, E., & Kinsella, J. E., (1993). Inhibition of oxidation of human low-density lipoprotein by phenolic substances in red wine. *The Lancet, 341,* 454–457.

Gerhauser, C., (2008). Cancer chemopreventive potential of apples, apple juice, and apple components. *Planta Medica, 74*(13), 1608–1624.

Gerster, H., (1997). The potential role of lycopene for human health. *Journal of the American College of Nutrition, 16*(2), 109–126.

Gibson, G. R., (2004). From probiotics to prebiotics and a healthy digestive system. *Journal of Food Science, 69*(5), M141–M143.

Giovannucci, E., Ascherio, A., Rimm, E. B., Stampfer, M. J., Colditz, G. A., & Willett, W. C., (1995). Intake of carotenoids and retinol in relation to risk of prostate cancer. *JNCI Journal of the National Cancer Institute, 87*(23), 1767–1776.

Gormley, T. R., (2006). Fish as a functional food. *Food Science and Technology, 20*(3), 25–28.

Ha, Y. L., Grimm, N. K., & Pariza, M. W., (1987). Anticarcinogens from fried ground beef: Heat-altered derivatives of linoleic acid. *Carcinogenesis, 8*(12), 1881–1887.

Haque, E., & Chand, R., (2008). Antihypertensive and antimicrobial bioactive peptides from milk proteins. *European Food Research and Technology, 227*(1), 7–15.

Harbowy, M. E., Balentine, D. A., Davies, A. P., & Cai, Y., (1997). Tea chemistry. *Critical Reviews in Plant Sciences, 16*(5), 415–480.

Hasegawa, S., & Miyake, M., (1996). Biochemistry and biological functions of citrus limonoids. *Food Reviews International, 12*(4), 413–435.

Hasler, C. M., (1998). Functional foods: Their role in disease prevention and health promotion. *Food Technology, 52,* 57–62.

Hasler, C. M., (2000). The changing face of functional foods. *Journal of the American College of Nutrition, 19*(5), 499S–506S.

Health Canada, (1998). *Policy Paper–Nutraceuticals/Functional Foods and Health Claims on Foods*.

Hecht, S. S., (1995). Chemoprevention by isothiocyanates. *Journal of Cellular Biochemistry*, *59*(22), 195–209.

Hemmingson, J. A., Falshaw, R., Furneaux, R. H., & Thompson, K., (2006). Structure and antiviral activity of the galactofucan sulfates extracted from *Undariapinnatifida* (Phaeophyta). *Journal of Applied Phycology, 18,* 185–193.

Hernández-Ledesma, B., Amigo, L., Recio, I., & Bartolomé, B., (2007). ACE-inhibitory and radical-scavenging activity of peptides derived from β-lactoglobulin f (19– 25). Interactions with ascorbic acid. *Journal of Agricultural and Food Chemistry*, *55*(9), 3392–3397.

Herrero, M., Del Pilar Sanchez-Camargo, A., Cifuentes, A., & Ibanez, E., (2015). Plants, seaweeds, microalgae and food by-products as natural sources of functional ingredients obtained using pressurized liquid extraction and supercritical fluid extraction. *Tr. AC. Trends in Analytical Chemistry, 71*, 26–38.

Hertog, M. G., Feskens, E. J., Kromhout, D., Hollman, P. C. H., & Katan, M. B., (1993). Dietary antioxidant flavonoids and risk of coronary heart disease: The Zutphen elderly study. *The Lancet*, *342*(8878), 1007–1011.

Hipkiss, A. R., & Brownson, C., (2000). A possible new role for the anti-aging peptide carnosine. *Cellular and Molecular Life Sciences CMLS*, *57*(5), 747–753.

Hosseini, T. A., & Shariati, M., (2009). Dunaliella biotechnology: Methods and applications. *Journal of Applied Microbiology*, *107*(1), 14–35.

Huang, T., Sinclair, A. J., Shen, L., Yang, B., & Li, D., (2009). Comparative effects of tuna oil and salmon oil on liver lipid metabolism and fatty acid concentrations in rats. *Journal of Food Lipids*, *16*(4), 436–451.

Ikeda, K., Kitamura, A., Machida, H., Watanabe, M., Negishi, H., Hiraoka, J., et al. (2003). Effect of Undariapinnatifida (Wakame) on the development of cerebrovascular diseases in stroke-prone spontaneously hypertensive rats. *Clinical and Experimental Pharmacology and Physiology, 30*, 44–48.

Ip, C., & Scimeca, J. A., (1997). Conjugated linoleic acid and linoleic acid are distinctive modulators of mammary carcinogenesis, *Nutrition and Cancer*, *27*, 131–135.

Iwasa, M., Kaito, M., Ikoma, J., Takeo, M., Imoto, I., Adachi, Y., Yamauchi, K., Koizumi, R., & Teraguchi, S., (2002). Lactoferrin inhibits hepatitis C virus viremia in chronic hepatitis C patients with high viral loads and HCV genotype 1b. *The American Journal of Gastroenterology*, *97*(3), 766.

Jaime, L., Rodríguez-Meizoso, I., Cifuentes, A., Santoyo, S., Suarez, S., Ibáñez, E., & Señorans, F. J., (2010). Pressurized liquids as an alternative process to antioxidant carotenoids' extraction from haematococcus pluvialis microalgae. *LWT-Food Science and Technology*, *43*(1), 105–112.

Jang, M., Cai, J., Udeani, G., Slowing, K. V., Thomas, C. F., Beecher, C. W. W., Fong, H. H. S., et al. (1997). Cancer chemopreventive activity of resveratrol, a natural product derived from grapes. *Science*, *275*, 218–220.

Jenssen, H., (2005). Anti-herpes simplex virus activity of lactoferrin/lactoferricin–an example of antiviral activity of antimicrobial protein/peptide. *Cellular and Molecular Life Sciences CMLS*, *62*(24), 3002–3013.

Kaito, M., Iwasa, M., Fujita, N., Kobayashi, Y., Kojima, Y., Ikoma, J., Imoto, I., Adachi, Y., Hamano, H., & Yamauchi, K., (2007). Effect of lactoferrin in patients with chronic hepatitis C: Combination therapy with interferon and ribavirin. *Journal of Gastroenterology and Hepatology*, *22*(11), 1894–1897.

Kannar, D., Mohandoss, P., Wattanapenpaiboon, N., & Wahlqvist, M. L., (1998). The prebiotic effect of garlic and inulin. *Proceedings of the Nutrition Society of Australia, 22,* 285.
Kelly, S. A., Summerbell, C. D., Brynes, A., Whittaker, V., & Frost, G., (2007). Wholegrain cereals for coronary heart disease. *The Cochrane Library.*
Khanal, R. C., & Olson, K. C., (2004). Factors affecting conjugated linoleic acid (CLA) content in milk, meat, and egg: A review. *Pakistan Journal of Nutrition, 3*(2), 82–98.
Kitts, D. D., & Weiler, K., (2003). Bioactive proteins and peptides from food sources: Applications of bioprocesses used in isolation and recovery. *Current Pharmaceutical Design, 9*(16), 1309–1323.
Klein, R., Klein, B. E., Jensen, S. C., & Meuer, S. M., (1997). The five-year incidence and progression of age-related maculopathy: The beaver dam eye study. *Ophthalmology, 104*(1), 7–21.
Kohlmeier, L., Kark, J. D., Gomez-Gracia, E., Martin, B. C., Steck, S. E., Kardinaal, A. F., Ringstad, J., Thamm, M., Masaev, V., Riemersma, R., & Martin-Moreno, J. M., (1997). Lycopene and myocardial infarction risk in the EURAMIC Study. *American Journal of Epidemiology, 146*(8), 618–626.
Kromhout, D., Bosschieter, E. B., & Coulander, C. D. L., (1985). The inverse relation between fish consumption and 20-year mortality from coronary heart disease. *New England Journal of Medicine, 312*(19), 1205–1209.
Kwon, M. J., & Nam, T. J., (2006). Porphyrin induces apoptosis-related signal pathway in AGS gastric cancer cell lines. *Life Sciences, 79,* 1956–1962.
Larsson, S. C., Bergkvist, L., & Wolk, A., (2005). High-fat dairy food and conjugated linoleic acid intakes in relation to colorectal cancer incidence in the Swedish Mammography Cohort–. *The American Journal of Clinical Nutrition, 82*(4), 894–900.
Lee, J. B., Hayashi, K., Hashimoto, M., Nakano, T., & Hayashi, T., (2004). Novel antiviral fucoidan from sporophyll of *Undariapinnatifida* (Mekabu). *Chemical and Pharmaceutical Bulletin, 52*(9), 1091–1094.
Li, Y., Elie, M., Blaner, W. S., Brandt-Rauf, P., & Ford, J., (1997). Lycopene, smoking and lung cancer. *Proceedings of the American Association for Cancer Research, 38,* 113.
MacDonald, H. B., (2000). Conjugated linoleic acid and disease prevention: A review of current knowledge. *Journal of the American College of Nutrition, 19*(2), 111S–118S.
Mann, N., (2000). Dietary lean red meat and human evolution. *European Journal of Nutrition, 39*(2), 71–79.
McIntosh, G. H., Royle, P. J., Le Leu, R. K., Regester, G. O., Johnson, M. A., Grinsted, R. L., Kenward, R. S., & Smithers, G. W., (1998). Whey proteins as functional food ingredients? *International Dairy Journal, 8*(5/6), 425–434.
Mehra, R., Marnila, P., & Korhonen, H., (2006). Milk immunoglobulins for health promotion. *International Dairy Journal, 16*(11), 1262–1271.
Meisel, H., (1998). Overview on milk protein-derived peptides. *International Dairy Journal, 8*(5/6), 363–373.
Messina, M., & Barnes, S., (1991). The role of soy products in reducing risk of cancer. *Journal of the National Cancer Institute, 83,* 541–546.
Mezzaroba, L. F. H., Carvalho, J. E., Ponezi, A. N., Antônio, M. A., Monteiro, K. M., Possenti, A., & Sgarbieri, V. C., (2006). Antiulcerative properties of bovine α-lactalbumin. *International Dairy Journal, 16*(9), 1005–1012.
Ministry of Health, Labor, and Welfare, Japan, (1991). *Foods for Specified Health Uses (FOSHU).*

Mistry, N., Drobni, P., Näslund, J., Sunkari, V. G., Jenssen, H., & Evander, M., (2007). The anti-papillomavirus activity of human and bovine lactoferricin. *Antiviral Research, 75*(3), 258–265.
Morris, M. C., Sacks, F., & Rosner, B., (1993). Does fish oil lower blood pressure? A meta-analysis of controlled trials. *Circulation, 88*(2), 523–533.
Murphy, M. G., Wright, V., Scott, J., Timmins, A., & Ackman, R. G., (1999). Dietary menhaden, seal, and corn oils differentially affect lipid and *ex vivo* eicosanoid and thiobarbituric acid-reactive substances generation in the guinea pig. *Lipids, 34*(2), 115–124.
Nagourney, R. A., (1998). Garlic: Medicinal food or nutritious medicine? *Journal of Medicinal Food, 1*(1), 13–28.
Nakagawa, T., & Yokozawa, T., (2002). Direct scavenging of nitric oxide and superoxide by green tea. *Food and Chemical Toxicology, 40*(12), 1745–1750.
National Institute of Nutrition, (2000). *Consumer Awareness of and Attitudes Towards Functional Foods, Highlights and Implications for Informing Consumer*. Leaflet.
Ness, A. R., Hughes, J., Elwood, P. C., Whitley, E., Smith, G. D., & Burr, M. L., (2002). The long-term effect of dietary advice in men with coronary disease: Follow-up of the diet and reinfarction trial (DART). *European Journal of Clinical Nutrition, 56*(6), 512–518.
Niva, M., (2007). All foods affect health: Understandings of functional foods and healthy eating among health-oriented Finns. *Appetite, 48*, 384–393.
Nys, M., & Debruyne, I., (2011). Lipids & Brain 2: A symposium on lipids and brain health. I*nform, 11,* 397–399.
Otles, S., & Cagindi, O., (2003). Kefir: A probiotic dairy-composition, nutritional and therapeutic aspects. *Pakistan Journal of Nutrition, 2*(2), 54–59.
Park, Y., K. J., Albrigh, W., Il, J. M., Storkson, M. E., & Cook, M. W., (1997). Pariza–Effect of conjugated linoleic acid on body composition in mice. *Lipids, 32*, 853–858.
Parodi, P. W., (2004). Milk fat in human nutrition. *Australian Journal of Dairy Technology, 59*(1), 3–59.
Pellegrini, A., (2003). Antimicrobial peptides from food proteins. *Current Pharmaceutical Design, 9*(16), 1225–1238.
Phipps, W. R., Martini, M. C., Lampe, J. W., Slavin, J. L., & Kurzer, M. S., (1993). Effect of flax seed ingestion on the menstrual cycle. *The Journal of Clinical Endocrinology & Metabolism, 77*(5), 1215–1219.
Plaza, M., Cifuentes, A., & Ibáñez, E., (2008). In the search of new functional food ingredients from algae. *Trends in Food Science & Technology, 19*(1), 31–39.
Potter, S. M., (1998). Soy protein and cardiovascular disease: The impact of bioactive components in soy. *Nutrition Reviews, 56*(8), 231–235.
Prates, J. M., & Mateus, C. M. R. P., (2002). Functional foods from animal sources and their physiologically active components. *Journal of Veterinary Medicine, 153*(3), 155–160.
Quirós, A., Ramos, M., Muguerza, B., Delgado, M. A., Miguel, M., Aleixandre, A., & Recio, I., (2007). Identification of novel antihypertensive peptides in milk fermented with Enterococcus faecalis. *International Dairy Journal, 17*(1), 33–41.
Rodriguez, E. B., Flavier, M. E., Rodriguez-Amaya, D. B., & Amaya-Farfán, J., (2006). Phytochemicals and functional foods. Current situation and prospect for developing countries. *Food and Nutrition Security, 13*(1), 1–22.
Rokka, T., Syväoja, E. L., Tuominen, J., & Korhonen, H. J., (1997). Release of bioactive peptides by enzymatic proteolysis of Lactobacillus GG fermented UHT-milk. *Milchwissenschaft, 52*, 675–678.

Sanchez-Machado, D. I., Lopez-Cervantes, J., Lopez-Hernandez, J., & Paseiro-Losada, P., (2004). Fatty acids, total lipid, protein and ash contents of processed edible seaweeds. *Food Chemistry, 85*(3), 439–444.

Schuster, G. S., Dirksen, T. R., Ciarlone, A. E., Burnett, G. W., Reynolds, M. T., & Lankford, M. T., (1980). Anticaries and antiplaque potential of free-fatty acids in vitro and *in vivo*. *Pharmacology and Therapeutics in Dentistry*, *5*(1/2), 25–33.

Setchell, K. D. R., Lawson, A. M., Borriello, S. P., Harkness, R., Gordon, H., Morgan, D. M. L., Kirk, D. N., Adlercreutz, H., Anderson, L. C., & Axelson, M., (1981). Lignan formation in man–microbial involvement and possible roles in relation to cancer. *The Lancet*, *318*(8236), 4–7.

Simopoulos, A. P., (1991). Omega-3 fatty acids in health and disease and in growth and development. *The American Journal of Clinical Nutrition, 54*(3), 438–463.

Simopoulos, A. P., (2004). Omega-3 essential fatty acid ratio and chronic diseases. *Food Review International, 20*, 77–90.

Singh, T. K., Fox, P. F., & Healy, Á., (1997). Isolation and identification of further peptides in the diafiltration retentate of the water-soluble fraction of cheddar cheese. *Journal of Dairy Research*, *64*(3), 433–443.

Srivastava, K. C., Bordia, A., & Verna, S. K., (1995). Garlic (*Allium sativum*) for disease prevention. *South African Journal of Science*, *91,* 68–68.

Steinmetz, K. A., & Potter, J. D., (1991). Vegetables, fruit and cancer II. Mechanisms. *Cancer Causes Control*, *2*, 427–442.

Tan, Z. S., Harris, W. S., Beiser, A. S., Au, R., Himali, J. J., Debette, S., & Robins, S. J., (2012). Red blood cell omega-3 fatty acid levels and markers of accelerated brain aging. *Neurology*, *78*(9), 658–664.

Thompson, K. D., & Dragar, C., (2004). Antiviral activity of *Undariapinnatifida* against herpes simplex virus. *Phytotherapy Research, 18*(7), 551–555.

Thompson, L. U., (1995). Flaxseed, lignans, and cancer. In: Cunnane, S., & Thompson, L. U., (eds.), *Flaxseed in Human Nutrition* (pp. 219–236). AOCS Press, Champaign, IL.

Thompson, L. U., Robb, P., Serraino, M., & Cheung, F., (1991). Mammalian lignan production from various foods. *Nutrition and Cancer*, *16*, 43–52.

Thormar, H., & Hilmarsson, H., (2007). The role of microbicidal lipids in host defense against pathogens and their potential as therapeutic agents. *Chemistry and Physics of Lipids*, *150*(1), 1–11.

Thormar, H., Isaacs, C. E., Kim, K., & Brown, H. R., (1994). Inactivation of visna virus and other enveloped viruses by free fatty acids and monoglycerides. *Annals of the New York Academy of Sciences*, *724*(1), 465–471.

U.S. Department of Agriculture, U.S. Department of Health and Human Services, (2010). *Dietary Guidelines for Americans* (7th edn.), U.S. Government Printing Office, Washington, DC.

Vercruysse, L., Van Camp, J., & Smagghe, G., (2005). ACE inhibitory peptides from enzymatic hydrolysates of animal muscle protein: A review. *Journal of Agricultural and Food Chemistry,* *53*(21), 2244–2245.

Verhoeven, D. T., Verhagen, H., Goldbohm, R. A., Van den Brandt, P. A., & Van Poppel, G., (1997). A review of mechanisms underlying anticarcinogenicity by brassica vegetables. *Chemical-Biological Interactions*, *103*(2), 79–129.

Vesper, H., Schmelz, E. M., Nikolova-Karakashian, M. N., Dillehay, D. L., Lynch, D. V., & Merrill Jr, A. H., (1999). Sphingolipids in food and the emerging importance of sphingolipids to nutrition. *The Journal of Nutrition*, *129*(7), 1239–1250.

Virtanen, J. K., Siscovick, D. S., Lemaitre, R. N., Longstreth, W. T., Spiegelman, D., Rimm, E. B., King, I. B., & Mozaffarian, D., (2013). Circulating omega-3 polyunsaturated fatty acids and subclinical brain abnormalities on MRI in older adults: The cardiovascular health study. *Journal of the American Heart Association, 2*(5), e000305.

Weisburger, J. H., (1998). Evaluation of the evidence on the role of tomato products in disease prevention. *Proceedings of the Society for Experimental Biology and Medicine, 218*(2), 140–143.

Wolk, A., Larsson, S. C., Johansson, J. E., & Ekman, P., (2006). Long-term fatty fish consumption and renal cell carcinoma incidence in women. *Jama, 296*(11), 1371–1376.

Yan, L., Yee, J. A., Li, D., McGuire, M. H., & Thompson, L. U., (1998). Dietary flaxseed supplementation and experimental metastasis of melanoma cells in mice. *Cancer Letters, 124*, 181–186.

Zafra-Stone, S., Yasmin, T., Bagchi, M., Chatterjee, A., Vinson, J. A., & Bagchi, D., (2007). Berry anthocyanins as novel antioxidants in human health and disease prevention. *Molecular Nutrition & Food Research, 51*(6), 675–683.

CHAPTER 3

Recent Trend on Dietary Natural Products for Prevention and Treatment of Cancer

MONOJ KUMAR DAS,[1] NEELU SINGH,[2] PITAMBAR BAISHYA,[1] PAULRAJ RAJAMANI,[2] SANKAR CHANDRA DEKA,[3] and ANAND RAMTEKE[1]

[1]*Cancer Genetics and Chemoprevention Research Group, Department of Molecular Biology and Biotechnology, Tezpur University, Napaam, Tezpur 784028, Assam, India*

[2]*School of Environmental Sciences, Jawaharlal Nehru University, New Delhi, 10067, India*

[3]*Department of Food Processing Technology, Tezpur University, Napaam, Tezpur 784028, Assam, India*

ABSTRACT

Cancer cells are abnormal cells, characterized by uncontrolled growth, their potential to escape the normal rules of cell division, where old and damaged cells continue to survive by altering cellular, genetic, and epigenetic machinery forming malignant mass. Autonomous cell division in cancer cells ignoring normal signals results in uncontrolled growth and proliferation that leads to tumor spreading and causes the cancer death. Statistical data reveals that 20% of the death in the world is due to cancer and rigorous efforts are being made by the researchers worldwide to discover drug to fight against the deadly disease. However, various types of drugs and therapies like radiotherapy, chemotherapy are already there but they end up with

---

Author Contribution: AR, MKD, SKD, contributed toward conceptualization, planning and writing the paper. NS contributed toward data collection and writing the paper.

Conflict of Interests: The authors declare no conflict of interests.

painful side effects and are also costly therefore, usage of natural herbal medicine is better alternative over the synthetic one. Cancer is a multifactorial disease, therefore a multitarget approach is needed to face the complex cancer biology which incorporate combined use of different natural anti-cancer agents able to target synergistically multiple signaling pathways involved in carcinogenesis, including angiogenesis and metastasis. Phytochemicals obtained from vegetables, fruits, spices, tea, herbs, and medicinal plants contain phenolic, flavonoids, carotenoids and terpenoids. These plants derived phytochemicals exhibit not only anti-cancer property but are also relatively less toxic/nontoxic offering minimal side effects. Therefore, the increasing interest is due to its easy availability and expensive. Also, past two decades have witnessed that 25% of crude drugs are derived from plants. Recent studies have reported that high intake of fruits, nuts, grains and cereals, which further confers there protective role against cancer. These herbal formulations target various major pathways by inducing apoptosis, cell cycle arrest, inhibition of cell proliferation and tumor angiogenesis, etc. However, there are certain major limitations which hinder the frequent usage of herbal medicines including lack of authentic sources of medicinal herbs, proper protocol for the herbal preparation, unknown effective dosages and lack of preclinical data.

3.1 INTRODUCTION

Across the world cancer has proven to be one of the leading cause of death (Fresco et al., 2006) which is further potentiated with the risk factors like obesity, smoking, reduced physical activity and altered reproductive behavior of the population (Torre et al., 2016). Although most of the chemotherapeutic agents which are cytotoxic to cancer cells are natural in origin and offer major limitations due to toxicity to the healthy cells (Lefranc et al., 2017). Apart from these, there are different mechanisms responsible for the tumor plasticity such as alteration in the target, hyperactivation of unconventional pathways, reactivation of the targeted pathway and cross-talk with the microenvironment (Ramos and Bentires, 2015). Heterogeneity of the cancer tissue is also one of the causal factors for the ineffectiveness of various chemotherapeutic agents (Ramos and Bentires, 2015; Lefranc et al., 2017). This heterogeneity in cancer cells introduces a complex network with various regulatory loops and superfluous pathways, responsible for the recurrent failure of the one-drug-one-target paradigm of treatment introducing drug resistance in patients (Dorel et al., 2015). In addition to this, poor prognosis and severe side effects of the conventional therapies like chemotherapy, radiotherapy and surgery offer limitations in their usage and results in surge for complementary and alternative drugs, which are not only toxic to the target cancer cells but offers negligible toxicity to the healthy cells. This further sails the boat of anti-cancer drug towards the island of natural plant derived products which are the reservoir of anti-cancer properties offering less toxicity to the normal cells (Wang et al., 2012). Various studies reported that, plants derived natural compounds play significant role to combat the robustness of the cell signaling pathway, and further combination therapeutic approach could be proposed in treatment of certain types of cancer (Dorel et al., 2015).

In addition to the conventional treatment, consumption of certain foods can further help in fighting against certain category of cancer (Lefranc et al., 2017). It has been already estimated that nearly 30–40% of all the cancer types can be forbidden by balanced diet and lifestyle alone (Donaldson, 2004). Normally diet rich with physiologically efficient components are found to be significantly efficient in preventing the recurrence of breast cancer (Pal et al., 2012). Therefore, etiologically role of diet in cancer prevention is widely accepted, a diet rich in saturated fat, alcohol intake <40 g/day shows strong correlations with various types of cancer (colon, breast, prostate, pharynx, oral cavity, etc.) incidences, whereas in those diet enriched with high intake of vegetables, fruits and grains showed reduced risk of cancer (Pal et al., 2012). This chapter is focused on those dietary compounds such as herbs, vegetables, fruits and spices that have demonstrated a promising potential as anticancer agents that includes different polyphenols and flavonoids and their role on different molecular targets in the field of cancer biology.

3.2 ROLE OF DIETARY PHYTOCHEMICALS OVER EXISTING DRUGS ON CANCER

The main target of the current available anticancer drugs are rapid dividing cells however, apart from cancer cells there are other cells also which undergo rapid division under normal conditions such as hair follicle cells, digestive tract cells and bone marrow cells this causes their further exposure of the introduced anticancer drugs to these non-targeted organs, which further requires large quantity of drugs to be administered leaving behind a series of side effects such as cardiotoxicity, hair loss, mucositis , myelosuppression, immunosuppression and neurotoxicity. Another encountered limitation of these drugs is the development of resistance of the cancer cells due to their ability to undergo mutations. These unsatisfactory outcomes of the drug used during chemotherapy introduce not only side effects, but are also inefficient in treatment and are also not economic. Therefore, there is an urgent need to search for the ideal drug which is cytotoxic to the cancer cells, whereas normal cells remains unaffected and are also cost effective. In this line bioactive compounds reported in various plants show an immense potential in anti-cancer effect and are biocompatible due to natural in origin, therefore act as a light house in the field of chemotherapeutics (Singh et al., 2016).

Recently, pre-clinical research have provided emphasis on the need for cancer prevention and new effective treatment strategies with minimal side effects, increased efficacy, less expensive and their ability to selectively target cancer cells on multiple cancer-related biological pathways. Plants are the integral part of the traditional indigenous healthcare system and are becoming concrete source of new drug discovery, there is scientific evidence that various dietary herbs and plant products prevents cancer and have anticancer potentials that fulfill the above-mentioned limitation of an effective drug for better management of cancer prevention and therapeutic use. Dietary phytochemicals are not limited to a particular part of plant but are reported to be found in other areas too, like leaves, roots, stem, bark and flower. This search is continuing to grow due to being blessed with highly rich flora and can be easily traced by the various research publications. Vegetables and fruits are excellent sources of cancer preventive substances. The most exciting findings have been achieved with antioxidant vitamins and their precursors, which are found in dark, green leafy vegetables and yellow/orange fruit and vegetables as mentioned in Table 3.1. And Table 3.2 shows the list of commonly occurring herbs and reported phytochemicals and their anticancer efficacy in the *in-vitro* and *in-vivo* system.

Spices have been extensively used as food flavoring agents and in traditional medicines for thousands of years as they include many bioactive compounds and have a lot of beneficial health effects (Zheng et al.). Studies have recognized the antioxidant, anti-inflammatory and immunomodulatory properties of spices, which might be related to effective management of several cancers. The biological effect of spices may occur from their ability to modulate a number of cellular processes, drug metabolism, cell division, differentiation, inducing apoptosis and inhibiting proliferation (Kaefer et al.). The low toxicity may make them predominantly useful as a subtle personal dietary change that may decrease the risk of numerous diseases. Table 3.3 shows the list of commonly occurring spices and reported phytochemicals and their anticancer efficacy in the *in-vitro* and *in-vivo* system.

TABLE 3.1 Examples of Herbs and Vegetables, Their Bioactive Compounds and Anticancer Activity

Active Constituent	Source	Active Against	References
PodophyllinPodophyllotoxin	*Podophyllum hexandrum* (Berberidaceae)	Leukemia, bronchogenic carcinoma, testis, and ovary cancer	Álvarez- et al., 2015
Nab-Paclitaxel	*Taxus brevifolia* (Taxaceae) Bark	Breast and ovary (both in vitro and in vivo)	Colombo et al., 2013
S-allylcysteine and S-allylmercapto-L-cysteine, alliin, allicin, methylallyltrisulfide	*Allium sativum* (Liliaceae)	Cancer of breast, skin, bladder, stomach, colon, esophagus, and the lung	Balasubramani et al., 2015
Quercetin, Doxorubicin, spinaninea, rutnine,	*Ziziphus spina-christi* (Rhamnaceae), leaves, and Flowers	Breast cancer and Lung cancer	Jaradat et al., 2016
Artemisinin	*Artemisia annua* (Asteraceae) whole plant	Breast, liver, and pancreatic cancer (both in vitro and in vivo)	Efferth, 2017
Gallic acid	*Leea indica* (Vitaceae) Leaves	Ehrlich ascites carcinoma (both in vitro and in vivo)	Raihan et al., 2016
Emodin, and Aloesin,	*Aloe vera* (Liliaceae) Whole plant	Anti-angiogenic activity (In vitro)	Rahman et al., 2017
Neferine	*Nelumbo nucifera* (Nelumbonaceae) Embryos	Liver cancer (in vitro)	Yoon et al., 2013
Charantin,	*Momordica charantia* (Cucurbitaceae) Roots and Leaves	Cancers of breast, and colon (in vitro)	Weng et al., 2013
Betulinic acid	*Betula utilis* (Betulaceae) Bark	Melanomas (in vitro)	Król et al., 2015
Isoliquiritigenin	*Glycyrrhiza uralensis* (Fabaceae) Roots	Lung cancer (in vitro)	Jung et al., 2014
(-)-epigallocatechin gallate	*Camellia sinensis* (Theaceae)	Prostate, brain, cervical, and bladder cancer (in vivo)	Ng et al., 2003

TABLE 3.1 *(Continued)*

Active Constituent	Source	Active Against	References
Luteolin	*Capsicum annuum* (Solanaceae) Peppe	Colorectal cancer (in vivo and in vitro)	Osman et al., 2015
Asiatic acid	*Centella asiatica* (Apiaceae) Leaves	Glioblastoma, Melanoma, and breast cancer (in vivo)	Arpita and Navneeta, 2017
Tylophorine	*Tylophora indica* (Asclepiadaceae) Leaves	Breast cancer (in vivo)	Gupta et al., 2010
Chrysoplenetin	Vitexnegundo (Lamiaceae) Fruits	Pancreatic cancer (in vitro)	Awale et al., 2011
Cardenolides, Asclepin	*Asclepias, Curassavica* (Asclepiadaceae) Aerial parts	Liver cancer (in vitro)	Li et al., 2009
5-hydroxy-6,7,8,30,40-penta methoxyflavone	Citrus limon (Rutaceae) Fruits	Human colon cancer (in vitro)	Hirata et al., 2009
Cannabisin, Berberine	*Berberis vulgaris* (Berberidaceae) stem bark., Root	Prostate, breast, and liver cancer (in vivo)	Pierpaoli et al., 2015
(þ)-higenamine,(þ)-argenaxine, argenaxine, angoline	Argemone Mexicana (Papaveraceae) Leaves	Cancer of Breast and gall bladder (in vivo)	Brahmachari et al., 2013
Solasonine, Solamargine,	*Solanum nigrum* (Solanaceae) leaves	Lung, breast, liver, and skin cancer	Al Sinani et al., 2016
Kaempferolgalactoside	*Bauhinia variegate* (Fabaceae) Flowers	Lung, liver, and breast (in vivo)	Tu et al., 2016
Butrin, (7,30,40-trihydroxyflavanone-7,30-diglucoside)	*Butea monosperma* (Fabaceae) Flower	Liver cancer (in vivo and in vitro)	Choedon et al., 2010
Saffron	*Saffron crocus* (Iridaceae) dry stigmas	Cancer if lung, liver, and pancreas (in vitro)	Ververidis et al., 2007
Bilobalide, EGb	*Ginkgo biloba* (Ginkgoaceae) Leaves	Colon cancer (in vivo)	Suzuki et al., 2004

TABLE 3.2 Examples of Spices, Their Bioactive Compounds, and Anticancer Activity

Active Constituent	Source	Cancer	References
Galangin	Galangal; rhizome (Zingiberaceae)	Liver, Colorectum	Su et al., 2013; Ha et al., 2013
Eugenol	Clove; flower buds (Myrtaceae)	Liver, Breast, Colorectum, Cervix	Iwano et al., 2013; Al-Sharif et al., 2013; Liu et al., 2014; Hussain et al., 2011
Carnosic acid	Rosemary; leaves (Lamiaceae)	Liver, Breast, Colorectum, Prostate	Gao et al., 2015; Gonzalez-Vallinas et al., 2014; Park et al., 2014; Petiwala et al., 2014
Curcumin	Turmeric; rhizome (Zingiberaceae)	Lung, Liver, Breast, Stomach, Colorectum, Cervix, Prostate	Yang et al., 2012; Kadasa et al., 2015; Strofer et al., 2011; Liu et al., 2014; Carroll et al., 2011; Lewinska et al., 2014; Shah et al., 2012
Thiacremonone, Diallyldisulfide, Diallylsulfide, Diallyltrisulfide, S-allyl mercaptocysteine	Garlic; bulb (Amaryllidaceae)	Lung, Breast, Stomach, Colorectum	Jo et al., 2014; Xiao et al., 2014; Ling et al., 2014; Tung et al., 2015
Capsaicin	Red chili Pepper; fruit (Solanaceae)	Lung, Breast, Stomach, Colorectum, Prostate	Chakraborty et al., 2014; Wu et al., 2014; Meral et al., 2014; Clark et al., 2015; Venier et al., 2015
Crocetin	Saffron; flower (Iridaceae)	Lung, Breast, Stomach, Colorectum, Prostate	Samarghandian et al., 2011; Chryssanthi et al., 2011; Bathaie et al., 2013; Amin et al., 2015; D'Alessandro et al., 2013
Thymoquinone	Black cumin; seed (Ranunculaceae)	Lung, Liver, Breast, Colorectum, Cervix	Al-Sheddi et al., 2014; Raghunandhakumar et al., 2013; Woo et al., 2013; Jrah-Harzallah et al., 2013; Ichwan et al., 2014

TABLE 3.2 *(Continued)*

Active Constituent	Source	Cancer	References
6-Shogaol, 6-gingerol, and 6-paradol	Ginger; rhizome (Zingiberaceae)	Lung, Breast, Colorectum, Prostate	Hsu et al., 2015; Lua et al., 2015; Radhakrishnan et al., 2014; Karna et al., 2012
Piperine	Black pepper; peppercorns (Piperaceae)	Breast, Colorectum, Prostate	Wu et al., 2014; Yaffe et al., 2013; Samykutty et al., 2013
Ethyl acetate extract	Coriander; Seed and leaf (Umbellifers)	Breast	Tang et al., 2013
6-MITC	Wasabi; root (Brassicaceae)	Breast	Fuke; et.al., 2014
Se-Methyl-L-selenocysteine	Onion; bulb (Amaryllidaceae)	Colorectum	Tung; et.al., 2015
Scallion extract	Scallion, leaf (Amaryllidaceae)	Colorectum	Arulselvan et al., 2012
Cinnamaldehyde	Cinnamon; bark (Lauraceae)	Colorectum	Yu et al., 2014
Carvacrol	Oregano; dried leaves, flowering tops (Lamiaceae)	Colorectum	Fan et al., 2015

TABLE 3.3 Anticancerous Phytochemicals Present in Fruits and Vegetables

Active Constituent	Source	Cancer	References
Procyanidins (B1 and B2), epicatechin, Catechin, phloretin, phloretin-2′-O-glycoside, quercetin, quercetin-3-O-glycoside, chlorogenic acids, and caffeic acid	Apple	Colon and Stomach	Serra et al., 2010
Kaempferol, p-coumaric acid, quercetin, and Artepillin C	Brazilian green propolis	Prostate	Szliszka el al., 2011
Allin	Garlic	Breast cancer	Sabnis, 2006
Quercetin 3-rhamnoside, quercetin 3-glucoside, Ellagic acid, cyanidin 3-glucoside, kaempferol 3-glucoside, cyanidin 3-xyloside, pelargonidin 3-glucoside	Blackberry	Leukemia and Lung cancer	Wang et al., 2008
Protocatechuic, p-hydroxybenzoic acids, Gallic, myricetin, kaempferol, quercetin	Cowpeas	Breast	Gutiérrez-Uribe et al., 2011
Caffeic, gallic, Catechin, and ferulic acids	Grape seeds	Colon	Lutterodt et al., 2011
Tannic acid	Black sesame	Colon	Kim et al., 2009
Bullatacin	Annona squamosa (Annonaceae) Seed	Liver (In vitro)	Biba et al., 2013
Cardol, Cardanol	Thai propolis	Breast, Liver, Colon, Stomach, Lung.	Tang, et al., 2013
Procyanidin A2, epicatechin,	Litchi (Sapindaceae) fruit pericarp	Breast cancer (both in vitro and in vivo)	Wang et al., 2006
Anthocyanins, monocaffeoylquinic, dicaffeoylquinic, tricaffeoylquinic acids	Sweet potato	Prostate cancer (both in vitro and in vivo)	Karna et al., 2011
Lupeol	Mango	Prostate cancer (both in vitro and in vivo)	Prasad et al., 2007

TABLE 3.3 *(Continued)*

Active Constituent	Source	Cancer	References
Cinnamic acid	Cinnamon oil, honeybee propolis		Lozynskyi et al., 2014
Chlorogenic and neo-chlorogenic acids, quercetin	Peach	Breast cancer (in vitro)	Noratto, 2009
Anthocyanins	Blueberry	Breast cancer (in vitro)	Faria et al., 2010
Skimmianine	Aegle marmelos (Rutaceae) Stem bark	Leukemia, breast cancer, malignant melanoma, malignant lymphoma	Mukhija et al., 2015

3.3 PHYTOCONSTITUENTS THAT KILL CANCER CELLS

Plants show a great diversity of phytocompounds. Phenolic compounds from medicinal herbs and dietary plants include phenolic acids, flavonoids, tannins, stilbenes, curcuminoids, coumarins, lignans, quinones, etc., that contribute to various pharmacological functions and have attracted considerable interest because of their potential in cancer chemo-prevention (e.g., anti-oxidant, anticarcinogenic or antimutagenic and anti-inflammatory effects) and also contribute to induction of apoptosis by arresting cell cycle, regulating carcinogen metabolism and oncogenesis expression, inhibiting DNA binding and cell adhesion, migration, proliferation or differentiation, and blocking signaling pathways (Wu-Yang Huang et al. 2009).

3.3.1 POLYPHENOLS

Phenolics are compounds possessing one or more aromatic rings bearing one or more hydroxyl groups with over 8,000 structural variants and are generally categorized as phenolic acids (Fresco et al., 2016). These polyphenolic compounds with antioxidants properties are very common, are widely distributed in the plant kingdom and belong to the important groups of a secondary metabolite of plants. The main sources of phenolics are dietary fruits, vegetables, herbs and spices.

3.3.2 FLAVONOIDS

Flavonoids are also from polyphenolic compounds and constitute a large family of secondary plant metabolites which are known for their high health benefits. The most common flavones are luteolin, apigenin, baicalein, chrysin, and their glycosides (e.g., apigetrin, vitexin, and baicalin), mainly distributed in many dietary plants (e.g., fruits, vegetables, grains, etc.). Flavonoids are diverse in chemical structure and are distributed to nearly all types of plants. Structurally flavonoids consist of two benzene rings bridged by three-carbon chain forming an oxygenated heterocycle (C6-C3-C6) and variations arises due to numerous arrangement of multiple hydroxyl, methoxyl, and O-glycoside groups substituents on the basic benzo-pyrone moiety (Hodek et al., 2002). However, they are present in food, usually as O-glycosides along with sugars bound at C3 site (Kühnau 1976; Chahar et al., 2011). Along with various properties like free radical scavenging activities, anti-oxidant and anti-inflammatory they also showed anticancerous activity (Manthey et al., 2001). It has already been reported that population consuming soy (active constituent is genistein) have shown a low risk of various types of cancers like colon, prostate, and breast cancer (Chahar et al., 2011) which is well documented in various studies such as 2-de-O-DMA exhibit growth inhibitory effect on various cancer cell lines such as human breast cancer cell lines (MCF-7 cells and MDA-MB-468), human prostate cancer (LNCaP and DU145) (Martin et al., 1978; Constantinou et al., 1998), etc. Due to these active constituents in dietary herbs, vegetables, fruits and spices have shown low risk of cancer.

3.4 MOLECULAR TARGETS OF PHYTOCHEMICALS FOR CANCER PREVENTION

These phytochemicals derived from dietary foods and vegetables have been reported to interfere with the 'biochemical' and 'molecular' mechanisms of carcinogenesis process and thereby reduce the risks. Carcinogenesis is a multistep and highly complex process, exhibiting distinct variations at both cellular as well as molecular level, which involves a series of epigenetics and genetics alteration affecting oncogenes and tumor suppressor genes. Cancer development occurs broadly at three steps, which begin with initiation followed by promotion and progression, where conversion of preneoplastic to neoplastic takes place with increased potential of metastasis, invasiveness, and angiogenesis (Chahar et al., 2011). Tumor initiation is a rapid and irreversible process, that involves an initial exposure to carcinogenic agents and finally its possible metabolic derivatives covalently interact with target-cell DNA leading to genotoxic damage. Phytochemical can prevent these multi step processes by exerting anti-proliferation, anti-inflammatory, anti-oxidative stress effects by modulating different molecular mechanism of action as well as molecular targets such as inhibition of cell cycle, induction of apoptosis, regulation of angiogenesis, inhibition of cyclooxygenase (COX)-2, signaling transducers and activator of transcription-3 (STAT3) inhibition, and down-regulation of nuclear factor-kappaB (NF-κB), activator protein-1, and mitogenic-activated protein kinases (MAPKs). The understanding of molecular mechanism of a specific plant derived compounds against a particular type of cancer will leads to invention of novel drugs and drugs targets for therapeutic intervention.

3.4.1 ANTI-PROLIFERATION ACTIVITY

Major catalysts of the tumor promotion and progression are ROS and growth promoting oxidants and therefore anti-proliferation may engross the hindrance of prooxidant process. It has been observed that Flavonoids are efficient in inhibiting COX or LOX55 (Mutoh et al., 2000), xanthine oxidase thus, inferring with tumor cell proliferation (Chang et al., 1993) additionally inhibition of polyamine biosynthesis can further contribute to the anti-proliferative activities of flavonoids.

3.4.2 CELL CYCLE ARREST

Cell growth and proliferation are highly regulated events in normal cells and dysregulation of cell cycle can leads to uncontrolled proliferation, contributes to the malignated phenotypes of tumor cells. Mitogenic signal initiation leads the cells to enter in organized and regulated cell cycle where cyclin-dependent kinases (CDKs) are the key regulators, therefore any alteration or deregulation in CDKs can further hamper the cell cycle progression. Due to mutation of CDK genes or CDK inhibitor genes leads to hyperactivation of CDKs or phosphorylated mediated inactivation of tumor suppressor gene Rb, further such modulation leads to aberrations in cell cycle control which are pathogenic trait of neoplasia. Therefore, modulators or inhibitors fetch more interest in therapeutic agents application in cancer treatment (Senderowicz, 1999, 2001) Various studies reported that cell cycle arrest have been observed at the boundaries of both the checkpoints G1/S and G2/M due to the modulatory effect of various flavonoids such as genistein, silymarin, quercetin, luteolin, daidzein, kaempferol, epigallocatechin 3-gallate and apigenin (Zi et al., 1998; Choi et al., 2001; Casagrande and Darbon, 2001).

3.4.3 ANTI-INFLAMMATORY EFFECT

Inflammation and cancer, exhibit a functional relationship, and chronic inflammation is the potential site at which cancer originates. During this, several classes of irritants along with tissue injury ensures inflammation resulting in augmentation of cell proliferation (Balkwill and Mantovani, 2001). Therefore, these proliferating cells with mutagenic assault/sustained damaged DNA and their prolong proliferation in microenvironments enriched with growth and inflammatory factors further continues their growth (Desmoulière, 2008). There are various plant polyphenols identified, which exhibit the potential to alter the activity and production of inflammatory molecules such as Curcumin can suppress the TNF expression by affecting the methylation of a TNF promoter. Similar observations has been observed in the subjects treated with resveratrol found in various fruits including red wine where suppressed level of TNF-α, IL-6, and C-reactive protein was shown, however, these changes were not observed in the placebo groups. Flavopiridol isolated from *Dysoxylum binectariferum* showed significant suppression of IL-1β, IL-6 and TNF-α whereas, expression of IL-10 was enhanced in astrocytes when administered in nanoparticulate form (Ren et al., 2014). Emodin obtained from *Rheum emodi* reduces the production of IL-6, IL-8, TNF-α, prostaglandin (PGE)-2, VEGF, MMP-1 and MMP-13 under hypoxic conditions in LPS-treated synoviocytes (Ha et al., 2011). There are various other polyphenols such as naringenin, apigenin, isoliquiritigenin, kaempferol, myricetin, catechins, fisetin, escin, vitexin, xanthohumol γ-mangostin identified which modulate inflammatory molecules in both in vivo and in-vitro (Gupta et al., 2014).

3.4.4 ANTI-ANGIOGENESIS

Development of new blood vessel known as Angiogenesis, significant during various stages like fetal development, wound healing, the menstrual cycle, pregnancy and skeletal growth, however found to be common in various pathological state also such as psoriasis, diabetic retinopathy, rheumatoid arthritis and in cancer (Carmeliet and Jain, 2000; Carmeliet, 2003; Khurana et al., 2005; Chung and Ferrara, 2011). During tumorigenesis the newly formed vasculature are abnormal which further limit the supply of not only oxygen but also drug. Therefore, anti-angiogenic agents normalize the vasculature transiently to continue the supply of oxygen as well as drug (Chung and Ferrara, 2011). Among the various factors like vascular endothelial growth factor (VEGF-A) is one of characteristic factor for the stimulation of angiogenesis via its receptor VEGFR-2, therefore being targeted by various anti-angiogenic agents (Hoeben et al., 2004; Weis and Cheresh, 2011). There are various plant-derived compound present in fruits and vegetables which exhibit anti-angiogenic property like resveratrol found to reduce VEGF expression through down-regulation of HIF-1alpha (Bishayee et al., 2010) other molecular mechanism like induction of the proteasomal degradation of HIF-1alpha, inhibition of Ras/MEK/ERK and phosphatidylinositol (PI)-3K/

Akt pathways and activation of forkhead box transcription factors are some of the commonly described pathways (Cao et al., 2004; Brå-Kenhielm et al., 2001; Liu et al., 2012; Srivastava et al., 2010; Morbidelli, 2016). Curcumin (isolated from *Curcuma longa*) suppresses cancerous neovascularization though down regulation of VEGF and COX-2 (Yoysungnoen et al., 2006), Quercetin commonly found in fruits and vegetables play role in anti-angiogenesis by inhibition of (endothelial cells) ECs growth and migration (Tan et al., 2003; Morbidelli, 2016), ellagic acid demonstrated to reduce angiogenesis though suppression of HDAC6 through impaired expression of HIF-1α and VEGF and obstructed Akt activation (Kowshik et al., 2014), Ursolic acid (active component of *Rosmarinus officinalis*) shows anti-angiogenic mediator by inhibiting MMP-2, MMP-9, VEGF and nitric oxide production (Kanjoormana and Kuttan, 2010; Morbidelli, 2016).

3.4.5 INDUCTION OF APOPTOSIS

Programmed cell death or Apoptosis, involves a series of event from cell development to death, different from necrosis which involves sudden demise of a cell due to accidental or lethal injury (Bansal et al., 2012). Apoptosis is essential and highly regulated as it involves the network of interacting protease and their inhibitor in response to stimuli from outside or inside of the cell which ultimately leads to the removal of unwanted damaged cell, further any dysregulation in apoptosis plays a significant role in oncogenesis. The molecular mechanism for flavonoids mediated apoptosis is not yet clear however, various examples reveals that it include decrease of ROS, inhibition of DNA topoisomerase I/II activity, regulation of heat shock protein expression, downregulation of nuclear transcription factor kappaB (NF-κB), modulation of signaling pathways, activation of endonuclease, and suppression of Mcl-1 protein. There are certain marker whose alteration serves as a key indicator of apoptosis such as Apopain Activity, Membrane permeability is the early marker of apoptosis whereas, late marker includes nuclear fragmentation. Further, there are certain apoptotic relevant genes such as Bax, bak and bcl-XL whose mRNA level changes frequently on receiving the apoptotic signal such as on exposure to flavones, increased expression of bak takes place in colon carcinoma cells (Wenzel et al., 2000), whereas bcl-2 mRNA and bcl-2 expression was downregulated on exposure to flavopiridol (König et al., 1997).

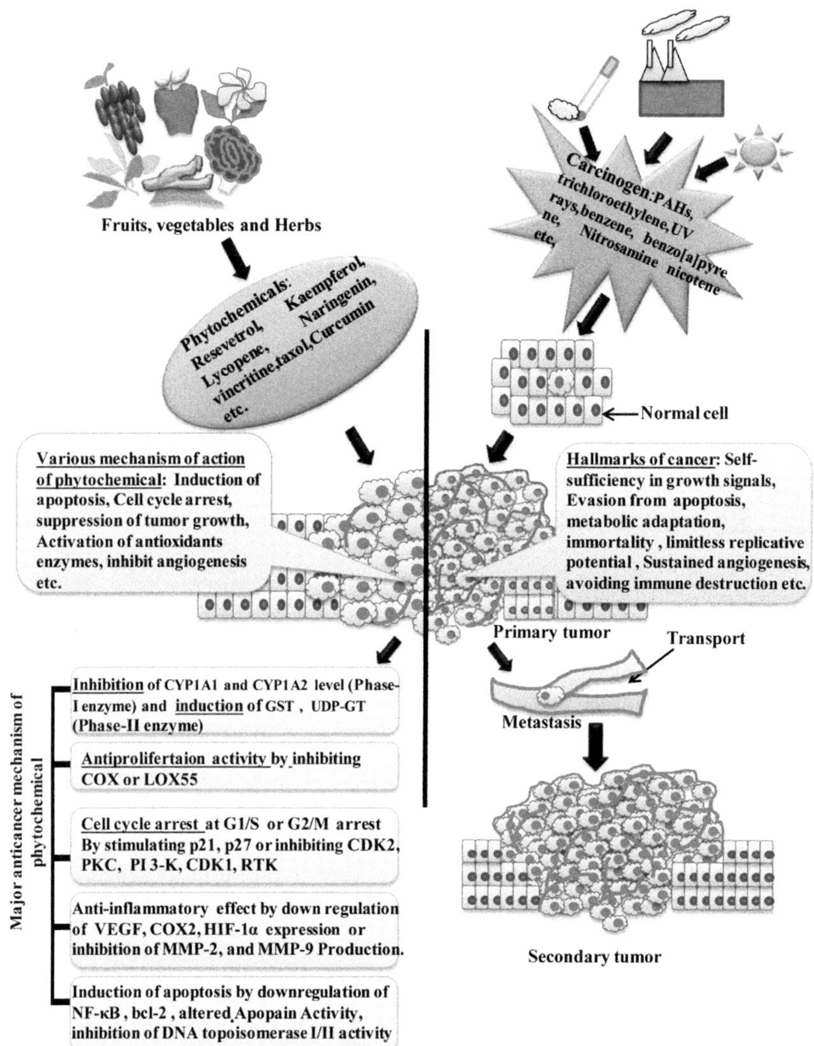

FIGURE 3.1 (See color insert.) Various environmental cues generating carcinogenic compound leads to the cancer incidence; however, in assistance with various phytochemicals exhibiting anti-cancer activity through several mechanisms among which some are shown.

3.5 HERBAL MEDICINE AND ITS IMPORTANCE

Herbal medicines have existed worldwide with long recorded history and they were used in ancient Chinese, Greek, Egyptian and Indian medicine for various therapeutic purposes. Modern medicine, despite so many achievements and progress remains inaccessible to the common mass to deal with various diseases and disorders. Till date, majority of world population mainly in developing/underdeveloped countries cannot access to allopath and therefore bound to rely on the time-tested traditional/alternative or complementary systems of medicine, among which many of these systems is much older compared to the allopathic medical wisdom (NCBS2015). In the 21st century, increasing pollution, unhealthy lifestyle and environmental toxins increases the risk of various diseases including cancer. Further, Irrational use of allopathic drugs generates numerous side effects, which is another major concern to be addressed. In 2013, WHO developed and lunched 'WHO Traditional Medicine Strategy 2014–2023' and emphasized to integrate traditional and complementary medicine to promote universal healthcare to ensure the quality, safety and effectiveness of such medicine. Therefore, the world is looking for cost effective, easily available, better physiological compatible traditional systems of medicine and holistic approach to avert such problem and provide the basic healthcare to all (WHO, 2013, Geneva).

3.5.1 INDIAN TRADITIONAL MEDICINE

Indian traditional medicinal system like Ayurveda, Siddha and Unani has a very rich history of effectiveness of drug, acknowledged by researcher working on modern medicine. Several evidences indicated that, in Indian subcontinent medical intervention like dentistry and trepanation were exercised as early as 7000 BCE. Current archeo-botanical excavations evidenced regarding the use of medicinal plants in the Middle Gangetic region, since the 2nd millennium BCE which are still found in Ayurvedic folk medicine. India inhabiting various group of people who follow their own religion, beliefs, culture, language and dialects leads to emergence of diverse medicinal systems (such as Ayurveda and homeopathy). However, number of medicinal systems introduced here from outside further contribute to enriched the Indian medicinal practices. Since ancient time, Indian society depends on traditional medicinal systems practiced. The herbal discovery is further complemented with the technique of isolation, purification, characterization of active ingredients and type of preparation. The term "herbal drug" determines the part/parts of a plant (leaves, flowers, seeds roots, barks, stems and etc.) used for preparing medicines (Frawley, 2000). Modern tools and technique have revolutionized the progression of drug discovery from medicinal plants and new approaches/concepts/technologies became a key apparatus in the development of traditional medicine further. Due to the scientific advancement today, more and more pharmacologically active ingredients of the *Ayurvedic* medicines as well as their usefulness in drug therapy have been identified. Basically, it is the phytochemical constituent such as saponins, tannins, alkaloids, alkenyl phenols, flavonoids, terpenoids, phorbol esters and sesquiterpenes lactones in the herbals, which lead to the desired healing effect. A single herb may even contain more than one of the aforementioned phytochemical constituents, which works synergistically with each other in producing pharmacological action.

3.5.2 HERBAL MEDICINE: LIMITATIONS NEEDS TO BE ADDRESSED

Drug discovery from medicinal plants used by national/international research is a burning approach of research area and therefore in past few years research on this aspect has been increased and lot more drug/formulations investigated by both public or private sector by different countries including India are in queue or in under clinical trial. India harbors a huge pool of diverse medicinal plant sources, among which many of them are already well characterized and enlisted in traditional medicinal system however, many of them remain unexplored and yet to be discovered, placing India at a unique place and confers its strong position in the research field of new drug discovery. However there is lot of limitation before the promotion of traditional herbal knowledge around the world, needed to be addressed before the promotion of traditional herbal knowledge around the world: Quality issues (such as adulteration, misidentification of plant, faulty collection and herbal preparation), processing and harvesting issues (includes poor pre and post-harvest practices, lack of processing techniques, poor agriculture and propagation method and Indiscriminate harvesting), Quality control related issues (like standardization, lack of good manufacturing practices also lack of awareness regarding the guideline and their poor implementation), Administrative issues (lack of proper monitoring and controlling authority), Pharmacogivilane (to investigate the Adverse reactions, contraindications, interactions with other drug or food). Therefore, all these factors work together to compromise the quality and quantity of herbal preparation that should be addressed wisely. Further, safety continues to be a foremost issue with the use of herbal remedies therefore clinical trials are necessary, to understand safe and efficacious use of drugs prior to introduce them in global market. Further to deal with Biopiracy proper documentation of folk knowledge of herbal medicine is required. Optimization of dosage of drug is equally important for the effective usage against the various disease also their mode of action, identification of active constituent therefore active participation of R&D group is required for better understanding of the mode of action and dosage composition for the treatment of any disease. There is a general believe that, herbal medicine has no side effect which is not true that may cause various health issues hindering the promotion of such drugs therefore, irrational usage of the herbal medicine should be checked. Other common issues like unethical practice of herbal medicine, exposure of unreliable and misleading information, absence of focused marketing and branding, lack of sufficient fund and knowledge sharing hold back the global promotion of herbal medicine. Lack of protection of biodiversity further inhibit the protection of the traditional medicinal plants are another major challenge to be addressed.

Therefore, to overcome various barriers like irrational use, quality control and standardization issues, high pharmacovigilance, etc. Further, strict implementation of rules, monitoring and periodic revision of regulations are mandatory to promote Indian traditional medicine. Altogether, to extract the maximum usage of herbal medicines, isolation of bioactive compounds and their mode of action at the effective dosage should be done prior to their clinical trial, so that major mass can taste the fruit of sacred herbs to uproot the various diseases.

3.6 CONCLUSION

Being blessed by nature, which offers a variety of plants, shows an immense potential to fight against the various disease including cancer, even the food including fruits and vegetables spices enlisted above have the ability to reduce the cancer risk. Therefore, routine intake of these fruits and vegetables helps to reduce the incidence of cancer also healthy lifestyle is equally important. Natural products are rich source of cancer chemotherapeutic drugs, and primarily target rapidly proliferating tumor cells by inhibition of cell cycle, induction of apoptosis, and regulation of angiogenesis. The understanding of molecular mechanism of a specific plant derived compound against a particular type of cancer will leads to the invention of novel drug and drug targets for therapeutic intervention. Chemoprevention by edible phytochemicals is of great interest and is considered to be an inexpensive, readily applicable, acceptable, and accessible approach for cancer control and management. But prior to this, in transfer of plants from the field to the lab, various factors plays an important role like irrational use, quality control and standardization issues, high pharmacovigilance, which varies with not species to species, but also on the above-mentioned variables. Therefore, a needful study is required to further harness the benefits of the herbs, fruit, and vegetables to fight against the diseases like cancer.

ACKNOWLEDGMENTS

This work was supported by the UGC DBT and DST India. The authors also thank Ms. Mandira Basumatary for paper editing.

KEYWORDS

- **cyclin-dependent kinases**
- **cyclooxygenase**
- **mitogenic-activated protein kinases**
- **nuclear factor-kappa B**
- **signaling transducers and activator of transcription 3**
- **vascular endothelial growth factor**

REFERENCES

Álvarez-Álvarez, R., Botas, A., Albillos, S. M., Rumbero, A., Martín, J. F., & Liras, P. (2015). Molecular genetics of naringenin biosynthesis, a typical plant secondary metabolite produced by Streptomyces clavuligerus. *Microbial Cell Factories*, *14*(1), 178.

Arpita, R., & Navneeta, B. (2017). Centella asiatica: a pharmaceutically important medicinal plant. *Curr Trends Biomed Eng Biosci*, *5*(3), 555661.

Awale, S., Linn, T. Z., Li, F., Tezuka, Y., Myint, A., Tomida, A., ... & Kadota, S. (2011). Identification of chrysoplenetin from *Vitex negundo* as a potential cytotoxic agent against PANC-1 and a panel of 39 human cancer cell lines (JFCR-39). *Phytotherapy Research*, *25*(12), 1770–1775.

Azmi, A. S., Bhat, S. H., Hanif, S., & Hadi, S. M. (2006). Plant polyphenols mobilize endogenous copper in human peripheral lymphocytes leading to oxidative DNA breakage: a putative mechanism for anticancer properties. *FEBS Letters*, *580*(2), 533–538.

Balasubramani, G., Ramkumar, R., Krishnaveni, N., Pazhanimuthu, A., Natarajan, T., Sowmiya, R., & Perumal, P. (2015). Structural characterization, antioxidant and anticancer properties of gold nanoparticles synthesized from leaf extract (decoction) of *Antigonon leptopus* Hook. & Arn. *Journal of Trace Elements in Medicine and Biology*, *30*, 83–89.

Balkwill, F., & Mantovani, A. (2001). Inflammation and cancer: back to Virchow? *The Lancet, 357*(9255), 539–545.

Bansal, P., Gupta V, V., Bansal, R. & Sapra, R. (2012). Dietary phytochemicals in cell cycle arrest and apoptosis-an insight. *Journal of Drug Delivery and Therapeutics, 2*, 2.

Biba, V. S., Jeba, M. P. W., & Remani, P. (2013). Differential effects of *Annona squamosa* seed extracts: antioxidant, antibacterial, cytotoxic and apoptotic study. *Int J Pharm Biol Sci, 4*, 899–907.

Bishayee, A., Politis, T., & Darvesh, A. S. (2010). Resveratrol in the chemoprevention and treatment of hepatocellular carcinoma. *Cancer Treatment Reviews, 36*(1), 43–53.

Brahmachari, G., Gorai, D., & Roy, R. (2013). *Argemone mexicana*: chemical and pharmacological aspects. *Revista Brasileira de Farmacognosia, 23*(3), 559–567.

BRÅKenhielm, E. B. B. A., Cao, R., & Cao, Y. (2001). Suppression of angiogenesis, tumor growth, and wound healing by resveratrol, a natural compound in red wine and grapes. *The FASEB Journal, 15*(10), 1798–1800.

Bu-Abbas, A., Clifford, M. N., Walker, R., & Ioannides, C. (1998). Contribution of caffeine and flavanols in the induction of hepatic phase II activities by green tea. *Food and Chemical Toxicology, 36*(8), 617–621.

Cao, Z., Fang, J., Xia, C., Shi, X., & Jiang, B. H. (2004). trans-3, 4, 5′-Trihydroxystibene inhibits hypoxia-inducible factor 1α and vascular endothelial growth factor expression in human ovarian cancer cells. *Clinical Cancer Research, 10*(15), 5253–5263.

Carmeliet, P. (2003). Angiogenesis in health and disease. *Nature Medicine, 9*(6), 653.

Carmeliet, P., & Jain, R. K. (2000). Angiogenesis in cancer and other diseases. *Nature, 407*(6801), 249.

Casagrande, F., & Darbon, J. M. (2001). Effects of structurally related flavonoids on cell cycle progression of human melanoma cells: regulation of cyclin-dependent kinases CDK2 and CDK11. *Biochemical Pharmacology, 61*(10), 1205–1215.

Casey, S. C., Amedei, A., Aquilano, K., Azmi, A. S., Benencia, F., Bhakta, D., ... & Crawford, S. (2015). Cancer prevention and therapy through the modulation of the tumor microenvironment. In *Seminars in Cancer Biology* (Vol. 35, pp. S199–S223). Academic Press.

Chahar, M. K., Sharma, N., Dobhal, M. P., & Joshi, Y. C. (2011). Flavonoids: A versatile source of anticancer drugs. *Pharmacognosy Reviews, 5*(9), 1.

Chang, W. S., Lee, Y. J., Lu, F. J., & Chiang, H. C. (1993). Inhibitory effects of flavonoids on xanthine oxidase. *Anticancer Research, 13*(6A), 2165–2170.

Choedon, T., Shukla, S. K., & Kumar, V. (2010). Chemopreventive and anti-cancer properties of the aqueous extract of flowers of *Butea monosperma*. *Journal of Ethnopharmacology, 129*(2), 208–213.

Choi, J. A., Kim, J. Y., Lee, J. Y., Kang, C. M., Kwon, H. J., Yoo, Y. D., ... & Lee, S. J. (2001). Induction of cell cycle arrest and apoptosis in human breast cancer cells by quercetin. *International Journal of Oncology, 19*(4), 837–844.

Chung, A. S., & Ferrara, N. (2011). Developmental and pathological angiogenesis. *Annual Review of Cell and Developmental Biology, 27*, 563–584.

Colombo, V., Lupi, M., Falcetta, F., Forestieri, D., D'Incalci, M., & Ubezio, P. (2011). Chemotherapeutic activity of silymarin combined with doxorubicin or paclitaxel in sensitive and multidrug-resistant colon cancer cells. *Cancer Chemotherapy and Pharmacology, 67*(2), 369–379.

Constantinou, A. I., Kamath, N., & Murley, J. S. (1998). Genistein inactivates bcl-2, delays the G2/M phase of the cell cycle, and induces apoptosis of human breast adenocarcinoma MCF-7 cells. *European Journal of Cancer*, *34*(12), 1927–1934.

Cornblatt, B. S., Ye, L., Dinkova-Kostova, A. T., Erb, M., Fahey, J. W., Singh, N. K., ... & Davidson, N. E. (2007). Preclinical and clinical evaluation of sulforaphane for chemoprevention in the breast. *Carcinogenesis*, *28*(7), 1485–1490.

Dasgupta, A., Biddle, D. A., Wells, A., & Datta, P. (2000). Positive and negative interference of the Chinese medicine Chan Su in serum digoxin measurement: Elimination of interference by using a monoclonal chemiluminescent digoxin assay or monitoring free digoxin concentration. *American Journal of Clinical Pathology*, *114*(2), 174–179.

Donaldson, M. S. (2004). Nutrition and cancer: a review of the evidence for an anti-cancer diet. *Nutrition Journal*, *3*(1), 19.

Dorel, M., Barillot, E., Zinovyev, A., & Kuperstein, I. (2015). Network-based approaches for drug response prediction and targeted therapy development in cancer. *Biochemical and Biophysical Research Communications*, *464*(2), 386–391.

Dvorak, H. F. (1986). Tumors: Wounds that do not heal. Similarities between tumor stroma generation and wound healing. *The New England Journal of Medicine*, *315*(26), 1650–1659.

Efferth, T., (2017). From ancient herb to modern drug: *Artemisia annua* and artemisinin for cancer therapy. Academic Press. *Seminars in Cancer Biology*, *46*, 65–83.

Faria, A., Pestana, D., Teixeira, D., De Freitas, V., Mateus, N., & Calhau, C. (2010). Blueberry anthocyanins and pyruvic acid adducts: anticancer properties in breast cancer cell lines. *Phytotherapy Research*, *24*(12), 1862–1869.

Fresco, P., Borges, F., Diniz, C., & Marques, M. P. M. (2006). New insights on the anticancer properties of dietary polyphenols. *Medicinal Research Reviews*, *26*(6), 747–766.

Gali-Muhtasib, H., Hmadi, R., Kareh, M., Tohme, R., & Darwiche, N. (2015). Cell death mechanisms of plant-derived anticancer drugs: beyond apoptosis. *Apoptosis*, *20*(12), 1531–1562.

Gali-Muhtasib, H., Hmadi, R., Kareh, M., Tohme, R., & Darwiche, N. (2015). Cell death mechanisms of plant-derived anticancer drugs: beyond apoptosis. *Apoptosis*, *20*(12), 1531–1562.

Greenwell, M., & Rahman, P. K. S. M. (2015). Medicinal plants: their use in anticancer treatment. *International Journal of Pharmaceutical Sciences and Research*, *6*(10), 4103.

Gupta, M., Mukhtar, H. M., & Ahmad, S. (2010). Phyto-pharmacological and plant tissue culture overview of *Tylophora indica* (burm f.) Merill. *J Pharm Sci Res*, *2*, 401–411.

Gupta, S. C., Kim, J. H., Prasad, S., & Aggarwal, B. B. (2010). Regulation of survival, proliferation, invasion, angiogenesis, and metastasis of tumor cells through modulation of inflammatory pathways by nutraceuticals. *Cancer and Metastasis Reviews*, *29*(3), 405–434.

Gupta, S. C., Tyagi, A. K., Deshmukh-Taskar, P., Hinojosa, M., Prasad, S., & Aggarwal, B. B. (2014). Downregulation of tumor necrosis factor and other proinflammatory biomarkers by polyphenols. *Archives of Biochemistry and Biophysics*, *559*, 91–99.

Gutiérrez-Uribe, J. A., Romo-Lopez, I., & Serna-Saldívar, S. O. (2011). Phenolic composition and mammary cancer cell inhibition of extracts of whole cowpeas (*Vigna unguiculata*) and its anatomical parts. *Journal of Functional Foods*, *3*(4), 290–297.

Ha, M. K., Song, Y. H., Jeong, S. J., Lee, H. J., Jung, J. H., Kim, B., ... & Kim, S. H. (2011). Emodin inhibits proinflammatory responses and inactivates histone deacetylase 1 in hypoxic rheumatoid synoviocytes. *Biological and Pharmaceutical Bulletin*, *34*(9), 1432–1437.

Hirata, T., Fujii, M., Akita, K., Yanaka, N., Ogawa, K., Kuroyanagi, M., & Hongo, D. (2009). Identification and physiological evaluation of the components from Citrus fruits as potential drugs for anti-corpulence and anticancer. *Bioorganic & Medicinal Chemistry*, *17*(1), 25–28.

Hodek, P., Trefil, P., & Stiborová, M. (2002). Flavonoids-potent and versatile biologically active compounds interacting with cytochromes P450. *Chemico-Biological Interactions*, *139*(1), 1–21.

Hoeben, A. N. N., Landuyt, B., Highley, M. S., Wildiers, H., Van Oosterom, A. T., & De Bruijn, E. A. (2004). Vascular endothelial growth factor and angiogenesis. *Pharmacological Reviews*, *56*(4), 549–580.

Hosbach, I., Neeb, G., Hager, S., Kirchhoff, S., & Kirschbaum, B. (2003). In defense of traditional Chinese herbal medicine. *Anaesthesia*, *58*(3), 282–283.

Huntley, A. L. (2009). The health benefits of berry flavonoids for menopausal women: cardiovascular disease, cancer and cognition. *Maturitas*, *63*(4), 297–301.

Igura, K., Ohta, T., Kuroda, Y., & Kaji, K. (2001). Resveratrol and quercetin inhibit angiogenesis in vitro. *Cancer Letters*, *171*(1), 11–16.

Iqbal, J., Abbasi, B. A., Mahmood, T., Kanwal, S., Ali, B., & Khalil, A. T. (2017). Plant-derived anticancer agents: A green anticancer approach. *Asian Pacific Journal of Tropical Biomedicine*. *7*(12), 1129–1150.

Jaradat, N. A., Al-Ramahi, R., Zaid, A. N., Ayesh, O. I., & Eid, A. M. (2016). Ethnopharmacological survey of herbal remedies used for treatment of various types of cancer and their methods of preparations in the West Bank-Palestine. *BMC Complementary and Alternative Medicine*, *16*(1), 93.

Jung, S. K., Lee, M. H., Kim, J. E., Singh, P., Lee, S. Y., Jeong, C. H., ... & Lee, N. H. (2014). Isoliquiritigenin induces apoptosis and inhibits xenograft tumor growth of human lung cancer cells by targeting both wild type and L858R/T790M mutant EGFR. *Journal of Biological Chemistry*, *289*(52), 35839–35848.

Kanjoormana, M., & Kuttan, G. (2010). Antiangiogenic activity of ursolic acid. *Integrative Cancer Therapies*, *9*(2), 224–235.

Karna, P., Gundala, S. R., Gupta, M. V., Shamsi, S. A., Pace, R. D., Yates, C., ... & Aneja, R. (2011). Polyphenol-rich sweet potato greens extract inhibits proliferation and induces apoptosis in prostate cancer cells in vitro and in vivo. *Carcinogenesis*, *32*(12), 1872–1880.

Khurana, R., Simons, M., Martin, J. F., & Zachary, I. C. (2005). Role of angiogenesis in cardiovascular disease: a critical appraisal. *Circulation*, *112*(12), 1813–1824.

Kim, M. J., Jeong, M. K., Chang, P. S., & Lee, J. (2009). Radical scavenging activity and apoptotic effects in HT-29 human colon cancer cells of black sesame seed extract. *International Journal of Food Science & Technology*, *44*(11), 2106–2112.

König, A., Schwartz, G. K., Mohammad, R. M., Al-Katib, A., & Gabrilove, J. L. (1997). The novel cyclin-dependent kinase inhibitor flavopiridol downregulates Bcl-2 and induces growth arrest and apoptosis in chronic B-cell leukemia lines. *Blood*, *90*(11), 4307–4312.

Kowshik, J., Giri, H., Kranthi Kiran Kishore, T., Kesavan, R., Naik Vankudavath, R., Bhanuprakash Reddy, G., ... & Nagini, S. (2014). Ellagic acid inhibits VEGF/VEGFR2, PI3K/Akt and MAPK signaling cascades in the hamster cheek pouch carcinogenesis model. *Anti-Cancer Agents in Medicinal Chemistry (Formerly Current Medicinal Chemistry-Anti-Cancer Agents)*, *14*(9), 1249–1260.

Kristo, A. S., Klimis-Zacas, D., & Sikalidis, A. K. (2016). Protective role of dietary berries in cancer. *Antioxidants*, *5*(4), 37.

Król, S. K., Kiełbus, M., Rivero-Müller, A., & Stepulak, A. (2017). Comprehensive review on betulin as a potent anticancer agent. *BioMed Research International*. 2015.

Kühnau, J. (1976). The flavonoids. A class of semi-essential food components: their role in human nutrition. In *World Review of Nutrition and Dietetics* (Vol. 24, pp. 117–191). Karger Publishers.
Kumar, S. R., Priyatharshni, S., Babu, V. N., Mangalaraj, D., Viswanathan, C., Kannan, S., & Ponpandian, N. (2014). Quercetin conjugated superparamagnetic magnetite nanoparticles for in-vitro analysis of breast cancer cell lines for chemotherapy applications. *Journal of Colloid and Interface Science*, *436*, 234–242.
Kumar, S., Pathania, A. S., Saxena, A. K., Vishwakarma, R. A., Ali, A., & Bhushan, S. (2013). The anticancer potential of flavonoids isolated from the stem bark of Erythrina suberosa through induction of apoptosis and inhibition of STAT signaling pathway in human leukemia HL-60 cells. *Chemico-Biological Interactions*, *205*(2), 128–137.
Le Marchand, L., Murphy, S. P., Hankin, J. H., Wilkens, L. R., & Kolonel, L. N. (2000). Intake of flavonoids and lung cancer. *Journal of the National Cancer Institute*, *92*(2), 154–160.
Lee, D., Kim, I. Y., Saha, S., & Choi, K. S. (2016). Paraptosis in the anti-cancer arsenal of natural products. *Pharmacology & Therapeutics*, *162*, 120–133.
Lefranc, F., Tabanca, N., & Kiss, R. (2017). Assessing the anticancer effects associated with food products and/or nutraceuticals using in vitro and in vivo preclinical development-related pharmacological tests. In *Seminars in Cancer Biology*. Academic Press. 46, pp. 14–32.
Li, J. Z., Qing, C., Chen, C. X., Hao, X. J., & Liu, H. Y. (2009). Cytotoxicity of cardenolides and cardenolide glycosides from *Asclepias curassavica*. *Bioorganic & Medicinal Chemistry Letters*, *19*(7), 1956–1959.
Liang, X. M., Jin, Y., Wang, Y. P., Jin, G. W., Fu, Q., & Xiao, Y. S. (2009). Qualitative and quantitative analysis in quality control of traditional Chinese medicines. *Journal of Chromatography A*, *1216*(11), 2033–2044.
Liu, Z., Li, Y., & Yang, R. (2012). Effects of resveratrol on vascular endothelial growth factor expression in osteosarcoma cells and cell proliferation. *Oncology Letters*, *4*(4), 837–839.
Lozynskyi, A., Zimenkovsky, B., & Lesyk, R. (2014). Synthesis and anticancer activity of new thiopyrano [2, 3-d] thiazoles based on cinnamic acid amides. *Scientia pharmaceutica*, *82*(4), 723–734
Lutterodt, H., Slavin, M., Whent, M., Turner, E., & Yu, L. L. (2011). Fatty acid composition, oxidative stability, antioxidant and antiproliferative properties of selected cold-pressed grape seed oils and flours. *Food Chemistry*, *128*(2), 391–399.
Manthey, J. A., Guthrie, N., & Grohmann, K. (2001). Biological properties of citrus flavonoids pertaining to cancer and inflammation. *Current Medicinal Chemistry*, *8*(2), 135–153.
Martin, P. M., Horwitz, K. B., Ryan, D. S., & Mcguire, W. L. (1978). Phytoestrogen interaction with estrogen receptors in human breast cancer cells. *Endocrinology*, *103*(5), 1860–1867.
Morbidelli, L. (2016). Polyphenol-based nutraceuticals for the control of angiogenesis: Analysis of the critical issues for human use. *Pharmacological Research*, *111*, 384–393.
Mukhija, M., Singh, M. P., Dhar, K. L., & Kalia, A. N. (2015). Cytotoxic and antioxidant activity of *Zanthoxylum alatum* stem bark and its flavonoid constituents. *Journal of Pharmacognosy and Phytochemistry*, *4*(4), 86.
Mutoh, M., Takahashi, M., Fukuda, K., Komatsu, H., Enya, T., Matsushima-Hibiya, Y., ... & Wakabayashi, K. (2000). Suppression by Flavonoids of Cyclooxygenase-2 Promoter-dependent Transcriptional Activity in Colon Cancer Cells: Structure-Activity Relationship. *Cancer Science*, *91*(7), 686–691.
National Centre for Biological Sciences (2015). Overview of Indian Healing Traditions. https://www.ncbs.res.in/HistoryScienceSociety/content/overview-indian-healing-traditions.

Ng, T. B., Lam, Y. W., & Wang, H. (2003). Calcaelin, a new protein with translation-inhibiting, antiproliferative and antimitogenic activities from the mosaic puffball mushroom *Calvatia caelata*. *Planta Medica*, *69*(03), 212–217.

Noratto, G., Porter, W., Byrne, D., & Cisneros-Zevallos, L. (2009). Identifying peach and plum polyphenols with chemopreventive potential against estrogen-independent breast cancer cells. *Journal of Agricultural and Food Chemistry, 57*(12), 5219–5226.

Okura, T., Ibe, M., Umegaki, K., Shinozuka, K., & Yamada, S. (2010). Effects of dietary ingredients on function and expression of P-glycoprotein in human intestinal epithelial cells. *Biological and Pharmaceutical Bulletin, 33*(2), 255–259.

Osman, N. H., Said, U. Z., El-Waseef, A. M., & Ahmed, E. S. (2015). Luteolin supplementation adjacent to aspirin treatment reduced dimethylhydrazine-induced experimental colon carcinogenesis in rats. *Tumor Biology*, *36*(2), 1179–1190.

Pal, D., Banerjee, S., & Ghosh, A. K. (2012). Dietary-induced cancer prevention: An expanding research arena of emerging diet related to healthcare system. *Journal of Advanced Pharmaceutical Technology & Research*, *3*(1), 16.

Parveen, S., Jan, U., & Kamili, A. (2013). Importance of Himalayan medicinal plants and their conservation strategies. *Australian Journal of Herbal Medicine*, *25*(2), 63.

Pierpaoli, E., Damiani, E., Orlando, F., Lucarini, G., Bartozzi, B., Lombardi, P., ... & Provinciali, M. (2015). Antiangiogenic and antitumor activities of berberine derivative NAX014 compound in a transgenic murine model of HER2/neu-positive mammary carcinoma. *Carcinogenesis*, *36*(10), 1169–1179.

Prasad, S., Kalra, N., & Shukla, Y. (2007). Induction of apoptosis by lupeol and mango extract in mouse prostate and LNCaP cells. *Nutrition and Cancer, 60*(1), 120–130.

Rahman, S., Carter, P., & Bhattarai, N. (2017). Aloe vera for tissue engineering applications. *Journal of Functional Biomaterials*, *8*(1), 6.

Raihan, M., Tareq, S. M., Brishti, A., Alam, M., Haque, A., & Ali, M. (2012). Evaluation of Antitumor Activity of *Leea indica* (Burm. f.) Merr. extract against Ehrlich Ascites Carcinoma (EAC) Bearing Mice. *American Journal of Biomedical Sciences*, *4*(2).

Ramos, P., & Bentires-Alj, M. (2015). Mechanism-based cancer therapy: resistance to therapy, therapy for resistance. *Oncogene*, *34*(28), 3617.

Ren, H., Han, M., Zhou, J., Zheng, Z. F., Lu, P., Wang, J. J., ... & Ouyang, H. W. (2014). Repair of spinal cord injury by inhibition of astrocyte growth and inflammatory factor synthesis through local delivery of flavopiridol in PLGA nanoparticles. *Biomaterials*, *35*(24), 6585–6594.

Sabnis, M. (2006). Chemistry and pharmacology of Ayurvedic medicinal plants. Chaukhambha Amarabharati Prakashan.

Senderowicz, A. M. (1999). Flavopiridol. The first cyclin-dependent kinase inhibitor in human clinical trials. *Invest New Drugs. 17,* 313–20.

Senderowicz, A. M. (2001). Development of cyclin-dependent kinase modulators as novel therapeutic approaches for hematological malignancies. *Leukemia*, *15*(1), 1.

Serra, A. T., Matias, A. A., Frade, R. F., Duarte, R. O., Feliciano, R. P., Bronze, M. R., ... & Duarte, C. M. (2010). Characterization of traditional and exotic apple varieties from Portugal. Part 2–antioxidant and antiproliferative activities. *Journal of Functional Foods*, *2*(1), 46–53.

Sinani, S. S., Eltayeb, E. A., Coomber, B. L., & Adham, S. A. (2016). Solamargine triggers cellular necrosis selectively in different types of human melanoma cancer cells through extrinsic lysosomal mitochondrial death pathway. *Cancer Cell International*, *16*(1), 11.

Singh, S., Sharma, B., Kanwar, S. S., & Kumar, A. (2016). Lead phytochemicals for anticancer drug development. *Frontiers in Plant Science*, *7*, 1667.

Singhuber, J., Zhu, M., Prinz, S., & Kopp, B. (2009). Aconitum in traditional Chinese medicine—a valuable drug or an unpredictable risk?. *Journal of Ethnopharmacology*, *126*(1), 18–30.

Sivaraj, R., Rahman, P. K., Rajiv, P., Narendhran, S., & Venckatesh, R. (2014). Biosynthesis and characterization of *Acalypha indica* mediated copper oxide nanoparticles and evaluation of its antimicrobial and anticancer activity. *Spectrochimica Acta Part A: Molecular and Biomolecular Spectroscopy*, *129*, 255–258.

Srivastava, R. K., Unterman, T. G., & Shankar, S. (2010). FOXO transcription factors and VEGF neutralizing antibody enhance antiangiogenic effects of resveratrol. *Molecular and cellular biochemistry*, *337*(1–2), 201–212.

Sun, X. Y., Plouzek, C. A., Henry, J. P., Wang, T. T., & Phang, J. M. (1998). Increased UDP-glucuronosyltransferase activity and decreased prostate specific antigen production by biochanin A in prostate cancer cells. *Cancer Research*, *58*(11), 2379–2384.

Suzuki, R., Kohno, H., Sugie, S., Sasaki, K., Yoshimura, T., Wada, K., & Tanaka, T. (2004). Preventive effects of extract of leaves of ginkgo (Ginkgo biloba) and its component bilobalide on azoxymethane-induced colonic aberrant crypt foci in rats. *Cancer Letters*, *210*(2), 159–169.

Szliszka, E., Zydowicz, G., Janoszka, B., Dobosz, C., Kowalczyk-Ziomek, G., & Krol, W. (2011). Ethanolic extract of Brazilian green propolis sensitizes prostate cancer cells to TRAIL-induced apoptosis. *International Journal of Oncology*, *38*(4), 941–953.

Tan, W. F., Lin, L. P., Li, M. H., Zhang, Y. X., Tong, Y. G., Xiao, D., & Ding, J. (2003). Quercetin, a dietary-derived flavonoid, possesses antiangiogenic potential. *European Journal of Pharmacology*, *459*(2–3), 255–262.

Torre, L. A., Siegel, R. L., Ward, E. M., & Jemal, A. (2016). Global cancer incidence and mortality rates and trends—an update. *Cancer Epidemiology and Prevention Biomarkers*, *25*(1), 16–27.

Tsyrlov, I. B., Mikhailenko, V. M., & Gelboin, H. V. (1994). Isozyme-and species-specific susceptibility of cDNA-expressed CYP1A P-450s to different flavonoids. *Biochimica et Biophysica Acta (BBA)-Protein Structure and Molecular Enzymology*, *1205*(2), 325–335.

Tu, L. Y., Pi, J., Jin, H., Cai, J. Y., & Deng, S. P. (2016). Synthesis, characterization and anticancer activity of kaempferol-zinc (II) complex. *Bioorganic & Medicinal Chemistry Letters*, *26*(11), 2730–2734.

Vallianou, N. G., Evangelopoulos, A., Schizas, N., & Kazazis, C. (2015). Potential anticancer properties and mechanisms of action of curcumin. *Anticancer Research*, *35*(2), 645–651.

Ververidis, F., Trantas, E., Douglas, C., Vollmer, G., Kretzschmar, G., & Panopoulos, N. (2007). Biotechnology of flavonoids and other phenylpropanoid-derived natural products. Part I: Chemical diversity, impacts on plant biology and human health. *Biotechnology Journal*, *2*(10), 1214–1234.

Wang, S. Y., Bowman, L., & Ding, M. (2008). Methyl jasmonate enhances antioxidant activity and flavonoid content in blackberries (*Rubus* sp.) and promotes antiproliferation of human cancer cells. *Food Chemistry*, *107*(3), 1261–1269.

Wang, S., Wu, X., Tan, M., Gong, J., Tan, W., Bian, B., ... & Wang, Y. (2012). Fighting fire with fire: poisonous Chinese herbal medicine for cancer therapy. *Journal of Ethnopharmacology*, *140*(1), 33–45.

Wang, X., Yuan, S., Wang, J., Lin, P., Liu, G., Lu, Y., ... & Wei, Y. (2006). Anticancer activity of litchi fruit pericarp extract against human breast cancer in vitro and in vivo. *Toxicology and Applied Pharmacology*, *215*(2), 168–178.

Wargovich, M. J. (2000). Anticancer properties of fruits and vegetables. *HortScience, 35*(4), 573–575.

Weis, S. M., & Cheresh, D. A. (2011). Tumor angiogenesis: molecular pathways and therapeutic targets. *Nature Medicine, 17*(11), 1359.

Wen, L., Wu, D., Jiang, Y., Prasad, K. N., Lin, S., Jiang, G., ... & Yang, B. (2014). Identification of flavonoids in litchi (*Litchi chinensis* Sonn.) leaf and evaluation of anticancer activities. *Journal of Functional Foods, 6*, 555–563.

Weng, C. J., & Yen, G. C. (2012). Chemopreventive effects of dietary phytochemicals against cancer invasion and metastasis: phenolic acids, monophenol, polyphenol, and their derivatives. *Cancer Treatment Reviews, 38*(1), 76–87.

Weng, J. R., Bai, L. Y., Chiu, C. F., Hu, J. L., Chiu, S. J., & Wu, C. Y. (2013). Cucurbitane triterpenoid from *Momordica charantia* induces apoptosis and autophagy in breast cancer cells, in part, through peroxisome proliferator-activated receptor γ activation. *Evidence-Based Complementary and Alternative Medicine.*

Wenzel, U., Kuntz, S., Brendel, M. D., & Daniel, H. (2000). Dietary flavone is a potent apoptosis inducer in human colon carcinoma cells. *Cancer Research, 60*(14), 3823–3831.

World Health Organization (2013). World Health Organization; Geneva. WHO Traditional Medicine Strategy, 2014–2023.

Yoon, J. S., Kim, H. M., Yadunandam, A. K., Kim, N. H., Jung, H. A., Choi, J. S., ... & Kim, G. D. (2013). Neferine isolated from Nelumbo nucifera enhances anti-cancer activities in Hep3B cells: molecular mechanisms of cell cycle arrest, ER stress induced apoptosis and anti-angiogenic response. *Phytomedicine, 20*(11), 1013–1022.

Yoysungnoen, P., Wirachwong, P., Bhattarakosol, P., Niimi, H., & Patumraj, S. (2006). Effects of curcumin on tumor angiogenesis and biomarkers, COX-2 and VEGF, in hepatocellular carcinoma cell-implanted nude mice. *Clinical Hemorheology and Microcirculation, 34*(1–2), 109–115.

Zi, X., Feyes, D. K., & Agarwal, R. (1998). Anticarcinogenic effect of a flavonoid antioxidant, silymarin, in human breast cancer cells MDA-MB 468: induction of G1 arrest through an increase in Cip1/p21 concomitant with a decrease in kinase activity of cyclin-dependent kinases and associated cyclins. *Clinical Cancer Research, 4*(4), 1055–1064.

CHAPTER 4

Probiotic Foods

ANANTA SAIKIA and FRANCIS DUTTA

Department of Horticulture, Assam Agricultural University, Jorhat–785013, Assam, India

ABSTRACT

Consumers, all over the world, nowadays intend to have a better and healthier lifestyle. This has led to a sharp increase in the demand for functional foods enriched with probiotic microorganisms. Very commonly, dairy-based products are considered ideal carrier food matrices for probiotics. However, in recent times the potential applications of probiotics in non-dairy matrices, like fruits juices, cereals, etc., are gaining importance because of some intrinsic benefits over the dairy-based ones. Probiotic meat and fish products have also been developed. Incorporation of prebiotics into probiotic food products to produce novel synbiotic foods has a growing scope for commercialization. A discussion on probiotic foods has been made in this chapter highlighting technological aspects of the incorporation of probiotic microorganisms in dairy, non-dairy, meat, and fish products.

4.1 INTRODUCTION

During the last decade, the inclination of modern-day consumers has changed significantly towards healthy foods mainly due to a steady increase in life expectancy, rising health care costs and intense desire for an improved life (Vasiljevic and Shah, 2008). Thus, demand for functional foods is increasing substantially opening up vast scope for economic benefits to food industries. Functional foods like fermented milk, yogurts, baby foods, drinks, and confectioneries include probiotics, prebiotics, vitamins, and minerals (Khan and Ansari, 2007). For centuries, probiotic microorganisms have been used in fermented dairy products like yogurt and cheese (Lourens-Hattingh and

Viljoen, 2001; Saarela et al., 2000). However, in recent times, the quest for newer food matrices for the delivery of probiotic strains have resulted in the development of novel non-dairy probiotic products like fruit juices, cereals, chocolates, etc. which were found better and superior carriers for delivery of probiotics (Sarao and Arora, 2017). However, several reports have also indicated poor survival of the probiotic organisms in these functional foods. Despite the strong scientific evidence associating these microorganisms to various health benefits, further investigation is required to confirm their stability and probiotic efficacy.

4.2 PROBIOTICS

The literal meaning of "probiotic" is "for life." Probiotics are live microorganisms, which improve the equilibrium of gut microbiota positively influencing the host (Marteau et al., 1995; Fuller, 1989). The term "probiotic" was first used by Lilly and Stillwell in 1965 to designate substances which were secreted by one microorganism and stimulated the growth of another, and clearly distinguishing it from "antibiotic" (Schrezenmeir and de-Vrese, 2001). The definition, as per the Expert Consultation of Food and Agriculture Organization and World Health Organization, identifies probiotics as *"live microorganisms that, when administered in adequate amounts, confer a health benefit on the host"* (FAO/WHO, 2006).

Various species of *Lactobacillus* and *Bifidobacterium* and some other species of microorganisms have been used as probiotics in foods (Boyle and Tang, 2006). Different strains of *Lactobacillus acidophilus, L. casei, L. crispatus, L. delbrueckii, L. fermentum, L. gasseri, L. helveticus, L. johnsonii, L. kefiranofaciens L. paracasei, L. plantarum, L. reuteri, L. rhamnous, L. lactis, L. salivarius, Bifidobacterium bifidum, B. breve, B. infantis, B. longum, B. lactis, B. adolescentis, B. essensis, B. laterosporus* and other species like *Candida kefir, Escherichia coli, Saccharomyces boulardii, Streptococcus thermophilus, Enterococcus francium, Propionibacterium, Pediococcus, and Leuconostoc* are considered as efficient probiotics (Martins et al., 2013; Shah, 2007; Senok et al., 2005). Species of *Lactobacillus* and *Bifidobacterium* are found safe for human, while many of the species of other genera are pathogenic and their safety is questionable although these are being used in yogurt preparation (Ranadheera et al., 2010).

A microorganism is considered probiotic when it fulfills the following conditions (Harish and Varghese, 2006; Salminen and von-Wright, 1998):

- The strain isolated should belong to the same species as in its intended host;
- The strain must have a demonstrable beneficial effect on the host;
- No pathogenic, toxic, allergic, mutagenic or carcinogenic reaction by the probiotic strain itself, its fermentation products or bacterial cell components;
- Ability to proliferate and/or colonize on the location where it is naturally active;
- Easy and reproducible production;
- Genetically stable without plasmid transfer;
- It should survive the transit through the gastrointestinal tract (GIT);
- It must be able to survive prolonged periods on storage;
- Viable during processing and storage.

A rigorous assessment of the microbial strain is needed to obtain the probiotic status. FAO/WHO (2002) has laid down the following requirements:

- Complete taxonomical assessment of the microorganism down to strain level;
- *In vitro* experimentation to assess the probiotic effects of the strain like tolerance/resistance to gastric acids, bile, and digestive enzymes;
- Confirmatory evidence of its antimicrobial activity against pathogenic bacteria;
- Evidence that the strain is safe in its delivered form and in pure form without contamination;
- *In-vivo* assessment to establish the health effects on the host.

Probiotic microorganisms, when administered as an active component of food, must be acid and bile tolerant to survive the journey through the digestive tract and able to flourish in the gut. The science of probiotics is in constant improvement, and recently, local or topical administration of probiotics is also proposed (Vandenplas et al., 2015). Probiotics must be supplied in numbers adequate to trigger the targeted effect on the host, which depend on strains, process, carrier food matrix, and targeted health effect. Most of the traditional probiotics exert the targeted benefits at a concentration around 10^7 to 10^8 probiotic cells per gram of food matrix, with a serving size around 100 to 200 mg per day (Rijkers et al., 2010).

Several evidences are available which establish the efficacy of probiotics in prevention and treatment of diseases of the gastrointestinal, respiratory, and urogenital tracts (Ringel et al., 2012; Parvez et al., 2006). Synbiotic therapy has shown fruitful results in treating critically ill patients with abdominal surgery, trauma, and those in intensive care units (Shimizu et al., 2013). Probiotics can maintain good balance and composition of intestinal microbiota, help increase body's resistance to pathogens and maintain the host's overall health (Plé et al., 2015). Maintenance of gut microflora (Velasquez-Manoff, 2015), protection against gastrointestinal pathogens (Ringel et al., 2012), improvement of the immune system (Kemgang et al., 2014), reduction of lactose intolerance (Gilliland, 1990), reduction in blood pressure (Rasic, 2003), anti-carcinogenic activity (Nagpal et al., 2012), anti-microbial property (Tharmaraj and Shah, 2009), improved utilization of food nutrients (Solis et al., 2002) are some of these proven health benefits. Health benefits derived from probiotics are shown in Figure 4.1.

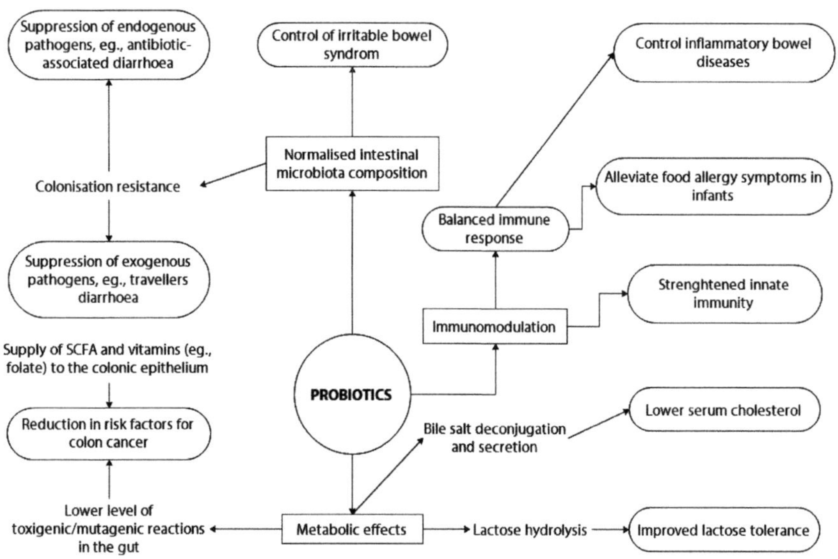

FIGURE 4.1 Health benefits of probiotics [redrawn from Parvez et al. (2006)].

The exact mechanism by which probiotics work is still not yet fully elucidated. Some studies reported that this might be due to production of bacteriocin and short chain fatty acid (SCFA), increase of acidity in gut, and nutrient competition to stimulate mucosal barrier function and immunomodulation (McNaught and MacFie, 2001).

4.3 PREBIOTICS AND SYNBIOTICS

A prebiotic was first defined by Gibson and Roberfroid (1995) as *"non-digestible food ingredient that positively influences the host by selectively encouraging the growth and/or activity of one or a limited number of bacteria in the colon, and thus improves host health."* The revised definition of prebiotics takes into consideration, not only the microbial changes in the colonic ecosystem of human beings, but also to other areas which may benefit from a selective targeting of particular microorganisms. The criteria which are needed to prove a food ingredient as prebiotic (Gibson et al., 2004) are:

- It should not be affected by gastric acids, hydrolyzed by digestive enzymes, and absorbed in GIT;
- It should fermentable by gut microflora;
- It selectively stimulates growth and performance of those gut bacteria that provide positive health benefits.

Prebiotics are food ingredients that help support growth of probiotic bacteria or non-digestible substances that act as food for the gut microbiota. Basically, prebiotics stimulate growth or activity of certain healthy bacteria that live in the gut (Hutkins et al., 2016). It is a selectively fermented component which favors certain changes in the gut bacteria which offers health benefits to the host (Roberfroid, 2007). Some of the most common prebiotics are oligosaccharides, inulin, fructo-oligosaccharides, isomalto-oligosaccharides, etc.

The 'synbiotic' is used to designate a product wherein both probiotics and prebiotics are present (Schrezenmeir and de-Vrese, 2001) with a synergism between them (Pandey et al., 2015). Synbiotics can be 'complementary,' when each prebiotic and probiotic components is individually chosen for overall health effect; and 'synergistic' if the prebiotic ingredient is selected to support performance of selected probiotic microorganism (Kolida and Gibson, 2011).

4.4 PROBIOTICS IN THE FOOD INDUSTRY

Probiotic foods fall into functional foods category. Recently, there has been an upsurge of interest among consumers in this area, which has propelled food companies to focus on the development of novel probiotic food products.

4.4.1 DAIRY-BASED PROBIOTIC FOODS

For thousands of years, dairy foods have played important roles in the diet of human beings all over the world. Milk products are good matrices of probiotic strains supporting their growth and viability (Gardiner et al., 1999). Typical examples of probiotic dairy products consumed worldwide are baby milk powder, cheese, fermented milk, ice-cream, and pasteurized milk (Awaisheh, 2012). But, the viability of probiotics in dairy matrices, their physical, chemical, and organoleptic properties, health effects, regulatory concerns, and labeling issues are to be taken into consideration while developing functional dairy probiotic products in commercial-scale (Song et al., 2012).

4.4.1.1 YOGURT

Yogurt is the most commonly chosen vehicle for the delivery of probiotics. Yogurt is produced from milk by the action of *L. delbrueckii* subsp. *bulgaricus* and *Streptococcus thermophilus* bacteria. Other *Lactobacilli* and *Bifidobacteria* are also used for yogurt preparation (Sfakianakis and Tzia, 2014). Yogurt can be prepared traditionally by allowing *Lactobacillus* and *Streptococcus* spp. to ferment milk at 42–45°C. Modern-day fermentation process is a well-controlled one wherein milk, milk powder, sugar, fruit, flavors, colorings, emulsifiers, stabilizers, and pure cultures of *S. thermophilus* and *L. bulgaricus* are used. These two organisms work symbiotically during the process, with their population changing frequently. The pH of a commercial yogurt is usually in the range of 3.7–4.3 (Lourens-Hattingh and Viljoen, 2001).

Nutrient availability, growth promoters and inhibitors, solute concentration, fermentation time, incubation, and storage temperature are some of the factors that affect viability of probiotic microorganisms in yogurt. When organic acids accumulate and pH of the medium decreases during the fermentation process, the viability of probiotics shall decline. The metabolic products of organic acids across storage may further affect the viability (Donkor et al., 2006). Fruits are commonly added to yogurts, but that should be done carefully. Fruits often contain some inherent antimicrobial components that may adversely affect probiotic viability. There is a tendency in food industries to add higher level of probiotics to counteract the antimicrobial effect of fruit components, which usually result in a poor quality product (Aryana and McGrew, 2007). However, adoption of strategies like in encapsulation in alginate beads, alginate starch, etc. may improve the viability of probiotics in yogurt (Song et al., 2012). Although, probiotic yogurt is believed to have positive effects on cardiovascular, immune, and metabolic health, sufficient scientific evidences are lacking to conclude that yogurt improves health or lowers the risks of aliment (Rijkers et al., 2011).

4.4.1.2 DRINKABLE FRESH AND FERMENTED MILKS

A wide range of dairy beverages containing probiotic bacteria are available in the market. Functional dairy beverages can be categorized into two categories: a) fortified dairy beverages, and b) whey-based beverages. The first category beverages are fortified with certain specific nutrients, such as vitamins and minerals, which lack in the beverage. Sometimes these beverages contain augmented levels of some of its natural components like protein or calcium to boost the natural health aspects of milk. During cheese making, whey is obtained as a by-product. These days, food industries are more oriented towards production of fermented whey beverages having probiotic bacteria to produce functional beverages with high nutritive value (Marsh et al., 2014). Magalhaes et al. (2010) formulated a fermented cheese whey-based beverage using kefir grains as starter culture which was identified as a probiotic one as yeasts and *Lactobacilli* were detected during analysis.

Although various probiotic strains are used in the commercial preparation of probiotic dairy beverages, *L. acidophilus*, *L. casei*, *L. rhamnosus*, and *L. plantarum* are most commonly employed. Additional ingredients are also added to milk to improve growth and viability of the probiotics. Addition of citrus fiber for making fermented milk is an example of such ingredients (Sendra et al., 2008). Some of the commercial probiotic dairy beverages are presented in Table 4.1.

4.4.1.3 KEFIR

Kefir or *kephir*, or *bulgaros*, a fermented milk beverage, originated in the Caucasus Mountains region in West Asia. The beverage is made with *kefir* "grains," a yeast/bacterial starter culture (Leite et al., 2013). It is not only produced with the traditional way, but also, with industrial processes, wherein, various types of milk are used (Kandylis et al., 2016). Probiotic bacteria found in kefir are *L. acidophilus, Bifidobacterium bifidum, Streptococcus thermophilus, L. delbrueckii* subsp. *bulgaricus, L. helveticus, L. kefiranofaciens, L. lactis* and *Leuconostoc* species. Acetic acid bacteria and non-lactose fermenting yeast are also found in *kefir* grains (Leite et al., 2013). Kefir is reported to protect the human body from toxins and inhibit the growth of *Helicobacter pylori* (Nielsen et al., 2014).

TABLE 4.1 Some Probiotic Dairy Beverages Available Commercially (Prado et al., 2008; Temmerman et al., 2003; Tamime and Marshall, 1997)

Product	Probiotic Strains
Acidophilus milk	*Lactobacillus acidophilus*
Acidophilin	*L. acidophilus, L. lactis* subsp. *lactis*, kefir yeasts
Sweet acidophilus milk	*L. acidophilus*
Acidophilus buttermilk	*L. acidophilus, L. lactis* subsp. *lactis,* subsp
Actimel	*L. acidophilus, Saccharomyces lactis*
AKTfit, Biola, BioAktiv, YOMO, LGG+	*L. casei* Immunitas™
Yoplait360°, KaikuActif	*L. rhamnosus* GG
Bifidus milk	*Bifidobacterium. bifidum* or B. *longum*
Bifighurt®	*B. longum* (CKL 1969) or *B. longum* (DSM 2054)
Biomild	*L. acidophilus, B. bifidum*
Chamyto	*L. johnsonii, L. helveticus*
Gaio	*L. casei* F19
ProViva	*L. plantarum* 299v
Verum	*L. rhamnosus* LB21
Yakult	*L. casei* Shirota
Yakult Miru-Miru	*L. casei, B. bifidum* or *B. bereve, L. acidophilus*

4.4.1.4 KOUMISS

Koumiss or *kumiss* is a fermented dairy product which is traditionally prepared from raw mare's milk. This drink is very popular among the nomadic cattle breeders in Asia and some regions of Russia (Yerlikaya, 2014). During the fermentation process, alcohol, and lactic acid are produced, which result in a sparkling, creamy, and acid milk drink. *Koumiss* is different from *kefir* because of the starter culture and the content of acid and alcohol in the product (Niamsiri and Batt, 2009). It is used to treat asthma, tuberculosis, pneumonitis, gynecological diseases, and cardiovascular diseases (CVD) (Marsh et al., 2014).

4.4.1.5 CHEESE

Compared to yogurt and fermented milks, cheese has a higher pH and buffering capacity, more solid consistency, longer shelf life, higher nutritional and energy values (Ong et al., 2007). Recently, mesophilic *lactobacilli,* as adjunct cultures, are commonly used in commercial cheese production (Georgieva et al., 2009). With the addition of these cultures, not only the quality is improved (Milesi et al., 2009), but also, the growth of undesirable microorganisms is inhibited (Banks and Williams, 2004). The first commercial probiotic cheese was developed by the Mills DA, Oslo, Norway. Till date, more than 200 commercial probiotic cheeses in various forms are found in the market worldwide (Song et al., 2012). Zhang et al. (2013), in their study, found that the probiotic cheese with *L. plantarum* K25 could cut the risk of CVD in a mice-model.

4.4.1.6 OTHER DAIRY BASED PROBIOTIC FOODS

During the last few decades, novel probiotic ice-cream products have been introduced to the markets. These products are developed by incorporating probiotic cultures into ice-cream products. The incorporation of probiotic cultures into ice-cream resulted in value-added functional product rich in dairy material, vitamins, and minerals (Awaisheh et al., 2005). Composition-wise ice-cream includes milk proteins, fat, and lactose, as well as other compounds, and they, make ice-cream a good vehicle for probiotic cultures. Moreover, ice-cream's relatively high pH values (5.5 to 6.5) lead to an increased survival of the probiotic bacteria across storage (Goff, 2008). Thus, fruits or their derivatives having acidic character should be avoided in ice-creams containing probiotic cultures which may impact, not only the sensory profile, but also negatively culture's viability (Favaro-Trindade et al., 2007). During an experimentation, *Quarg* (a traditional form of cheese) was prepared from partially skimmed milk with two probiotic cultures to improve its nutrition characteristics, and the results showed that probiotics could ensure the highest level of utilization of fat, protein, lactose, and phosphorus (Durić et al., 2007). Alegro et al. (2007) produced a probiotic chocolate mousse using *L. paracasei* subsp. *Paracasei* LBC 82, solely, and together with inulin, which turned out to be a good matrix for delivery of the used probiotic strain.

4.4.2 NON-DAIRY-BASED PROBIOTIC FOODS

Dairy products are commonly and conveniently used as the carriers of probiotic bacteria, as they offer suitable environment to support growth and viability of the bacteria. However, dairy-based probiotic foods can exert certain negative effects on people having lactose intolerance and allergy to milk proteins; and on people with obesity and blood pressure due to high fat and cholesterol content of milk (Kumar et al., 2015). Dairy-based products require cold chain for storage and distribution which compelled R&D food scientists to look for new non-dairy based probiotic product lines (Heenan et al., 2004). Additionally, there is a constant increase in the consumer vegetarianism throughout the developed countries, which have created an increasing demand for vegetarian probiotic products (Prado et al., 2008). In the last two decades, several non-dairy probiotic products are developed and market. Some of the traditional and recently developed non-dairy probiotic products around the world are presented in Table 4.2.

4.4.2.1 CEREAL-BASED PROBIOTIC FOODS

Cereals contain prebiotic fibers which can protect the probiotic microbiota from the adverse conditions of the gastrointestinal tract. These provide good matrices for the production of probiotic products (Kandylis et al., 2016). Whole grains are sources of phytochemicals, including phytoestrogens, phenolic compounds, antioxidants, phytic acid and sterols (Katina et al., 2007), and can reduce Type 2 diabetes, CVD, obesity, and certain type of cancers (Clemens and Pressman, 2006). Fermentation of cereals increases the bioavailability of minerals (Sankara and Deosthale, 1983) due to the action of microbial enzymes and the organic acids produced during the fermentation process (Hotz and Gibson, 2007; Teucher et al., 2004). Although a large number of fermented cereal products have long been produced, the probiotic characteristics of microorganisms involved in cereal foods fermentation have recently been reported. Due to the complexity of cereals, a systematic approach is required to identify the factors that enhance the growth of probiotics in the cereals (Kedia et al., 2007).

Boza, an acidic beverage with low level of alcohol produced in the Balkan peninsula from cereals, mostly wheat, is natural source of probiotic LAB (Todorov et al., 2008). *Ogi*, a fermented maize product is a popular weaning and breakfast cereal in sub-Saharan Africa. Traditionally it is prepared by natural fermentation beginning with steeping maize grains in water for 2–4 days at room temperature, followed by wet milling, sieving, and souring of slurry for 2–3 days at room condition. The duration of fermentation and souring determine the degree of sourness and the nutrient status of *ogi*. *Lactobacillus. plantarum*, *L. fermentum,* and *Streptococcus lactis* are the predominant microorganisms in *ogi* fermentation (Enujiugha, 2006).

TABLE 4.2 Non-Dairy Probiotic Products (Cruz et al., 2010)

Category	Probiotic Products
Cereal products	Yogurt-based on rice,
	Puddings from cereals
	Oat drinks
	Yosa (pudding prepared from oat-bran)
	Mahewu (drink from slightly fermented maize-meal porridge)
	Drinks from wheat, millet, rye, maize, and other cereals
	Malt-based drink
	Boza (made from fermented cereals like wheat, maize, millets, etc.)
	Fermented beverage from millet or sorghum flour
Fruit and vegetable products	Banana pulp
	Banana puree
	Beet-root drink
	Tomato drink
	Juice of pineapple, cranberry, and orange
	Juice of passion fruit and grape
	Juice of cabbage
	Juice of carrot
	Juice of blackcurrant
Soy products	Fermented soymilk drink
	Non-fermented soy-based frozen desserts
Other non-dairy products	Meat products
	Dosa
	Probiotic cassava-flour product
	Starch-saccharified probiotic drink

Sobia, a traditional cereal-based sweet-sour fermented beverage made from malt and wheat flour, is very popular in the Western region of Saudi Arabia (Gassem, 2002). *Yosa* is another popular snack food in Finland and other Scandinavian countries. It is made from oat bran pudding cooked in water. Its texture is similar to yogurt but free from milk products (Wood, 1997).

Angelov et al. (2006) developed a probiotic drink, in which, a whole-grain oat substrate was fermented with *L. plantarum* B28. It had a shelf life of 21 days under refrigerated condition, and it combined the health benefits of a probiotic culture with the oat prebiotic beta-glucan. Saman et al. (2011) investigated the fermentability of brown rice and rice bran for the growth of human *L. plantarum* NCIMB 8826 isolated from the human gut and found that rice bran or rice bran extracts could be used in new probiotic food developments.

4.4.2.2 SOY-BASED PROBIOTIC FOODS

Soybean is rich in high-quality proteins. Relatively high concentrations of bioactive compounds like isoflavones and saponins are also present in soy, which contribute to reduction of the risk of various diseases (Tseng and Xiong, 2009). Soymilk fermentation with probiotic bacteria can increase the antioxidative activities of the fermented soymilk, which further increases the potential of developing a probiotic diet adjunct with probiotic fermented soymilk (Heenan et al., 2004). Sanni et al. (2013) developed a sorghum flour-based yogurt like product '*sorghurt*' which had a viable count of 5.5 $\log_{10}$cfu/ml of the product and had acceptable sensory scores. Recently, a probiotic soy-based product, similar to *petit-suisse* cheese was developed by Matias et al. (2014).

4.4.2.3 FRUITS AND VEGETABLES-BASED PROBIOTIC FOODS

Fruits and vegetables are considered healthy foods, and unlike dairy products, they lack allergens, lactose, and cholesterol, which adversely affect some portions of the population (Sheehan et al., 2007). Research is being continued in developing functional beverages from fruits, fortified with the probiotic and prebiotic components. The fruit juices have been recommended as an efficient matrix for the functional health ingredients since they intrinsically contain beneficial nutrients, and most importantly have taste profiles that are agreeable to all age groups (Tuorila and Cardello, 2002). Fruit juices naturally contain sugars which support the growth of probiotics (Ding and Shah, 2008).

Zheng et al. (2014) studied the suitability of litchi juice, treated with high hydrostatic pressure, for the production of a probiotic beverage by *L. casei*. They reported the production of a probiotic beverage with good Color, flavor, and overall acceptance. The beverage also retained total phenolics, and antioxidant capacity and the viability count of *L. casei* was more than 8.0 log cfu/mL after 28 days of storage at 4°C. Pakbin et al. (2014) studied the feasibility of producing a probiotic peach juice by inoculating with 24-hour old lactic acid bacteria (LAB) culture incubated at 30°C. They found that after 28 days of cold storage at 4°C, the viable cell counts of *L. delbrueckii* were 1.72×10^7 cfu/ml in fermented peach juice, which ascertained that peach juice could be a suitable matrix for the delivery of probiotic bacteria. Perricone et al. (2014) evaluated the viability of *L. reuteri* DSM 20016, in a different type of fruit juices in both refrigerated and thermal abuse conditions. They found that fruit juices could be a good matrix for delivery of the probiotic strain, which showed a high viability in apple, pineapple, and orange juices compared to red-fruit juice (a commercial formulation, having pH, 2.97; soluble solids, 10.3Bx; sugar 1.05 g/l, 20% red-orange, 20% blueberry, 10% pomegranate). Thus, the viability of the bacteria is strongly affected by the type of juice, and this can be improved by induction of resistance through exposure to sub-lethal stress. Minimally processed pineapple is also a suitable vehicle for probiotic bacteria like *L. plantarum* and *L. fermentum (*Russo et al., 2014). The *Lactobacilli* can also counteract the growth of *Listeria monocytogens* and *Escherichia coli* O157:H7 on pineapple plugs. Noni juice is also identified as a suitable carrier for probiotic *B. longum* and *L. plantarum* (Wang et al., 2009).

To counteract the harsh acidic environment in certain fruit juices, microencapsulation technologies have been designed and successfully applied using various matrices (Perricone et al., 2015). Ca-alginate can carry out a fermentation of banana puree, which can give a novel probiotic banana product with higher number of viable cells (Tsen et al., 2003). Gaanappriya et al. (2013) evaluated the viability of encapsulated *L. plantarum* in sapodilla, grapes, orange, and watermelon juices, and maintained the probiotic at 7 $\log_{10}$ CFU/ml or more for one month.

Vegetables are rich sources of the biologically active compounds, which have beneficial effects in prevention of some diseases and certain types of cancer. Due to the presence of carbohydrates, vegetable juices can be a suitable substrate for lactic acid fermentation. Fermentation of these juices cannot only help in their preservation but also can improve its nutritive and protective properties (Campbell-Platt, 1994). Beetroot juice is a good carrier of *L. acidophilus* and *L. plantarum*. The probiotic bacterial count in the fermented juice is around 10^9 cfu/ml after storage for 28 days at 4°C (Yoon et al., 2005). Several other substrates like cabbage, watermelon, tomato juice have been tested with positive results (Sivudu et al., 2014; Tamminen et al., 2013; Yoon et al., 2006).

Vegetable juices enriched with brewer's yeast autolysate before lactic acid fermentation can increase the number of LAB during fermentation (Rakin et al., 2007). Fermented fruits and vegetables are a potential source of probiotics as they harbor several LAB. Some of the fermented fruits and vegetables of the Asian region that have been indicated as a potential source of probiotic cultures is given in Table 4.3.

TABLE 4.3 Fermented Foods Made from Fruits and Vegetables of Asia (Swain et al., 2014)

Fermented Product	Country	Fruit and Vegetables Used	Microorganisms
Ca muoi	Vietnam	Eggplant	*L. fermentum L. pentosus L. brevis*
Dakguadong	Thailand	Mustard leaf	*L. plantarum*
Gundruk	Nepal, India	Cabbage, radish, mustard, cauliflower	*Pediococcus*spp. *Lactobacillus* spp.
Inziangsang	India	Mustard leaf	*L. plantarum L. brevis, Pediococcusacidilactici*
Khalpi	Nepal	Cucumber	*L. plantarum P. pentosaceus*
Nozawana-Zuke	Japan	Turnip	*L. curvatus*
Soidon	India	Bamboo Shoot	*L. brevis L. fallax L. lactis*
Tempoyak	Malaysia	Durian	*L. brevis L. mesenteroides L. mali L. fermentum*
Yan-taozih	China, Taiwan	Peaches	*L. mesenteroides Weissellacibaria L. lactis* subsp. *Lactis W. paramesenteroides Enterococcus faecalis W. minor L. brevis*

4.4.2.4 MEAT-BASED PROBIOTIC PRODUCTS

Because of its unique composition and structure, meat is considered as an excellent vehicle for probiotics. Meat exhibits a protective nature on LAB's against the harmful action of bile (Awaisheh, 2012). Probiotic applications in meat are limited to fermented meats, such as dry sausages (Song et al., 2012). But, the production of probiotic meat products necessitates overcoming certain technological constrains native microflora of meat, use of additives like nitrites and salt, low water activity, and low content or absence of natural sugars. Moreover, the probiotic bacteria used in the manufacturing process should be capable of surviving in conditions found in fermented products, and the sensory profile of the product must be intact (Krajewska and Dolatowski, 2012). The popular LAB strains most employed in meat starter cultures include *L. casei, L. curvatus, L. pentosus, L. plantarum, L. sakei, Pediococcus acidilactici* and *P. pentosaceus (*Tyopponen et al., 2003).

4.4.2.5 FISH-BASED PROBIOTICS

The literature on fish-based probiotics is meager. People of North East India have some traditional recipe for fermenting fish for long-term storage and use. Thapa et al. (2004) studied the microbial diversity of some traditional North-east Indian fermented fish products like *Hentak, Ngari, and Tungtap* and found the existence of potentially probiotic LAB such as *L. plantarum, L. lactis* subsp. *cremoris, L. amylophilus, L. fructosus, L. coryniformis* subsp. *torquens*, and *Enterococcus faecium.*

4.5 FUTURE SCOPE

As of now, consumers are oriented towards functional products, and there is an indication that probiotic foods are going to have a major share in the food market in the near future. There is a need for new studies to check the combination of dairy and non-dairy substrates, which may result in novel products with synergistic traits. Also, the delivery techniques like microencapsulation need to be fine-tuned in the near future. Probiotic foods have attained quite a lot of popularity in the developed nations, the market penetration of these products in developing and third world countries is low. So, efficient marketing of these products, highlighting the benefits is a dire necessity. In the quest for developing novel probiotic strains, gene technology, and relative genomics will play a key role. Moreover, with gene sequencing, an increased thoughtful mechanisms and functionality of probiotics will be known.

4.6 CONCLUSION

In the last two decades, there has been a tremendous increase in the use of probiotics and their applications in the food industry. Although it has been proved that dairy-based fermented products are the best matrices for delivering probiotics, there is growing evidence of the possibility of obtaining probiotic foods from non-dairy matrices. Several raw materials, such as cereals, fruits, and vegetables, meat, etc., have recently been assessed as suitable substrate for designing new, non-dairy probiotic foods. The development of novel, economic, and technological matrices is an utmost necessity to bring the non-dairy probiotic foods *at par* with dairy probiotic foods. While developing novel probiotic products for commercialization, certain aspects like stability, sensory acceptance, and price are to be taken into consideration. With technological advances in this area, the use of prebiotics in combination with non-dairy probiotic products to produce synbiotics on a commercial-scale at reasonable prices can be done.

KEYWORDS

- **cereal-based probiotic foods**
- **dairy-based fermented products**
- **fish-based probiotics**
- **fruits and vegetables-based probiotic foods**
- **meat-based probiotics**
- **soy-based probiotic foods**

REFERENCES

Alegro, L. C. A., Alegro, J. H. A., Cardarelli, H. R., Chiu, M. C., & Saad, S. M. I., (2007). Potentially probiotic and synbiotic chocolate mousse. *LWT- Food Sci. Technol., 40*, 669–675.

Angelov, A., Gotcheva, V., Kuncheva, R., & Hristozova, T., (2006). Development of a new oat-based probiotic drink. *Int. J. Food Microbiol., 112*, 75–80.

Aryana, K. J., & McGrew, P., (2007). Quality attributes of yogurt with *Lactobacillus casei* and various prebiotics. *LWT: Food Sci. Technol., 40*, 1808–1814.

Awaisheh, S. S., (2012). Probiotic food products classes, types, and processing. In: *Probiotics* (pp. 551–582). In Tech, Rijeka, Croatia.

Awaisheh, S. S., Hadaddin, M. S., & Robinson, R. K., (2005). Incorporation of selected nutraceuticals and probiotic bacteria into fermented dairy product. *Int. Dairy Technol., 10*, 1189–1195.

Banks, J. M., & Williams, A. G., (2004). The role of the nonstarter lactic acid bacteria in Cheddar cheese ripening. *Int. J. Dairy Technol., 57*, 145–152.

Boyle, R. J., & Tang, M. L., (2006). The role of probiotics in the management of allergic disease. *Clin. Exp. Allerg., 36*, 568–576.
Campbell-Platt, G., (1994). Fermented foods–a world perspective. *Food Res. Int., 27*, 253–257.
Clemens, R., & Pressman, P., (2006). Heyday in grain land. *Food Technol., 60*, 18.
Cruz, A. G., Cadena, R. S., Walter, E. H. M., Mortazavian, A. M., Granato, D., Faria, J. A. F., & Bolini, H. M. A., (2010). Sensory analysis: Relevance for prebiotic, probiotic, and synbiotic product development. *Compr. Rev. Food Sci. Food Saf., 9*, 358–373.
Ding, W. K., & Shah, N. P., (2008). Survival of free and microencapsulated probiotic bacteria in orange and apple juices. *Int. Food Res. J., 15*, 219–232.
Donkor, O., Henriksson, A., Vasiljevic, T., & Shah, N., (2006). Effect of acidification on the activity of probiotics in yogurt during cold storage. *Int. Dairy J., 16*, 1181–1189.
Durić, M. S., Iličić, M. D., Milanović, S. D., Carić, M. D., & Tekić, M. N., (2007). Nutritive characteristics of probiotic quark as influenced by type of starter. *Acta Period. Technol., 38*, 11–19.
Enujiugha, V. N., (2006). Supplementation of ogi, a maize-based infant weaning food, with African oil bean seed (*Pentaclethra macrophylla* Benth). *Intern. J. Postharv. Technol. Innov., 1*, 202–211.
FAO/WHO, (2002). *Guidelines for the Evaluation of Probiotics in Food.* Food and Agriculture Organization of the United Nations, and World Health Organization, Report of a joint FAO/WHO working group on drafting guidelines for the evaluation of probiotics in food, London Ontario, Canada.
FAO/WHO, (2006). *Probiotics in Food: Health and Nutritional Properties and Guidelines for Evaluation.* Report of a joint FAO/WHO expert consultation on evaluation of health and nutritional properties of probiotics in food including powder milk and live lactic acid bacteria, World Health Organization and Food and Agriculture Organization of the United Nations, Rome.
Favaro-Trindade, C. S., Carvalho-Balieiro, J. C. D., Dias, P. F., Sanino, F. A., & Boschini, C., (2007). Effects of culture pH and fat concentration on melting rate and sensory characteristics of probiotic fermented yellow mombin (*Spondias mombin* L) ice creams. *Food Sci. Technol. Int., 13*, 285–291.
Fuller, R., (1989). Probiotics in man and animals. *J. Appl. Bacteriol., 66*, 365–378.
Gaanappriya, M., Guhankumar, P., Kiruththica, V., Santhiya, N., & Anita, S., (2013). Probiotic action of fruit juices by *Lactobacillus acidophilus*. *Int. J. Adv. Biotechnol. Res., 4*, 72–77.
Gardiner, G., Stanton, C., Lynch, P., Collins, J., Fitzgerald, G., & Ross, R., (1999). Evaluation of cheddar cheese as a food carrier for delivery of a probiotic strain to the gastrointestinal tract. *J. Dairy Sci., 82*, 1379–1387.
Gassem, M. A. A., (2002). A microbiological study of Sobia: A fermented beverage in the western province of Saudi Arabia. *World J. Microbiol. Biotechnol., 18*, 173–177.
Georgieva, R., Iliev, I., Haertlé, T., Chobert, J. M., Ivanova, I., & Danova, S., (2009). Technological properties of candidate probiotic *Lactobacillus plantarum* strains. *Int. Dairy J., 19*, 696–702.
Gibson, G. R., & Roberfroid, M. B., (1995). Dietary modulation of the human colonic microbiota: Introducing the concept of prebiotics. *J. Nutr., 125*, 1401–1412.
Gibson, G. R., Probert, H. M., Loo, J. V., Rastall, R. A., & Roberfroid, M. B., (2004). Dietary modulation of the human colonic microbiota: Updating the concept of prebiotics. *Nutr. Res. Rev., 17*, 259–275.
Gilliland, S. E., (1990). Health and nutritional benefits from lactic acid bacteria. *FEMS Microbiol. Lett., 87*, 175–188.

Goff, D., (2008). Sixty-five years of ice-cream science. *Int. Dairy J., 7*, 754–758.

Harish, K., & Varghese, T., (2006). Probiotics in humans: Evidence-based review. *Calicut Medl. J., 4*(4), e3.

Heenan, C. N., Adams, C., Hoskena, R. W., & Fleet, H., (2004). Survival and sensory acceptability of probiotic microorganisms in a nonfermented frozen vegetarian dessert. *LWT- Food Sci. Technol., 37*, 461–466.

Hotz, C., & Gibson, R. S., (2007). Traditional food-processing and preparation practices to enhance the bioavailability of micronutrients in plant-based diets. *J. Nutr., 137*, 1097–1100.

Hutkins, R. W., Krumbeck, J. A., Bindels, L. B., Cani, P. D., Fahey, G., Goh, Y. J., et al. (2016). Prebiotics: Why definitions matter. *Curr. Opin. Biotechnol., 37*, 1–17.

Kandylis, P., Pissaridi, K., Bekatorou, A., Kanellaki, M., & Koutinas, A. A., (2016). Dairy and non-dairy probiotic beverages. *Curr. Opin. Food Sci., 7*, 58–63.

Katina, K., Liukkonen, K. H., Kaukovirta-Norja, A., Adlercreutz, H., Heinonen, S. M., Lampi, A. M., Pihlava, J. M., & Poutanen, K., (2007). Fermentation-induced changes in the nutritional value of native or germinated rye. *J. Cereal Sci., 46*, 348–355.

Kedia, G., Wang, R., Patel, H., & Pandiella, S. S., (2007). Use of mixed cultures for the fermentation of cereal-based substrates with potential probiotic properties. *Process Biochem., 42*, 65–70.

Kemgang, T. S., Kapila, S., Shanmugam, V. P., & Kapila, R., (2014). Cross-talk between probiotic lactobacilli and host immune system. *J. Appl. Microbiol., 117*, 303–319.

Khan, S. H., & Ansari, F. A., (2007). Probiotics: The friendly bacteria with market potential in global market. *Pak. J. Pharm. Sci., 20*, 71–76.

Kolida, S., & Gibson, G. R., (2011). Synbiotics in health and disease. *Ann. Rev. Food Sci. Technol., 2*, 373–393.

Krajewska, D. K., & Dolatowski, Z. J., (2012). Probiotic meat products and human nutrition. *Process Biochem., 47*, 1761–1772.

Kumar, B. V., Venkata, S., Vijayendra, N., Vijaya, O., & Reddy, S., (2015). Trends in dairy and non-dairy probiotic products - a review. *J. Food Sci. Technol., 52*, 6112–6124.

Leite, A. M. D. O., Miguel, M. A., Peixoto, R. S., Rosado, A. S., Silva, J. T., & Paschoalin, V. M., (2013). Microbiological, technological and therapeutic properties of kefir: A natural probiotic beverage. *Braz. J. Microbiol., 44*, 341–349.

Lourens-Hattingh, A., & Viljoen, B. C., (2001). Yogurt as probiotic carrier food. *Int. Dairy J., 11*, 1–17.

Magalhaes, K. T., Pereira, M. A., Nicolau, A., Dragone, G., Domingues, L., Teixeira, J. A., Silva, J. B. D. A., & Schwan, R. F., (2010). Production of fermented cheese whey-based beverage using kefir grains as starter culture: Evaluation of morphological and microbial variations. *Bioresour. Technol., 101*, 8843–8850.

Marsh, A. J., Hill, C., Ross, R. P., & Cotter, P. D., (2014). Fermented beverages with health-promoting potential: Past and future perspectives. *Trends Food Sci. Technol., 38*, 113–124.

Marteau, P., Gerhardt, M. F., Myara, A., Bouvier, E., Trivin, F., & Rambaud, J. C., (1995). Metabolism of bile salts by alimentary bacteria during transit in the human small intestine. *Microbiol. Ecol. Health Dis., 8*, 151–157.

Martins, E. M. F., Ramos, A. M., Vanzela, E. S. L., Stringheta, P. C., Pinto, C. L. D. O., & Martins, J. M., (2013). Products of vegetable origin: A new alternative for the consumption of probiotic bacteria. *Food Res. Int., 51*, 764–770.

Matias, N. S., Bedanin, R., Castro, A., & Saad, S. M. I., (2014). A probiotic soy-based innovative product as an alternative to petit-suisse cheese. *LWT- Food Sci. Technol., 59*, 411–417.

McNaught, C. E., & MacFie, J., (2001). Probiotics in clinical practice: A critical review of the evidence. *Nutr. Res., 21*, 343–353.

Milesi, M. M., Vinderola, G., Sabbag, N., Meinardi, C. A., & Hynes, E., (2009). Influence on cheese proteolysis and sensory characteristics of non-starter *Lactobacilli* strains with probiotic potential. *Food Res. Int., 42*, 1186–1196.

Nagpal, R., Kumar, A., Kumar, M., Behare, P. V., Jain, S., & Yadav, H., (2012). Probiotics, their health benefits and applications for developing healthier foods: A review. *FEMS Microbiol. Lett., 334*, 1–15.

Niamsiri, N., & Batt, C. A., (2009). Dairy products. In: Schaechter, M., (ed.), *Encyclopedia of Microbiology* (pp. 34–44). Academic Press, Oxford.

Nielsen, B., Gürakan, G. C., & Ünlü, G., (2014). Kefir: A multifaceted fermented dairy product. *Probiot. Antimicrob. Proteins, 6*, 123–135.

Ong, L., Henriksson, A., & Shah, N. P., (2007). Chemical analysis and sensory evaluation of cheddar cheese produced with *Lactobacillus acidophilus*, *Lb. casei*, *Lb. paracasei* or *Bifidobacterium*. *Int. Dairy J., 17*, 937–945.

Pakbin, B., Razavi, S. H., Mahmoudi, R., & Gajarbeygi, P., (2014). Producing probiotic peach juice. *Biotechnol. Health Sci., 1*, 1–5.

Pandey, K. R., Naik, S. R., & Vakil, B. V., (2015). Probiotics, prebiotics and synbiotics- a review. *J. Food Sci. Technol., 52*, 7577–7587.

Parvez, S., Malik, K. A., Kang, S. A., & Kim, H. Y., (2006). Probiotics and their fermented food products are beneficial for health. *J. Appl. Microbiol., 100*, 1171–1185.

Perricone, M., Bevilacqua, A., Altieri, C., Sinigaglia, M., & Corbo, M. R., (2015). Challenges for the production of probiotic fruit juices. *Beverages, 1*, 95–103.

Perricone, M., Corbo, M. R., Sinigaglia, M., Speranza, B., & Bevilacqua, A., (2014). Viability of *Lactobacillus reuteri* in fruit juices. *J. Funct. Foods, 10*, 421–426.

Plé, C., Breton, J., Daniel, C., & Foligné, B., (2015). Maintaining gut ecosystems for health: Are transitory food bugs stowaways or part of the crew? *Int. J. Food Microbiol., 213*, 139–143.

Prado, F. C., Parada, J. L., Pandey, A., & Soccol, C. R., (2008). Trends in non-dairy probiotic beverages. *Food Res. Int., 41*, 111–123.

Rakin, M., Vukasinovic, M., Siler-Marinkovic, S., & Maksimovic, M., (2007). Contribution of lactic acid fermentation to improved nutritive quality vegetable juices enriched with brewer's yeast autolysate. *Food Chem., 100*, 599–602.

Ranadheera, R. D. C. S., Baines, S. K., & Adams, M. C., (2010). Importance of food in probiotic efficacy. *Food Res. Int., 43*, 1–7.

Rasic, J. L., (2003). Microflora of the intestine probiotics. In: Caballero, B., Trugo, L., & Finglas, P., (eds.), *Encyclopedia of Food Sciences and Nutrition* (pp. 3911–3916). Academic Press, Oxford.

Rijkers, G. T., Bengmark, S., Enck, P., Haller, D., Herz, U., Kalliomaki, M., et al. (2010). Guidance for substantiating the evidence for beneficial effects of probiotics: Current status and recommendations for future research. *J. Nutr., 140*, 671S–676S.

Rijkers, G. T., Vos, W. M. D., Brummer, R. J., Morelli, L., Corthier, G., & Marteau, P., (2011). Health benefits and health claims of probiotics: Bridging science and marketing. *Br. J. Nutr., 106*, 1291–1296.

Ringel, Y., Quigley, E. M. M., & Lin, H. C., (2012). Using probiotics in gastrointestinal disorders. *Am. J. Gastroenterol., 1*, 34–40.

Roberfroid, M., (2007). Prebiotics: The concept revisited. *J. Nutr., 137*, 830S–837S.

Russo, P., Chiara, M. L. V. D., Vernile, A., Amodio, M. L., Arena, M. P., Capozzi, V., Massa, S., & Spano, G., (2014). Fresh-cut pineapple as a new carrier of probiotic lactic acid bacteria. *Biomed. Res. Int.,* 1–9.

Saarela, M., Mogensen, G., Fondén, R., Mättö, J., & Mattila-Sandholm, T., (2000). Probiotic bacteria: Safety, functional and technological properties. *J. Biotechnol., 84,* 197–215.

Salminen, S., & Von-Wright, A., (1998). Current probiotics-safety assured? *Microb. Ecol. Health Dis., 10,* 68–77.

Saman, P., Fuciños, P., Vázquez, J. A., & Pandiella, S. S., (2011). Fermentability of brown rice and rice bran for growth of human *Lactobacillus plantarum* NCIMB 8826. *Food Technol. Biotechnol., 49,* 128–132.

Sankara, R. D. S., & Deosthale, Y. G., (1983). Mineral composition, ionizable iron and soluble zinc in malted grains of pearl millet and ragi. *Food Chem., 11,* 217–223.

Sanni, A., Franz, C., Schillinger, U., Huch, M., Guigas, C., & Holzapfel, W., (2013). Characterization and technological properties of LAB in the production of "sorghurt," a cereal-based product. *Food Biotechnol., 27,* 178–198.

Sarao, L. K., & Arora, M., (2017). Probiotics, prebiotics and microencapsulation - a review. *CRC Crit. Rev. Food Sci. Nutr., 57,* 344–371.

Schrezenmeir, J., & De-Vrese, M., (2001). Probiotics, prebiotics, and synbiotics—approaching a definition. *Am. J. Clin. Nutr., 73,* 361S–364S.

Sendra, E., Fayos, P., Lario, Y., Fernández-López, J., Sayas-Barberá, E., & Pérez-Alvarez, J. A., (2008). Incorporation of citrus fibers in fermented milk containing probiotic bacteria. *Food Microbiol., 25,* 13–21.

Senok, A. C., Ismaeel, A. Y., & Botta, G. A., (2005). Probiotics: Facts and myths. *Clin. Microbiol. Infect., 11,* 958–966.

Sfakianakis, P., & Tzia, C., (2014). Conventional and innovative processing of milk for yogurt manufacture, development of texture and flavor: A review. *Foods, 3,* 176–193.

Shah, N. P., (2007). Functional cultures and health benefits. *Int. Dairy J., 17,* 1262–1277.

Sheehan, V. M., Ross, P., & Fitzgerald, G. F., (2007). Assessing the acid tolerance and the technological robustness of probiotic cultures for fortification in fruit juices. *Innov. Food Sci. Emerg. Technol., 8,* 279–284.

Shimizu, K., Nomoto, K., Ogura, H., Morotomi, M., Kuwagata, Y., Asahara, T., Tasaki, O., Shimazu, T., Matsushima, A., & Sugimoto, H., (2013). Probiotic/synbiotic therapy for treating critically ill patients from a gut microbiota perspective. *Dig. Dis. Sci., 58,* 23–32.

Sivudu, S. N., Umamahesh, K., & Reddy, O. V. S., (2014). A comparative study on probiotication of mixed watermelon and tomato juice by using probiotic strains of *Lactobacilli. Int. J. Curr. Mibriol. Appl. Sci., 3,* 977–984.

Solis, B., Samartín, S., Gómez, S., Nova, E., Rosa, B. D. L., & Marcos, A., (2002). Probiotics as a help in children suffering from malnutrition and diarrhoea. *Eur. J. Clin. Nutr., 56*(3), S57–S59.

Song, D., Ibrahim, S., & Hayek, S., (2012). Recent application of probiotics in food and agricultural science. In: *Probiotics* (pp. 3–36). In Tech, Rijeka, Croatia.

Swain, M. R., Anandharaj, M., Ray, R. C., & Rani, R. P., (2014). Fermented fruits and vegetables of Asia: A potential source of probiotics. *Biotechnol. Res. Int.,* 1–19.

Tamime, A. Y., & Marshall, V. M. E., (1997). Microbiology and technology of fermented milks. In: *Microbiology and Biochemistry of Cheese and Fermented Milk* (pp. 57–152). Blackie Academic and Professionals, London.

Tamminen, M., Salminen, S., & Ouwehand, A. C., (2013). Fermentation of carrot juice by probiotics: Viability and preservation of adhesion. *Int. J. Biotechnol. Wellness Ind., 2,* 10–15.

Temmerman, R., Pot, B., Huys, G., & Swings, J., (2003). Identification and antibiotic susceptibility of bacterial isolates from probiotic products. *Int. J. Food Microbiol., 81*, 1–10.

Teucher, B., Olivares, M., & Cori, H., (2004). Enhancers of iron absorption: Ascorbic acid and other organic acids. *Int. J. Vitam. Nutr. Res., 74*, 403–419.

Thapa, N., Pal, J., & Tamang, J. P., (2004). Microbial diversity in ngari, hentak and tungtap, fermented fish products of north-east India. *World J. Microbiol. Biotechnol., 20*, 599–607.

Tharmaraj, N., & Shah, N. P., (2009). Antimicrobial effects of probiotics against selected pathogenic and spoilage bacteria in cheese-based dips. *Int. Food Res. J., 16*, 261–276.

Todorov, S. D., Botes, M., Guigas, C., Schillinger, U., Wiid, I., Wachsman, M. B., Holzapfel, W. H., & Dicks, L. M. T., (2008). Boza, a natural source of probiotic lactic acid bacteria. *J. Appl. Microbiol., 104*, 465–477.

Tsen, J. H., Lin, Y. P., & King, V. A. E., (2003). Banana puree fermentation by *Lactobacillus acidophilus* immobilized in Ca-alginate. *J. Gen. Appl. Microbiol., 49*, 357–361.

Tseng, Y., & Xiong, Y. L., (2009). Effect of inulin on the rheological properties of silken tofu coagulated with glucono-d-lactone. *J. Food Eng., 90*, 511–516.

Tuorila, H., & Cardello, A. V., (2002). Consumer responses to an off-flavor in juice in the presence of specific health claims. *Food Qual. Prefer., 13*, 561–569.

Tyopponen, S., Petaja, E., & Mattila-Sandholm, T., (2003). Bioprotectives and probiotics for dry sausages. *Int. J. Food Microbiol., 83*, 233–244.

Vandenplas, Y., Huys, G., & Daube, G., (2015). Probiotics: An update. *J. Pediatr., 91*, 6–21.

Vasiljevic, T., & Shah, N. P., (2008). Probiotics-from Metchnikoff to bioactives. *Int. Dairy J., 18*, 714–728.

Velasquez-Manoff, M., (2015). Gut microbiome: The peace-keepers. *Nat., 518*, S3–S11.

Wang, C. Y., Ng, C. C., Su, H., Tzeng, W. S., & Shyu, Y. T., (2009). Probiotic potential of noni juice fermented with lactic acid bacteria and *Bifidobacteria. Int. J. Food Sci. Nutr., 60*, 98–106.

Wood, P. J., (1997). Functional foods for health: Opportunities for novel cereal processes and products. *Cereal, 8*, 233–238.

Yerlikaya, O., (2014). Starter cultures used in probiotic dairy product preparation and popular probiotic dairy drinks. *Food Sci. Technol., 34*, 221–229.

Yoon, K. Y., Woodams, E. E., & Hang, Y. D., (2005). Fermentation of beet juice by beneficial lactic acid bacteria. *LWT, 38*, 73–75.

Yoon, K. Y., Woodams, E. E., & Hang, Y. D., (2006). Production of probiotic cabbage juice by lactic acid bacteria. *Bioresour. Technol., 97*, 1427–1430.

Zhang, L., Zhang, X., Liu, C., Li, C., Li, S., Li, T., Li, D., Zhao, Y., & Yang, Z., (2013). Manufacture of cheddar cheese using probiotic *Lactobacillus plantarum* K25 and its cholesterol-lowering effects in a mice model. *World J. Microbiol. Biotechnol., 29*, 127–135.

Zheng, X., Yu, Y., Xiao, G., Xu, Y., Wu, J., Tang, D., & Zhang, Y., (2014). Comparing product stability of probiotic beverages using litchi juice treated by high hydrostatic pressure and heat as substrates. *Innov. Food Sci. Emerg. Technol., 23*, 61–67.

CHAPTER 5

Bioactive Compounds and Health Benefits of Phalsa: An Underutilized Fruit

ARADHITA BARMAN RAY and KIRAN BALA

Department of Food Technology, Guru Jambheshwar University of Science and Technology, Hisar (Haryana)–125001, India,
E-mail: dhitaray@gmail.com

ABSTRACT

In recent years, more attention has been focused on darkly pigmented fruits and vegetables which are considered very rich dietary sources of bioactive compounds. Owing to their numerous health benefits, natural bioactive compounds are in great demand today being non-toxic in nature and for providing protection against various oxidative damages as well as various chronic diseases inside our body. These compounds also possess superb health benefits like antioxidant potential, free radical scavenging activity (SA), and reducing potential. Phalsa (*Grewia asiatica* L.), is a very small sized fruit belonging to the family *Tiliaceae*, is an underutilized indigenous minor berry, which possesses very high nutritional as well as medicinal components. It serves as a good source of proteins, amino acids, vitamins, minerals, vitamin A and C, phosphorus, iron along with various bioactive compounds including anthocyanins, tannins, phenolics, and flavonoids. This multipurpose plant possesses antioxidant, anti-hyperglycemic, radioprotective, hepato-protective, antifungal, and antiviral activity and also utilized for treating lack of appetite, typhus, acidity, giddiness, diarrhea, hypertension, and anorexia. Henceforth, this fruit is establishing importance as a prospective source of natural bioactive compounds due to the presence of various phytochemicals (flavonoids, phenolics, anthocyanins, and tannins) and consequently attracted the attention of food scientists and researchers.

This unexplored fruit possesses a higher scope to be recognized by the food industries as its study on multiple health benefits and value addition by processing is still scanty. Therefore, an in-depth study regarding the importance of the phalsa bioactive compounds and their probable health benefits has been discussed and presented in this chapter.

5.1 INTRODUCTION

Fruits are the basic part of our diet, which plays a preventive and curative role against health-related problems, since time immemorial. Due to varying environmental conditions, India possesses the huge potential of a growing variety of major and minor fruits (Ravani and Joshi, 2014). *Grewia asiatica* L., locally known as phalsa, is counted among the least-utilized edible fruit-bearing medicinal plant possess high nutritional as well as medicinal potential (Zia-Ul-Haq et al., 2015). India is considered the home of phalsa, which is one of the most popular sub-tropical and tropical berry type fruit (Gupta et al., 2006). This fruit is reported to be native to South Asia, cultivated mainly for its edible fruit which are a very rich source of nutrients like proteins, amino acids, vitamins, minerals, and various bioactive compounds such as anthocyanins, tannins, phenolics, and flavonoids (Zia-Ul-Haq et al., 2013). Ahaskar et al. (2007) reported that the ripe fruit is a good source of anthocyanin type cyanidin 3-glucoside, minerals, and dietary fibers, etc. The phalsa plant belongs to the family *Tiliaceae* is a scraggly shrub (dwarf type) and small tree (tall type), which grows to 4 m or more in height and mainly found in India, Pakistan, Southeast Asia and the USA, etc. (Gupta et al., 2007). In India, it is commercially cultivated in states of Punjab, Haryana, Uttar Pradesh, and Andhra Pradesh (Abid et al., 2012).

Phalsa plant possesses antioxidant, antihyperglycaemic, radioprotective, hepatoprotective, antifungal, and antiviral activity and also utilized for treating lack of appetite, typhus, acidity, giddiness, diarrhea, hypertension, and anorexia as mentioned by Bhangale et al. (2013). The fruit is also termed as 'star apple' which is small round, 1.0–1.9 cm in diameter, 0.8–1.6 cm in height, 0.5–2.2 g in weight and bears clusters of 2–8 in numbers on leaf axils (Tiwari et al., 2014). Gupta et al. (2006) mentioned that these fruits ripen in stages on bushes when the color of fruit skin changes from light green to cherry red or purplish red during ripening and finally to dark purple or nearly black in color. Ripened fruits are found covered with a very thin, whitish blush. Kumar et al. (2014) reported that these berry-like deep reddish brown colored fruits are a good source of vitamin A and C, along with a fair source of phosphorus and iron, sub-acidic in nature and exert a cooling effect. Goyal (2012) described that phalsa fruits are astringent and stomachic, unripe fruits alleviates inflammation, administered in respiratory, cardiac, blood disorders, and treating fever. Two varieties of phalsa, i.e., tall (Figure 5.1a, b) and dwarf (Figure 5.2a, b) are locally grown in dry, arid regions, but both of these differ with respect to various physical and chemical characteristics (Pareek and Panwar, 1981).

FIGURE 5.1 (See color insert.) (a) Tall plant of phalsa; (b) Tall variety phalsa fruit.

FIGURE 5.2 (See color insert.) (a) Dwarf plant; (b) Dwarf variety phalsa fruit.

5.2 BIOACTIVE COMPOUNDS AND HEALTH BENEFITS

Bioactive compounds from natural plant sources are gaining more attention in recent times and emerging as natural antioxidants. These are the extra-nutritional constituents found in a very small amount in some foods (Kris-Etherton et al., 2002). Antioxidant potential of fruits and vegetables is of great interest from consumer's health point of view. Phytochemical antioxidants are the secondary plant metabolites and carotenoids, flavonoids, cinnamic acids, benzoic acids, folic acid, ascorbic acid, tocopherols, tocotrienols, etc., are some of the examples of these compounds (Ramamoorthy and Bono, 2007). Loganayaki and Manian (2010) reported that disease-preventing potential is due to many antioxidants found in fruits and vegetables, which include phytochemicals like polyphenols, carotenoids, and vitamins. Akhtar et al. (2015) mentioned that antioxidants could be synthetic also but owing to toxic and carcinogenic effects, synthetic antioxidants, like butyl hydroxyanisole and butylhydroxytoluene are being replaced with natural antioxidants. Natural bioactive compounds are in great demand for food and pharmaceutical industries due to their health improving potential. As reported by Gungor and Sengul (2008), especially, deep-colored fruits are rich sources of various bioactive compounds like phenols, flavonoids, anthocyanins, carotenoids, and thiols as well as vitamins (ascorbic acid and tocopherols).

Bioactive compounds have attracted the attention of Food scientists, Technologists, and Researchers, as health consciousness among people has been increased during recent years. These are the biologically active compounds/molecules from renewable resources and applications to be used as additives in a number of food and other industries. Amount of these compounds in plants varies among species, plant-part, specific tissues, and plant maturity (Julkunen-Tiitto and Sorsa, 2001).

Natural bioactive compounds are also called phytochemicals, found in fruits, vegetables, grains, legumes, tea, and wine, which also play a significant role for the prevention of many diseases like chronic and degenerative diseases, acute inflammation and cancer (Shashirekha et al., 2015). Flavonoids are the most important bioactive molecules from plant sources among polyphenolic compounds and are used as nutraceuticals, which possess health-beneficial influences for the treatment or prevention of many diseases (Routray and Orsat, 2011). Foods rich in natural antioxidants were found responsible for reducing the chances of ailments like cancers, stroke, and heart diseases, as well as help in boosting the antioxidant potential of the blood plasma and this potential, is mainly due to the presence of minerals, vitamins, fiber, and a large number of phytochemicals as reported by Maron (2004). Baba and Malik, (2015) also described that presence of antioxidants (phenolics, flavonoids, tannins, and proanthocyanidins) in plants might offer protection against various diseases such as ingestion of natural antioxidants has been inversely associated with morbidity and mortality from degenerative disorders. Srivastava et al. (2012) correlated phytochemical profile and antioxidant activity in phalsa and reported that the presence of flavonoids and tannins was found responsible for the therapeutic potential of the fruit.

5.2.1 PHALSA: SOURCE OF BIOACTIVE COMPOUNDS

Studies about the bioactive composition of this fruit are limited as indicated by the literature reviewed till now. However, few studies regarding the phytochemical composition of leaves, stem, and bark are available, but literature pertaining to phalsa fruit is scanty. Most of the existing phytochemical studies on this fruit are based on qualitative screening, and very less focus was given on quantitative analysis. Presence of vitamin C, carbohydrates, phenols, flavonoids, saponins, acids, glycosides, alkaloids, steroids, tannin, and mucilaginous compounds in phalsa fruit was confirmed by Gupta et al. (2006), Mukhtar et al. (2012) and Joshi et al. (2013) in separate research studies. Different qualitative and quantitative researches reported that berry fruits, including phalsa, are good sources of bioactive compounds (Wani et al., 2016). Siddiqi et al. (2011) observed that the polyphenolic fractions of the fruit possess quercetin, quercetin-O-β-d-glucoside, pelargonidin-3,5-glucoside, delphinidin-3-glucoside, naringenin-7-O-β-d- glucoside, whereas, anthocyanin type cyanidin 3-glucoside, cyanidin-3-glucoside, tannins, gallic acid, ellagic acid, coumaric acid, chlorogenic acid, quercetin, myricetin, rutin, catechin, and ascorbic acid were confirmed by Sharma and Gupta (2013) from phalsa pomace. Singh et al. (2009) reported that dark purple color of the fruit is mainly attributed to the presence of anthocyanins, which are important bioactive compounds. Two anthocyanin pigments, including cyanidin-3-glucoside and delphinidin-3-glucoside, were also observed by Khurdiya and Anand (1981) with the aid of paper chromatography. In a preliminary phytochemical analysis, Khattab et al. (2015) found the presence of phenolics acids, flavonoids, alkaloids, glycosides, and steroids whereas, tannins, saponins, terpenoids, and resins were absent in ethanolic extract of this fruit. In a quantitative phytochemical analysis on seven tropical fruit residues left after juice extraction was carried out by Gupta et al. (2014) and the presence of 12.42 mg/g flavonoid (as catechin equivalent), 1.05 g/100 g saponin, 1.56 g/100 g alkaloid and 0.52 g/100 g tannins was confirmed. Srivastava et al. (2012) also studied the phytochemical profile of crude methanolic extracts of this fruit and 4.608 QE (quercetin equivalent) mg/g total flavonoids, 144.11 mg GAE (gallic acid equivalent)/g total phenols and 4.882 mg/kg total anthocyanin were observed. However, 55–87 mg/100 g total phenols were observed by Kaur and Kapoor (2005). Phytochemical and biological properties of *G. asiatica and Grewia optiva* which belongs to family *Tiliaceae,* were noticed by Khanal et al. (2016) and the presence of tannins, flavonoids, glycosides, terpenoids, and steroids was confirmed wheres, alkaloids, and saponins were not observed. A comparative study on antioxidant activity, total flavonoids and quercetin content of *in vitro* and *in vivo* parts of *G. asiatica* mast was conducted by Sharma and Patni (2013a) conducted and a significant amount of flavonoids were observed in leaves, stem, and callus. Ethanolic extract of leaves possessed the highest quercetin content (4.28 ng (nanogram)/μl (microliter)), whereas, maximum antioxidant activity (98.2%) was showed by fruit extract. Mucilaginous extract of phalsa showed the presence of quercetin, quercetin-3-o-β-D-glucoside, pelargonidin-3, 5-diglucoside, delphinidin 3-glucoside, and cyanidin-3-glucoside, naringenin 7-o-β-D-glucoside, glutaric acid, proline, phenylalanine, lysine, glucose, xylose, arabinose, catechin, tannins, and leucoanthocyanins, as reported by Sateesha et al. (2013). Four fractions of crude methanolic extract of the fruit were analyzed by Siddiqi et al. (2011) separately and a significant amount of flavonoids, total phenolics and anthocyanins were observed in each fraction. Thus, this fruit can be a future source of bioactive compounds.

During recent times, synthetic therapeutic agents are replaced by the natural additives. Qualitative analysis or preliminary screening is an initiative step for further quantitative estimation of plant phytochemicals. *In vivo* and *in vitro* phytochemical screening of *Grewia* species was carried out by Sharma and Patni (2013b) and bioactive compounds including flavonoids were observed in methanol, ethanol, ethyl acetate, distilled water extracts of stem, leaf, and callus of *G. asiatica* mast. Musa (2015) also reported the presence of such compounds from N-hexane extract (tannin, glycoside, saponin, resin, anthraquinone, steroid, and flavonoid) as well as in methanolic extract (saponin, resins, and flavonoids). A large number of bioactive compounds (carbohydrates, proteins, phenols, flavonoids, anthocyanins, tannins, and alkaloids) were also observed by Islary et al. (2016) in a wild edible fruit (*G. sapida*), found in Assam (North-East) India. Along with Pawar et al. (2015)

confirmed the presence of carbohydrates, flavonoids, alkaloids, and steroids in Indian medicinal plants, including phalsa during a preliminary qualitative analysis for phytochemicals.

Phytochemicals isolated from various plant parts of phalsa are represented in Table 5.1.

TABLE 5.1 Bioactive Compounds Isolated from Various Plant Parts of Phalsa

SL. No.	Plant parts	Compounds separated	References
1.	Fruit	Anthocyanins: delphinidin-3 glucoside, pelargonidin-3,5-diglucoside and cyanidin-3-glucoside; Flavonoids: quercetin, quercetin-3-O-β-D-glucoside and naringenin-7-O-β-D-glucoside; catechin	Goyal, 2012
2.	Leaves	Fats, phenols, triterpenoids, alkaloids, sterols, tannins, flavonoids, kaempferol, quercetin, and mixture of their glycosides	Ali et al., 1982; Patil et al., 2011
3.	Seeds	Stearic, palmitic, sitosterol, oleic, and linoleic acid	Morton, 1987
4.	Seed oil	Palmitic, oleic, Stearic, linoleic acids, β-sitosterol, and tetrabromostearic acid	Ahmad et al., 1964
5.	Flower	Flavonoids: naringenin and naringenin-7-O-β-D-glucoside; quercetin, quercetin-3-O-β-D-glucoside, grewinol 3, 21, 24-trimethyl-5, 7 dihydroxyhentriacontanoic acid, β-sitosterol, and δ-lactone	Lakshmi and Chauhan, 1976a, b; Lakshmi et al., 1976
6.	Bark	β-sitosterol, betulin, β-amyrin and erythrodiol taraxasterol	Tripathi et al., 1973; Joshi et al., 1974

5.2.2 EXTRACTION OF BIOACTIVE COMPOUNDS

For obtaining bioactive compounds from natural sources, extraction is the basic and most useful step. Various extraction methods as solvent and heating reflux have been widely used as traditional extraction methods since last few years. To overcome the problems of high solvent use and more time consumption, many novel techniques like ultrasound-assisted and microwave-assisted extraction have been introduced in recent times, as reported by Wang et al. (2011). Prasad et al. (2012) explained that extraction efficiency of bioactive compounds depends on various factors including solvent and its composition, solvent to solid ratio, pH, and temperature.

Gao and Liu (2005) described that ultrasonic methods which use high-frequency sound to disrupt the target compound from the plant materials are most commonly used for the extraction of various phytochemicals. They further added that microwave-assisted extraction is also having wide applications for the extraction of numerous biologically active compounds. Bhadoriya et al. (2011) reported that in microwave heating, dipole rotation of the solvent, which causes a rise in temperature and enhances the solubility of the compound of interest. In addition to, other extraction techniques like supercritical fluid (SFC) extraction, sub-critical water extraction and pressure assisted extraction are also employed for the extraction of phenolic compounds from fruits and vegetable sources (Sharma and Gupta, 2013). Do et al. (2014) found that polar solvents are most commonly used for obtaining polyphenols from plant sources, as compounds having different chemical characteristics and polarities may or may not dissolve in a particular solvent. They also reported that aqueous mixtures of methanol, ethanol, ethyl acetate, and acetone are suitable solvents for extraction of bioactive compounds from natural sources.

5.2.3 QUANTIFICATION OF BIOACTIVE COMPOUNDS

In recent years, food scientists and technologists are attracted towards bioactive compounds as consciousness towards health is increasing among people. Phenols and flavonoids are very valuable phytochemicals; hence, their quantification is a necessary step. Due to the inherent diversity of flavonoids, analytical chemists and biochemists faced many challenges for quantification in a fast and efficient way (Routray and Orsat, 2011). Cote et al. (2010) described that clean-up or purification is the primary step for further characterization and quantification of the crude extracts and solid-phase extraction (SPE) is most commonly used the technique for this purpose. Recently, for characterization and quantification of purified extracts, a chromatographic technique such as high-performance liquid chromatography (HPLC) is widely used which is considered most suitable as reported by Abidin et al. (2014). Chawla and Ranjan (2016) explained that during these days, Ultra Performance Liquid Chromatography (UPLC) which is the unique version of HPLC, is used for identification and quantification purpose of phytochemicals, as this technique showed sensitivity, resolution, and speed superior to HPLC. An in-depth study about extraction, identification, and quantification of bioactive compounds from this underutilized fruit would be of great significance which will possibly explore the hidden potential of this miracle fruit.

5.3 NUTRITIONAL COMPOSITION AND MEDICINAL IMPORTANCE OF PHALSA FRUIT

5.3.1 NUTRITIONAL COMPOSITION

Phalsa fruits are a good source of nutrients such as proteins, amino acids, carotenes, and dietary fibers as well as vitamins (A & C) and minerals (phosphorus & iron) (Dave et al., 2015). These fruits contain 50–60% juice, 2.0–2.5% acid and 10–11% sugar (mainly in the form of reducing sugar) as described by Kacha et al. (2012). Vitamin C of phalsa varied from 4.38 to 32.10 mg/100 g, tannin from 1.13 to 2.46%, anthocyanin from 70 to 72 mg/100 g, phosphorous 24.2 mg/100 g and iron 1.08 mg/100 g, which are necessary for maintaining good health (Pangotra, 2016). Phalsa fruit was found a fruit nutritionally rich in proximate, minerals, and vitamins by Yadav (1999). This fruit is a good source of micronutrients including Cu, Zn, Co, Ni, Fe, and Cr which are necessary to sound health and human body required these micronutrients in microgram (Khan et al., 2006). According to Morton (1987), this fruit contains 81.13% moisture, 1.58% protein, 1.77% crude fiber, 1.82% fat, 10.27% sugar, and 725 /kg edible fruit calories.

Underutilized fruit crops including phalsa were found rich sources of iron (Arivalagan et al., 2012) thus; the deficiency of micronutrients of Indian population could be overcome by including such fruits into the government meal campaign. Grover and Samson (2014) reported that phalsa was found good source of vitamin C (4.4 mg/100 g), β- carotene (16.11 mg/100 g), Cu (0.48 mg/100 g) and Zn (144 mg/100 g) which acts as antioxidants. They also correlated vision health with antioxidant intake and concluded that antioxidants play a powerful role in boosting eye health. Srivastava (1953) reported that phalsa juice contained maltose, fructose, and glucose, but sucrose was not found. Phalsa beverage blended with pear was found a good source of reducing sugars, total sugars, tannin, iron, phosphorus, and anthocyanins, as reported by Pangotra (2016) while working on the blended beverage of pear and phalsa. Table 5.2 shows the nutritional composition of phalsa.

Study on the amino acid profile of seed and pulp (hydrolyzed and unhydrolyzed state) by Hasnain and Ali (1992) showed that in pulp threonine, in seed methionine and in juice phosphoserine, serine, and taurine were the major amino acids. Moreover, phosphoserine was found in higher quantity in pulp whereas, aspartic acid, glycine, and tyrosine in hydrolyzed products. Thus, the above study clearly indicates that this fruit is a good source of proteins and amino acids. Along with the fruits, seeds are also equally important and usually eaten with the fruits. Many scientific studied have been carried out on seed's nutritional composition. Maury et al. (2012) studied about the fatty acid profile of phalsa seed oil and reported that oleic acid (13.5%), stearic acid (11%), linoleic acid (64.5%) and palmitic acid (8%) were the major amino acids in the oil. Madaan and Khurdiya (1987) also observed linoleic acid as main fatty acid, whereas palmitic, palmitoleic, linolenic, arachidic, oleic, heptadecanoic, myristic, and stearic acids were present in lower amounts. Intake of fatty acids and amino acids is directly connected to good health as suggested by many epidemiological studies.

TABLE 5.2 Nutritional Value of Phalsa (Yadav, 1999; Khan et al., 2006)

Nutritional Parameters	Values/100 g
Proximate	
Carbohydrate (g)	21.1
Protein (g)	1.57
Total lipid (fat) (g)	< 0.1
Fiber (g)	5.53
Ash (g)	1.1
Vitamins	
Vitamin B_1 (mg)	0.02
Vitamin B_2 (mg)	0.264
Vitamin B_3 (mg)	0.825
Vitamin A (μg)	16.11
Vitamin C (mg)	4.385
Minerals	
Calcium (mg)	136
Phosphorus (mg)	24.2
Iron (mg)	1.08
Sodium (mg)	17.3
Potassium (mg)	372
Micronutrients	
Iron (mg/100 g FW)	140.8
Zinc (mg/100 g FW)	1.44
Copper (mg/100 g FW)	0.48
Nickel (mg/100 g FW)	2.61
Chromium (mg/100 g FW)	1.08
Cobalt (mg/100 g FW)	0.99

FW = Fresh weight basis

5.3.2 MEDICINAL IMPORTANCE

All the plant parts of this underutilized phalsa plant are important and provide many health benefits. As evident by many scientific studies, phalsa plays a significant role in preventing and curing of health-related problems. Phalsa fruit is considered as phytomedicine and also act as a source for phytochemicals, oil, and hydrocarbon as reported by Tripathi et al. (2010). Phalsa also acts as a blood purifier and multiple health benefits provided by all the plant parts (leaves, bark, pulp, flowers, and seeds) of plant are familiar since historical time (Sharma et al., 2014). Fruits are well known for curing many diseases such as throat trouble, cardiac, respiratory, and blood disorders, fever, and also possess antioxidant, antidiabetic, anticancer, antirheumatic, antipyretic, hypoglycemic, radioprotective, anti-inflammatory (Shrimanker et al., 2013), anti-ulcer, antimalarial (Paviaya et al., 2013), antihyperglycemic, and hepatoprotective activities (Khattab et al., 2015). Srivastava et al. (2012) reported that ripe fruits showed astringent, stomachic, and aphrodisiac potential, whereas inflammation is alleviated with the use of unripe fruits. According to Paul (2015), unripe fruits are also helpful in eliminating the kapha, vata, and biliousness. Phalsa also acts as a cooling agent, used as a refreshing drink during hot summers and for treating urological disorders (Shukla et al., 2016). According to Rahman and Rahman (2014), these fruits can be used for treating diarrhea, weakness, anoxia, and worm. Moreover, Orwa et al. (2009) described that these are beneficial in thirst quenching and for burning sensations due to heat. A large number of reports in the literature explained the role of this fruit as antimicrobial, antifungal, antiviral, antioxidant, antidiabetic, anticancer, antihyperglycemic, radioprotective, chemopreventive, antiplatelet, analgesic, antiemetic, antipyretic activities and also act as immunomodulatory agent as mentioned by Paul (2015). This fruit possesses well-known therapeutic potential owing to various phytochemicals, minerals, and vitamins present and also facilitate the dead fetus ejection (Gupta et al., 2006). Traditionally, the extract obtained from fruits has been utilized in polyherbal ayurvedic medicines (Gupta et al., 2010; Mesaik et al., 2013). Crude ethanolic extracts of the stem bark, leaves, and fruit were analyzed by Parveen et al. (2012) it was concluded that significant anti-hyperglycemic activity was shown by all the plant parts. As reported by Khanal et al. (2016), n-hexane extract of phalsa possessed anti-inflammatory potential. Debajyoti et al. (2012) described that phalsa juice was found good for treating alcoholism, gynecological disorders, and heart disease; roots for curing rheumatism, whereas anti-tubercular property was shown in leaves.

An *in vitro* study on cytotoxic activity against breast cancer cells, cervical cancer cells, and hepatocellular carcinoma cells was conducted by Gupta et al. (2014) and among all the tested cell lines, pomace extract of phalsa was found effective against breast cancer cell. Marya et al. (2011) also studied anticancer activity of phalsa fruits and leaves (aqueous extracts), and these extracts were found effective against breast and liver cancer. In non-diabetic humans, influence of the fruit on phagocytosis and glycemic index (GI) were examined by Mesaik et al. (2013) and the fruit was observed with low GI, i.e., 5.34 with modest hypoglycemic activity. Low GI of phalsa juice was also reported by Saddozai et al. (2015) and the juice was found good for diabetic people as well as was helpful in decreasing the risk of coronary heart diseases (CHDs) and obesity-related problems. Preventive role against radiation-induced alterations to brain (Ahaskar and Sisodia, 2006; Ahaskar et al., 2007), testis (Sharma and Sisodia, 2010b), liver (Sharma and Sisodia, 2010a), cerebrum (Sisodia et al., 2008) and blood (Singh et al., 2007) was also shown by fruit extract. According to Sinha et al. (2015) biofunctional and chemo-preventive compounds present in plant parts, might be responsible for health-boosting potential of the plant and this is the main reason behind their overall acceptability and increased consumption. Table 5.3 represents the medicinal importance of phalsa plant parts.

5.4 CONCLUSION REMARKS

Phalsa fruit is not well known as potential source of antioxidants till now. In spite of high therapeutic and nutritive potential, market position of this fruit is unsatisfactory owing to under exploitation and utilization as a food source. Phalsa fruit is very rich in various bioactive compounds like flavonoids, anthocyanins, vitamins, and phenolic compounds but is underutilized because of various limitations like extremely perishable nature, seasonal availability, smaller size, repeated harvesting, and low yield. Bioactive compounds possess countless health benefits; therefore, knowledge about their functions and uses is very necessary. Processed foods can be fortified with bioactive compounds rich extract for boosting the immune system in human beings. To achieve this objective phytochemicals rich fresh fruits, vegetables, and food industry wastes can be used for the extraction of these compounds. Thus, by considering the above-mentioned aspects and importance of this fruit, there is vast need of complete investigation and exploration of this fruit, for better utilization, increased acceptability, and value addition.

TABLE 5.3 Medicinal Importance of Phalsa Plant Parts

Plant Part	Medicinal Value
Fruits	Fruits have antitumor, antipyretic, antidiabetic, immunomodulatory, analgesic, cytotoxic, and aphrodisiac potential. Unripe fruits used in fever, cardiac, respiratory, and blood disorders. Ripe fruits provide cooling effect in thirst and burning sensation, cure throat troubles, inflammation, and remove dead fetus. Ethanolic extract of fruit possesses hepatoprotective efficacy, antihyperglycemic activity, helpful in purification of blood and curing anemia. Aqueous extract of the fruit showed *in vitro* cytotoxic. Activity.
Pomace	Pomace possesses antimicrobial activity.
Leaves	Leaves are recommended for curing leprosy, fever, gout, arthritis, elephantiasis, diarrhea, rheumatoid, inflammations, leucoderma, and bronchitis. These possess phytotoxic, antihyperglycaemic, insecticidal, cytotoxic, larvicidal, nematicidal, antibiotic, anti-hyperlipidemic, antibacterial, anti-diabetic, natural antiemetics and antifungal properties as well as applied on skin eruptions. Aqueous leaf extract have *in vitro* cytotoxic activity, and methanolic extract have anti-platelet activity.
Seeds	Seeds are used for the remediation of Pb (II) contaminated water and acts as anti-fertility agent.
Root Bark	It is useful for biliousness and treats rheumatism as well as burning in vagina. Anti-inflammatory and analgesic activities are shown by root bark. Ethanol extract is useful for diabetes and its related complications.

KEYWORDS

- **bioactive compounds**
- **medicinal importance**
- **multiple uses**
- **nutritional composition**
- **phalsa**

REFERENCES

Abid, M., Muzamil, S., Kirmani, S. N., Khan, I., & Hassan, A., (2012). Effect of different levels of nitrogen and severity of pruning on growth, yield and quality of phalsa (*Grewia subinaequalis L.*). *African Journal of Agricultural Research, 7*(35), 4905–4910.

Abidin, L., Mujeeb, M., Mir, S. R., Khan, S. A., & Ahmad, A., (2014). Comparative assessment of extraction methods and quantitative estimation of luteolin in the leaves of Vitex negundo Linn. by HPLC. *Asian Pacific Journal of Tropical Medicine, 7*(1), S289–S293.

Ahaskar, M., & Sisodia, R., (2006). Modulation of radiation-induced biochemical changes in brain of Swiss albino mice by *Grewia asiatica. Asian Journal of Experimental Science, 20,* 399–404.

Ahaskar, M., Sharma, K. V., Singh, S., & Sisodia, R., (2007). Post-treatment effect of *Grewia asiatica* against radiation-induced biochemical changes in brain of Swiss albino mice. *Iranian Journal of Radiation Research, 5*(3), 105–112.

Ahmad, M. N., Zahid, N. D., Rafiq, M., & Ahmad, I., (1964). Composition of the oil of *Grewia asiatca* seeds. *Pakistan Journal of Scientific and Industrial Research, 7*(2), 145–146.

Akhtar, N., Haq, I., & Mirza, B., (2015). Phytochemical analysis and comprehensive evaluation of antimicrobial and antioxidant properties of 61 medicinal plant species. *Arabian Journal of Chemistry*, (doi: org/10.1016/j.arabjc.2015.01.013).

Ali, S. I., Khan, N. A., & Husain, I., (1982). Flavonoid constituents of *Grewia asiatica. Journal of Scientific Research (Bhopal, India), 4*(1), 55–56.

Arivalagan, M., Prasad, T. V., & Bag, M. K., (2012). Role of underutilized crops for combating iron deficiency in Indian population. *Current Science, 103*(2), 137.

Baba, S. A., & Malik, S. A., (2015). Determination of total phenolic and flavonoid content, antimicrobial and antioxidant activity of a root extract of *Arisaema jacquemontii* Blume. *Journal of Taibah University for Science, 9,* 449–454.

Bhadoriya, U., Tiwari, S., Mourya, M., & Ghule, S., (2011). Microwave-assisted extraction of flavonoids from *Zanthoxylum budrunga* W. optimization of extraction process. *Asian Journal of Pharmacy and Life Science, 1*(1), 81–86.

Bhangale, J., Acharya, S., & Deshmukh, T., (2013). Antihyperglycaemic activity of ethanolic extract of *Grewia asiatica* (L.) leaves in alloxan-induced diabetic mice. *World Journal of Pharmaceutical Research, 2*(5), 1486–1500.

Chawla, & Ranjan, C., (2016). Principle, instrumentation, and applications of UPLC: A novel technique of liquid chromatography. *Open Chemistry Journal, 3,* 1–16.

Cote, J., Caillet, S., Doyon, G., Sylvain, J. F., & Lacroix, M., (2010). Analyzing cranberry bioactive compounds. *Critical Reviews in Food Science and Nutrition, 50*(9), 872–888.

Dave, R., Rao, T. V. R., & Nandane, A. S., (2015). RSM- based optimization of edible-coating formulations for preserving post-harvest quality and enhancing storability of phalsa (*Grewia asiatica* L.). *Journal of Food Processing and Preservation*, (doi: 10.1111/ jfpp.12630).

Debajyoti, D., Achintya, M., Debdas, D., Achintya, S., & Jayram, H., (2012). Evaluation of antipyretic and analgesic activity of Parusaka (*Grewia asiatica Linn*): An indigenous Indian plant. *International Journal of Research in Ayurveda and Pharmacy, 3*(4), 519–523.

Do, Q. D., Angkawijaya, A. E., Tran-Nguyen, P. L., Huynh, L. H., Soetaredjo, F. E., Ismadji, S., & Ju, Y., (2014). Effect of extraction solvent on total phenol content, total flavonoid content and antioxidant activity of *Limnophila aromatica. Journal of Food and Drug Analysis, 2*(2), 296–302.

Gao, M., & Liu, C. Z., (2005). Comparison of techniques for the extraction of flavonoids from cultured cells of *Saussurea medusa* Maxim. *World Journal of Microbiology & Biotechnology, 21,* 1461–1463.

Goyal, P. K., (2012). Phytochemical and pharmacological properties of the genus *Grewia*: A review. *International Journal of Pharmacy and Pharmaceutical Sciences, 4*(4), 72–78.

Grover, A. K., & Samson, S. E., (2014). Antioxidants and vision health: Facts and fiction. *Molecular and Cellular Biochemistry, 388,* 173–183.

Gungor, N., & Sengul, M., (2008). Antioxidant activity, total phenolic content and selected physicochemical properties of white mulberry (*Morus alba* L.) Fruits. *International Journal of Food Properties, 11*(1), 44–52. doi: 10.1080/10942910701558652.

Gupta, M. K., Lagarkha, R., Sharma, D. K., Sharma, P. K., Singh, R., & Ansari, H. S., (2007). Antioxidant activity of the successive extracts of *Grewia asiatica* leaves. *Asian Journal of Chemistry, 19*(5), 3417–3420.

Gupta, M. K., Sharma, P. K., Ansari, S. H., & Lagarkha, R., (2006). Pharmacognostical evaluation of *Grewia asiatica* fruits. *International Journal of Plant Sciences, 1*(2), 249–251.

Gupta, M., Shaw, B. P., & Mukherjee, A., (2010). A new glycosidic flavonoid from *Jwarhar mahakashy* (antipyretyic) ayurvedic preparation. *International Journal of Ayurveda Research, 2*(1), 106–111.

Gupta, P., Bhatnagar, I., Kim, S., Verma, A. K., & Sharma, A., (2014). *In-vitro* cancer cell cytotoxicity and alpha-amylase inhibition effect of seven tropical fruit residues. *Asian Pacific Journal of Tropical Biomedicine, 4*(2), S665–S671.

Hasnain, A., & Ali, R., (1992). Amino acid composition of *Grewia asiatica* (Phalsa) as index of juice quality. *Pakistan Journal of Scientific of Industrial Research, 35,* 514–515.

Islary, A., Sarmah, J., & Basumatary, S., (2016). Proximate composition, mineral content, phytochemical analysis and *in vitro* antioxidant activities of a wild edible fruit (*Grewia Sapida* Roxb. Ex DC.) found in Assam of North-East India. *Journal of Investigational Biochemistry, 5*(1), 21–31.

Joshi, K. C., Prakash, L., & Shah, R., (1974). Chemical investigation of the bark and heartwood of *Grewia asiatica. Journal of the Indian Chemical Society, 51*(9), 830.

Joshi, P., Preeti, P., & Priya, D. L., (2013). Pharmacognostical and phytochemical evaluation of *Grewia asiatica* Linn (*Tiliaceae)* fruit pulp and seed. *International Journal of Pharmaceutical & Biological Archives, 4*(2), 333–336.

Julkunen-Tiitto, R., & Sorsa, S., (2001). Testing the effects of drying methods on willow flavonoids, tannins, and salicylates. *Journal of Chemical Ecology, 27*(4), 779–789.

Kacha, H. L., Viradia, R. R., Leua, H. N., Jat, G., & Tank, A. K., (2012). Effect of NAA, GA3 and ethrel on yield and quality of phalsa (*Grewia asiatica* L.) under South- Saurashtra condition. *The Asian Journal of Horticulture*, *7*(2), 242–245.

Kaur, C., & Kapoor, H. C., (2005). Antioxidant activity of some fruits in Indian diet. *Acta Horticulture, 696,* 563–565.

Khan, A. S., Hussain, A., & Khan, F., (2006). Nutritional importance of micronutrients in some edible wild and unconventional fruits. *Journal of Chemical Society of Pakistan, 28*(6), 576–582.

Khanal, D. P., Raut, B., & Kafle, M., (2016). A comparative study on phytochemical and biological activities of two *Grewia* species. *Journal of Manmohan Memorial Institute of Health Sciences, 2091–1041*(2), 53–60.

Khattab, H. A. H., El-Shitany, N. A., Abdallah, I. Z. A., Yousef, F. M., & Alkreathy, H. M., (2015). *Antihyperglycemic Potential of Grewia Asiatica Fruit Extract Against Streptozotocin-Induced Hyperglycemia in Rats: Anti-Inflammatory and Antioxidant Mechanisms*. Hindawi Publishing Corporation, Oxidative Medicine and Cellular Longevity doi: org/10.1155/2015/549743.

Khurdiya, D. S., & Anand, J. C., (1981). Anthocyanins in Phalsa (*Grewia subinaequalis* L.) fruits. *Journal of Food Science and Technology, 18*(3), 112–114.

Kris-Etherton, P. M., Hecker, K. D., Bonanome, A., Coval, S. M., Binkoski, A. E., & Hilpert, K. F., (2002). Bioactive compounds in foods: Their role in the prevention of cardiovascular disease and cancer. *The American Journal of Medicine, 113*(9), 71–88.

Kumar, M., Dwivedi, R., Anand, A. K., & Kumar, A., (2014). Effect of nutrient on vegetative growth, fruit maturity and yield attributes of phalsa *(Grewia Subinaequalis* D. C.). *Global Journal of BioSciences and Biotechnology, 3*(3), 264–268.

Lakshmi, V., & Chauhan, J. S., (1976a). Grewinol, a keto-alcohol from the flowers of *Grewia asiatica*. *Lloydia*, *39*(5), 372–374.

Lakshmi, V., & Chauhan, J. S., (1976b). Chemical examination of the flowers of *Grewia asiatica* Linn. *Journal of the Indian Chemical Society*, *53*(6), 632–633.

Lakshmi, V., Agarwal, S. K., & Chaijhan, J. S., (1976). A new δ- lactone from the flowers of *Grewia asiatica. Phytochemistry, 15,* 1397–1399.

Loganayaki, N., & Manian, S., (2010). *In vitro* antioxidant properties of indigenous underutilized fruits. *Food Science and Biotechnology, 19*(3), 725–734.

Madaan, T. R., & Khurdiya, D. S., (1987). A study on phalsa (*Grewia subinaequalis* L.) seed oil. *Journal of the Oil Technologist's Association of India, 19*(1), 23.

Maron, D. J., (2004). Flavonoids for reduction of atherosclerotic risk. *Current Atherosclerosis Report, 6,* 73–78.

Marya, B., Dattani, K. H., Patel, D. D., Patel, P. D., Patel, D., Suthar, M. P., Patel, V. P., & Bothara, S. B., (2011). *In-vitro* cytotoxicity evaluation of aqueous fruit and leaf extracts of *Grewia asiatica* using MTT assay. *Pharma Chemica, 3*(3), 282–287.

Maury, P. K., Jain, S. K., & Lal, N., (2012). *Grewia asiatica*: An overview. *International Journal of Pharmaceutical Research and Development*, *4*(4), 154–158.

Mesaik, M. A., Ahmed, A., Khalid, A. S., Jan, S., Siddiqui, A. A., Perveen, S., & Azim, M. K., (2013). Effect of *Grewia asiatica* fruit on Glycemic index and phagocytosis tested in healthy human subjects. *Pakistan Journal of Pharmaceutical Sciences*, *26*(1), 85–89.

Morton, J. F., (1987). Phalsa. In: *Fruits of Warm Climates* (pp. 276–277). USA, ISBN: 0-9610184-1-0.

Mukhtar, H. M., Kaur, H., Singh, S., & Singh, M., (2012). Standardization and preliminary phytochemical investigation of the fruits of *Grewia asiatica* Linn. *Research Journal of Pharmacognosy and Phytochemistry*, *4*(4), 212–214.

Musa, Y. M., (2015). Isolation and purification of flavonoids from the leaves of locally produced *Carica papaya*. *International Journal of Scientific & Technology Research, 4*(12), 282–285.
Orwa, C., Mutua, A., Kindt, R., Jamnadass, R., & Anthony, S., (2009). *Agroforestree Database: A Tree Reference and Selection Guide Version 4.0*. http://www.worldagroforestry (Accessed on 12 June 2019).
Pangotra, B. B., (2016). *Development and Evaluation of Phalsa Blended Beverage*. MSc. Thesis. Sher-e-Kashmir University of Agricultural Sciences & Technology of Jammu Main Campus, Chatha, Jammu.
Pareek, O. P., & Panwar, H. S., (1981). Vegetative, floral and fruit: Characteristics of two phalsa (*Grewia subinequalis* DC) types. *Annals of Arid Zone, 20*(4), 281–290.
Parveen, A., Irfan, M., & Mohammad, F., (2012). Antihyperglycemic activity in *Grewia asiatica*, a comparative investigation. *International Journal of Pharmacy and Pharmaceutical Sciences, 4*(1), 210–213.
Patil, P., Patel, M. M., & Bhavsar, C. J., (2011). Preliminary phytochemical and hypoglycemic activity of leaves of *Grewia asiatica* L. *Research Journal of Pharmaceutical, Biological and Chemical Sciences, 2*(1), 516–520.
Paul, A., (2013). Minor and uncultivated fruits of Eastern India. *2nd International Symposium on Minor Fruits and Medicinal Plants for Better Lives*, 54–67.
Paul, S., (2015). Pharmacological actions and potential uses of *Grewia asiatica*: A review. *International Journal of Applied Research, 1*(9), 222–228.
Paviaya, U. S., Kumar, P., Wanjari, M. M., Thenmozhi, S., & Balakrishnan, B. R., (2013). Analgesic and anti-inflammatory activity of root bark of *Grewia asiatica* Linn. in rodents. *Ancient Science of Life, 32*(3), 150–155.
Pawar, M. M., Patil, S. D., Jadhav, A. P., & Kadam, V. J., (2015). *In-vitro* antimicrobial activity and phytochemical screening of selected Indian medicinal plants. *Indo American Journal of Pharmaceutical Research, 5*(01), 593–600.
Prasad, N. K., Kong, K. W., Ramanan, R. N., Azlan, A., & Ismail, A., (2012). Selection of experimental domain using two-level factorial design to determine extract yield, antioxidant capacity, phenolics, and flavonoids from *Mangifera pajang* Kosterm. *Separation Science and Technology, 47*, 2417–2423.
Rahman, M., & Rahman, J., (2014). Medicinal value and nutrient status of indigenous fruits in Bangladesh. *Nova Journal of Medical and Biological Sciences, 3*(4), 1–9.
Ramamoorthy, P. K., & Bono, A., (2007). Antioxidant activity, total phenolic and flavonoid content of *Morinda citrifolia* fruit extracts from various extraction processes. *Journal of Engineering Science and Technology, 2*(1), 70–80.
Ravani, A., & Joshi, D. C., (2014). Processing for value addition of underutilized fruit crops. *Trends in Post Harvest Technology, 2*(2), 15–21.
Routray, W., & Orsat, V., (2011). Microwave-assisted extraction of flavonoids: A review. *Food Bioprocess Technol.* doi: 10.1007/s11947-011-0573-z.
Saddozai, A. A., Mumtaz, A., Raza, S., & Saleem, S. A., (2015). Microbial count and shelf life of phalsa (*Grewia asiatica*) juice. *Pakistan Journal of Agricultural Research, 28*(4), 395–399.
Sateesha, S. B., Balaji, S., Rajamma, A. J., Shekar, H. S., & Chandan, K., (2013). Prospective of *Grewia* fruit mucilage as gastro retentive drug delivery system: Statistical optimization of formulation variables. *RGUHS Journal of Pharmaceutical Sciences, 3*(3), 11–19.
Sharma, A., & Gupta, P., (2013). Evaluation of antioxidant activity and validated method for analysis of polyphenols from non-edible parts of Indian tropical fruits by using

microwave-assisted extraction and LC-MS/MS. *International Journal of Pharma and Bio Sciences*, *4*(1), 227–241.
Sharma, K. V., & Sisodia, R., (2010a). Hepatoprotective efficacy of *Grewia asiatica* fruit against oxidative stress in Swiss albino mice. *Iranian Journal of Radiation Research*, *8*(2), 75–85.
Sharma, K. V., & Sisodia, R., (2010b). Radioprotective potential of *Grewia asiatica* fruit extract in mice testis. *Pharmacologyonline, 1,* 487–495.
Sharma, N., & Patni, V., (2013a). Comparative analysis of total flavonoids, quercetin content and antioxidant activity of *in vivo* and *in vitro* plant parts of *Grewia asiatica* mast. *International Journal of Pharmacy and Pharmaceutical Sciences, 5*(2), 464–469.
Sharma, N., & Patni, V., (2013b). *In vivo* and *in vitro* qualitative phytochemical screening of *Grewia* species. *International Journal of Biological & Pharmaceutical Research, 4*(9), 634–639.
Sharma, N., Meena, R., Arya, D., & Patni, V., (2014). *In vivo* and *in vitro* GC-MS characterization of *Grewia asiatica* mast. *International Journal of Pharmaceutical Research & Development*, *6*(4), 96–102.
Shashirekha, M. N., Mallikarjuna, S. E., & Rajarathnam, S., (2015). Status of bioactive compounds in foods, with focus on fruits and vegetables. *Critical Reviews in Food Science and Nutrition, 55,* 1324–1339.
Shrimanker, M. V., Patel, D. D., Modi, H. K., Patel, K. P., & Dave, R. M., (2013). *In-vitro* antibacterial and antifungal activity of *Grewia asiatica* Linn. Leaves. *American Journal of Pharm. Tech Research, 3*(1), 382–385.
Shukla, R., Sharma, D. C., Baig, M. H., Bano, S., Roy, S., Provazník, I., & Kamal, M. A., (2016). Antioxidant, antimicrobial activity and medicinal properties of *Grewia asiatica* L. *Medicinal Chemistry*, *12*(3), 211–216, DOI: 10.2174/1573406411666151030110530.
Siddiqi, R., Naz, S., Ahmad, S., & Sayeed, S. A., (2011). Antimicrobial activity of the polyphenolic fractions derived from *Grewia asiatica, Eugenia jambolana* and *Carissa carandas*. *International Journal of Food Science and Technology*, *46,* 250–256.
Singh, D., Chaudhary, M., & Chauhan, P. S., (2009). Value addition to forest produce for nutrition and livelihood. *The Indian Foreste.*, 1271–1284.
Singh, S., Sharma, K. V., & Sisodia, R., (2007). Radioprotective role of *Grewia asiatica* in mice blood. *Pharmacologyonline*, *2,* 32–43.
Sisodia, R., Ahaskar, M., Sharma, K. V., & Singh, S., (2008). Modulation of radiation-induced biochemical changes in cerebrum of Swiss albino mice by *Grewia asiatica. Acta Neurobiologiae Experimentalis*, *68,* 32–38.
Srivastava, I. H. C., (1953). Paper chromatography of fruit juices. *Journal of Scientific & Industrial Research, 12B,* 363–365.
Srivastava, J., Kumar, S., & Vankar, P. S., (2012). Correlation of antioxidant activity and phytochemical profile in native plants. *Nutrition & Food Science*, *42*(2), 71–79.
Tiwari, D. K., Singh, D., Barman, K., & Patel, V. B., (2014). Bioactive compounds and processed products of phalsa (*Grewia subinaequalis* L.) fruit. *Popular Kheti*, *2*(4), 128–132.
Tripathi, S., Chaurey, M., Balasubramaniam, A., & Balakrishnan, N., (2010). *Grewia asiatica* Linn. as a phytomedicine: A review. *Research Journal of Pharmacy and Technology, 3*(1), 1–3.
Tripathi, V. J., Ray, A. B., & Dasgupta, B., (1973). Triterpenoid constituents of *Grewia asiatica. Current Science, 42*(23), 820–821.
Wang, Z., Shang, Q., Wang, W., & Feng, X., (2011). Microwave-assisted extraction and liquid chromatography/mass spectrometry analysis of flavonoids from grapefruit peel. *Journal of Food Process Engineering, 34,* 844–859.

Wani, T. A., Rana, S., Bhat, W. W., Pandith, S. A., Dhar, N., Razdan, S., Chandra, S., Sharma, N., & Lattoo, S. K., (2016). Efficient *in vitro* regeneration, analysis of molecular fidelity and *Agrobacterium tumifaciens*-mediated genetic transformation of *Grewia asiatica* L. *Journal of Plant Biochemistry & Physiology*, *4*(167), 1–7.

Yadav, A. K., (1999). Phalsa: A potential new small fruit for Georgia. In: Janick, J., (ed.), *Perspectives on New Crops and New Uses* (pp. 348–352). ASHS Press: Alexandria, VA, USA.

Zia-Ul-Haq, M., Ahmad, S., Imran, I., Ercisli, S., & Moga, M., (2015). Compositional study and antioxidant capacity of *Grewia asiatica* L. seeds grown in Pakistan. *Proceeding of the Bulgarian Academy of Sciences*, *68*(2), 191–200.

Zia-Ul-Haq, M., Milan, S. K. R., & Feo, V. D., (2013). *Grewia asiatica* L., a food plant with multiple uses. *Molecules*, *18,* 2663–2682.

CHAPTER 6

Pseudocereals: Nutritional Composition, Functional Properties, and Food Applications

RITU SINDHU[1] and B. S. KHATKAR[2]

[1]*Center of Food Science and Technology, Chaudhary Charan Singh Haryana Agricultural University, Hisar–125001 (Haryana), India*

[2]*Department of Food Technology, Guru Jambheshwar University of Science and Technology, Hisar–125001 (Haryana), India,*
E-mail: bskhatkar@yahoo.co.in

ABSTRACT

Plants producing seeds or fruits that are used and consumed as grains are pseudocereals, despite the fact that botanically they are neither grasses nor true cereal grains. Best-known pseudocereals are amaranth, buckwheat, and quinoa. Although production of the pseudocereals is comparatively minor on a global scale of grain production, they add considerably to the human diet in certain regions and cultures since ancient times. Recently, there is a resurgence of interest in pseudocereals due to their excellent nutritional and biological value, gluten-free composition, and presence of some health-promoting compounds. The pseudocereals can serve as an important source of energy in the diet due to high starch content of seeds, and their excellent quality protein, high dietary fiber and sufficient lipid content rich in unsaturated fatty acids can contribute to the improvement of nutrition all over the world. Sufficient amount of vitamins, minerals, and a good level of other bioactive compounds like saponins, fagopyritols, squalene, polyphenols, and phytosterols in pseudocereals depicts their applicability as supplements or major cereal replacers in the human diet. Amaranth, buckwheat, and quinoa are potentially valuable for people allergic to traditional cereals as these grains do not have any prolamins toxic to celiac disease and can be used as ingredients in gluten-free diets. Mostly pseudocereals are able to grow in poor soils and conditions not inappropriate for other cereals.

Therefore, strong resistance of pseudo-cereals crops represents a sustainable food source, even if potential food shortage happens for various reasons. The present chapter focuses on the main aspects, including composition, properties, importance, and applications of different pseudocereals.

6.1 INTRODUCTION

In order to ensure global food security for future, exploration of sustainable plant resources is important. Therefore, researchers have been focused on valorizations of some underutilized, forgotten or neglected crops during the last few decades worldwide. Pseudocereals are fruits or seeds producing plants, which are neither grasses nor true cereal, although these are used and consumed as grains. Three plants namely amaranth, quinoa, and buckwheat are known as pseudocereals because their seeds are similar in composition and utilization to the true cereals while botanically their plants are dicotyledonous different from monocotyledonous plants of true cereals like wheat and rice. The word Amaranthus came from "Anthor" a Greek word, which stands for everlasting or unwilting. Amaranthus is an ancient crop that belongs to family Amaranthaceae, which is considered to have originated from central and southern America. Out of more than 60 species, only three species (*A. caudatus, A. cruentus,* and *A. hypochondriacus*) present remarkable agricultural characteristics, such as fast establishment, tolerance to water deficit, biomass production and nutrient cycling. Amaranth is a multi-use crop gives highly nutritious seeds and leaves used as human food and animal feed; due to eye-catching colored inflorescence, it is also grown as an ornamental plant (Figure 6.1). In addition to nutritional characteristics, amaranth plants have agronomic characteristics making out it as an alternative crop where cereals and vegetables cannot be grown (dry soils, high altitudes, and high temperatures). In India, about 20 species are found wild/cultivated mainly in the Himalayas from Kashmir to Bhutan, besides South Indian hills. Buckwheat (*Fagopryum esculentum* Monch) work is originated from a combination of two words–"boc" which is the Anglo-Saxon word meaning beech and "whoet" meaning wheat since it bears a resemblance to the beechnut. Buckwheat is an annual crop, which belongs to family Polygonaceae. Amongst the various varieties of buckwheat, only nine comprise desirable agricultural as well as nutritional quality. Buckwheat is mainly produced in China, the Russian Federation, Ukraine, and Kazakhstan. It is also cultivated in some other countries like Slovenia, Poland, Hungary, and Brazil. Mainly two species of buckwheat are cultivated around the world:

tartary buckwheat (*Fagopyrum tataricum*) and common buckwheat (*Fagopyrum esculentum* Moench). Images of the inflorescence of buckwheat plants are given in Figure 6.2. The seed of buckwheat is basically a fruit (achene) enclosed by a hull (pericarp), and the hull can be shiny or dull with color may be brown or black. Buckwheat is handled and processed like other cereals due to its starchy seeds, which resemble cereal grains in composition as well as uses. For many years, buckwheat cultivation had dropped off, however, in recent times, there has been a revival of interest in the cultivation of this pseudocereal for the reason that buckwheat is very nutritious having a good quality of protein, starch, dietary fiber, some minerals, vitamins, flavonoids, and other bioactive compounds in its seed. In India, tartary buckwheat is cultivated in hilly regions where the climate is cold, dry, and harsh, has a higher resistance to stress than common buckwheat. Buckwheat has been known by different names during the history of its development and names of buckwheat play major role in tracing its journey in Europe and Asia. Tartary buckwheat is known as phapar in India. Its name is tite phapar in Nepal whereas in Bhutan it is called as bjo and brow in Pakistan. It is interesting that common and tartary buckwheat is known as sweet buckwheat and bitter buckwheat, respectively in India and China.

FIGURE 6.1 (See color insert.) Images of the inflorescence of amaranth plants.

Quinoa (*Chenopodium quinoa Willd.*) plant belongs to the *Chenopodiaceae* family. Approximately 250 species both cultivated and weedy forms are included in this family worldwide. Though, before thousands of years, it was cultivated in the Andes, mainly in Peru and Bolivia. Like amaranth and buckwheat, quinoa draws attention due to its high nutritional value and its high resistance for climatic and weather and soil conditions. Quinoa plant constitutes leaves as well as grains as edible parts while grains studied more for their economic and scientific characteristics.

F. esculentum *F. tataricum*

FIGURE 6.2 (See color insert.) Images of flowers of buckwheat plants.

6.2 SEED CHARACTERISTICS OF PSEUDOCEREALS

The physical properties of seed play a significant role in the selection of proper separating and cleaning equipments. Physical properties like seed weight, seed volume, true density, and bulk density are the detrimental factor for post-harvest treatment of grains. Seed weight symbolizes the grain size, and big grain size means trouble-free post-harvest processing. Images of seeds of pseudocereals are represented in Figure 6.3. Large sized grain has more endosperm indicates more edible portion. Data on bulk density, angle of repose, and friction coefficients against bin wall materials are required to calculate the pressures and loads on storage structures. Seeds of amaranths have different Colors that may be white, cremish yellow, gold, red, and dark depending on plant species. Generally seed color of A. hypochondriacus is cremish yellow, and A. caudatus gives seeds red color. Lenticular shaped seeds of amaranth are small in size (1.0–1.5 mm diameter) and weight of a single kernel is 0.6–1.3 mg. Seeds of quinoa are usually larger in size as compared to seeds of amaranth. Quinoa seed is 1–2.6 mm in diameter with 1 g weight for 350 kernels. Shape of quinoa seeds may be round, flattened, and oval with Colors varying from pale yellow to pink or black (Taylor and Parker, 2002). The structure of seeds of quinoa and amaranth are considerably different from maize and wheat like cereals. In the case of amaranth and quinoa seeds, the germ is campylotropous, which surrounds the starchy perisperm like a hoop (Bressani, 2003). Seed of buckwheat is actually a fruit or achene-comprising hull, spermoderm, germ, and endosperm (Mazza and Oomah, 2003). The seeds of buckwheat are 4 to 9mm long; triangular in shape and color of seeds depends on the color of the hull, which may be glossy or dull, light or dark brown, black, or grey. Similar to common cereals, starch is reserved in the endosperm of buckwheat seed and embryo with its two cotyledons is extended through the endosperm. Tartary buckwheat seeds are round in cross-section, whereas common buckwheat seeds are typically three-angled and a bit larger in size than tartary buckwheat (Marshall and Pomeranz, 1982).

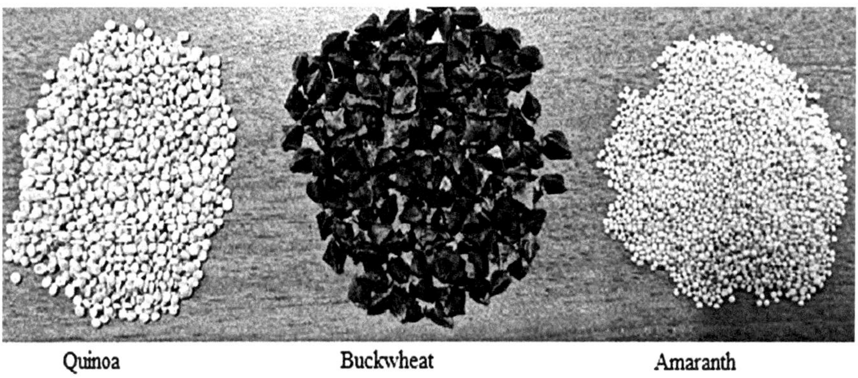

FIGURE 6.3 **(See color insert.)** Images of grains of pseudocereals.

6.3 NUTRITIONAL COMPOSITION OF PSEUDOCEREALS

Pseudocereal grains are well known for their good nutritional value containing energy-rich carbohydrates in the form of sugar, starch, and dietary fibers; well-balanced amino acids and sufficient levels of vitamins and minerals. Nutrients of amaranth, buckwheat, and quinoa seeds are discussed in the following subsections and comparison of the chemical composition of flours of pseudocereal grains with some common cereal flours is represented in Table 6.1.

6.3.1 CARBOHYDRATES (STARCH AND DIETARY FIBER)

Important components of human died are the carbohydrates, which contribute 50–70% of dietary energy. Carbohydrates are categorized into three groups based on their degree of polymerization- sugars (monosaccharides, disaccharides, polyols), oligosaccharide, and polysaccharides (starch and non-starch). In seeds of pseudocereals, starch comprises the major and most important component of carbohydrate present in the form of granules with different size and shape. Starch is an important energy source in diet and also associated with the gut microflora. Studies on starches from amaranth, buckwheat, and quinoa have been carried out from the last few decades with respect to isolation, physico-chemical, thermal, and morphological characteristics. Starch granules of quinoa and amaranth are smaller in size than buckwheat starch granules. Starch of amaranth starch comprises 88.9–99.9% amylopectin and therefore classified as "waxy type" starch with some unique characteristics such as high viscosity and gelatinization at a higher temperature in comparison to normal starches with amylose contents between 17 and 24%. Amaranth starch has higher swelling power and greater water-binding capacity due to an extremely small size (0.8 to 2.5 μm) of granules as compared to the starch granules of other grains as rice (3 to 8 μm), wheat (3 to 34 μm) and maize (5 to 25 μm). Starch is the main carbohydrate in the quinoa seed and constitutes 52 to 69% of it. Quinoa starch contains amylose content in the range of 3 to 22%, which is similar to that of basic rice types while lower than that of wheat and corn. Similar to amaranth starch granules quinoa starch shows high viscosity, great water absorption capacity, and high swelling capacity. Quinoa starch has great freeze-thaw stability. Starch content in the buckwheat grains ranges from 59 to 70% of the dry mass, depending on the geographical conditions. Granules of buckwheat starch are 5.5 to 10.7 μm in size and shape may be spherical or polygonal or oval with the visible flat surface due to dense packing in the endosperm (Sindhu and Khatkar, 2016). Generally, in the starch of buckwheat, amylose content varies from 15 to 52% depending on the varieties and climatic conditions (Table 6.1).

Dietary fiber is the edible portion of a plant or corresponding carbohydrates which are not digested and absorbed in the small intestine of the human digestive system. However dietary fibers can be fermented by intestinal microflora. Dietary fibers are generally existing within thick cell wall tissues, aleurone layer, hulls, and seed coat. Dietary fiber is the important component of diet helps in food digestion, appropriate functioning, and good health of the digestive tract. It also presents a feeling of fullness and supports the weight loss of the body. Studies revealed that pseudocereals are a good source of dietary fibers. Amaranth contains 8–16% dietary fiber with soluble fibers ranging from 18 to 40% fluctuating according to varieties and climatic conditions. Share of dietary fibers also varies with processing such as milling, sifting, and pneumatic classification procedures used. Buckwheat grains contain dietary fiber in the range of 50–110 mg/g, the soluble fiber in the range 30–70 mg/g, and the concentration of the insoluble fiber varied from 20–40 mg/g. The dietary fiber content of quinoa (12.88 to 14.20%) exists, particularly in the embryo.

TABLE 6.1 Comparison of Pseudocereal Flours Composition (Dry Basis) with Other Commonly Used Cereals

Crop	Protein (%)	Ash (%)	Lipid (%)	Soluble fiber (%)	Insoluble fiber (%)	Total fiber (%)
Amaranth	13.9	2.1	7.3	6.3	8.2	14.5
Quinoa	16.5	3.8	6.1	4.3	9.6	13.9
Common buckwheat	11.0	2.6	3.4	1.2	5.3	6.5
Tartary buckwheat	10.3	1.8	2.5	0.5	5.8	6.3
Oat	12.6	1.8	7.1	3.3	4.9	8.2
Rye	11.7	1.5	1.8	3.6	10.0	13.6
Wheat	11.5	1.7	1.0	1.0	1.5	2.4

Adapted from Ahmed et al. (2014).

6.3.2 PROTEINS

Protein content in pseudocereals is generally higher than in common cereals, and highest protein content in amaranth grains is followed by quinoa and buckwheat. Protein mainly constitutes the globulins and albumins in amaranth, quinoa, and buckwheat. Contrary to most common grains, pseudocereals comprise very less or lack of prolamin, which are the chief storage proteins in major cereals, (Drzewiecki et al., 2003). Comparison of amino acid levels of pseudocereal grains with some common cereal grains is summarized in Table 6.2. The protein of amaranth has high digestibility and rich in lysine content ranging 4.9 to 6.1% of total protein, whereas cereal proteins are generally deficient in this amino acid. Dissimilar to cereals grains containing protein mainly in the endosperm, protein in amaranth grains distributed as 65% in the embryo and only 35% in perisperm (Senft, 1979; Betschart et al., 1981). Amaranth proteins are deficient in some amino acids such as leucine, isoleucine, and valine, whereas they are present in excess in mostly cereals; therefore, amaranth is well matched with cereals for blending purpose. In some studies it was found that nutrition score of mixture of maize and amaranth flour (50:50) is close to 100 and amaranth flour in combination with wheat flour enhances the nutritional score of baked products (Academy of Sciences, 1984; Saunders and Becker, 1984; Bressani, 1989; Joshi and Rana, 1991; Segura-Nieto et al., 1994).

Buckwheat seeds contain protein in the range from 8.5 to 18.5% depending on the variety. Mainly albumin, globulin, and glutelin constitute the protein in buckwheat, but the concentration of each individual protein fraction varies widely depending on the variety. Buckwheat seeds contain albumin and globulin as the main storage proteins while a very lesser amount of prolamin and glutelin is present. Amino acid profile of buckwheat protein with higher lysine content than cereals proteins contributes to its high biological value. The amino acid score of buckwheat flour is 100 that is the maximum score among proteins of plant sources. Buckwheat flour can be easily used in the preparation of gluten-free food products. However, buckwheat products cause or encourage some allergic reactions in the human body due to disulfide bonds in the proteins of albumins. Also, the low digestibility is the main drawback of buckwheat protein. Anti-nutritional factors such as protease inhibitors (trypsin inhibitors) and tannins present in buckwheat are responsible for low digestibility. Also, buckwheat products induce allergic reactions due to disulfide bonds in proteins of albumin family (Table 6.2).

TABLE 6.2 Comparison of Composition of Essential Amino Acid (g/100 g Protein) of Pseudocereals with Other Common Cereals

Amino acid	Amaranth	Buckwheat	Quinoa	Barley	Wheat	Maize
Lysine	6.2	5.1	6.1	3.7	2.5	2.8
Methionine	2.0	1.9	4.8	1.8	1.8	2.4
Cystine	2.0	2.2	4.8	2.3	1.8	2.2
Threonine	3.3	3.5	3.8	3.6	2.8	3.9
Valine	4.4	4.7	4.5	5.3	4.5	5.0
Isoleucine	3.7	3.5	4.4	3.7	3.4	3.8
Leucine	6.1	6.1	6.6	7.1	6.8	1.05
Phenylalanine	4.6	4.2	7.3	4.9	4.4	4.5
Histidine	2.6	2.2	3.2	2.2	2.3	2.4
Tryptophan	3.3	1.6	1.2	1.1	1.0	0.6
BV (%)	-	93.1	-	76.3	62.5	64.3
NPU (%)	-	74.4	-	64.3	57.8	59.9

BV: biological value (based on amino acid composition); NPU: net protein utilization. Adapted from Ahmed et al. (2014).

The quantity and quality of quinoa proteins are generally greater than those of major cereal grains, at the same time having a gluten-free property and high digestibility. Quinoa has protein level in the range of 12.9 to 16.5%, which is comparatively more than that of oat, rice, rye, barley, and maize and near to the protein content of wheat. Like amaranth and buckwheat, main storage proteins are globulin (37%) and albumin (35%) with absence or low content of prolamins. The protein content of quinoa and composition of quinoa varies in different species. Quinoa has a wonderful amino acid balance with a high content of lysine and thionic amino acids. It is considered as one of the exceptional plants providing all essential amino acids for the human body, and quinoa protein is acknowledged as high-quality proteins contrary to grain proteins having lysine as limiting amino acid. The protein value of quinoa protein is similar to casein in milk. Along with the high protein content and perfect amino acid balance, quinoa also contains sufficient concentration of tryptophan, which is another limiting amino acid. Digestibility of protein varies depending on the species and generally increased with cooking.

6.3.3 LIPIDS

Generally, the level of lipid in quinoa and amaranth is 2 to 3 times more as compared to buckwheat and other major cereals. Lipids of pseudocereals are recognized by a high level of unsaturation that is nutritionally important. Most abundant fatty acid in pseudocereals is linoleic acid, which is approximately 50% of total fatty acids in both quinoa as well as amaranth, and in buckwheat, it is about 35%. Oleic acid is another important fatty acid in pseudocereals reported nearly 25% in case of both quinoa and amaranth while 35% in buckwheat (Alvarez-Jubete et al., 2009; Ruales and Nair, 1993). In amaranth grains, approximately 76% of fatty acids are unsaturated, and the ratio of saturated to unsaturated fatty acid ranges from 0.26 to 0.31. Amaranth oil is characterized by a high amount of squalene ranges from 2 to 8% in refined oil. Squalene is a highly unsaturated hydrocarbon, a triterpene, recognized as the requisite biochemical precursor of sterols. Commercially, squalene is derived from liver of sharks, whales, and some other marine species and synthesized industrially. A little quantity of squalene is also obtained from plant sources like wheat germ, olive, and rice bran oils. It is generally utilized as a lubricant in fine electronic instruments and pharmaceutical industry. Utilization of squalene in the food industry is contradictory as results of some studies have indicated that high squalene diet raises the level of LDL-cholesterol in blood plasma while other investigations showed no change in triacylglycerols and cholesterol level of blood serum while significant increment appeared in bile acids and fecal cholesterol along with its non-polar derivatives. However, the antioxidative, antibacterial, and anti-tumor effects of squalene are scientifically confirmed and accepted (Kopicová and Vavreinová, 2007). Amaranth oils contain phospholipids in relatively low amount (around 5%), and tocols content ranges from 191 mg/ kg to 2000 mg/kg in oil (Becker, 1994; Berghofer and Schoenlechner, 2002). Lipid content of buckwheat seeds varies from 15 to 37mg/g depending on varieties. The highest concentration of lipids is in the embryo and during milling of buckwheat bran is produced as a highly lipid-rich fraction. The linoleic, palmitic, and oleic acids are the major fatty acids of buckwheat lipids. Lipid content of buckwheat is similar to that of wheat and rye. However, in per gram of total lipids, neutrals lipids varied from 810 to 850 mg in buckwheat and about 350mg in wheat and rye. In quinoa seeds, lipids vary from 2 to 9.5% depending on the species. Quinoa is accepted as alternative oily seed due to quality and amount of lipid. Lipid content of quinoa is higher than common cereals such as maize while lower than soybean. Quinoa oil is rich in essential fatty acids like α-linolenic, and linoleic acids, and these fatty acids along with the oleic, constitute about 88% of total fatty acids of quinoa grains. The main saturated fatty acid is palmitic fatty acid, which constitutes 10% of its total fatty acid. Compared to amaranth and buckwheat, quinoa seeds contain a higher content of α-linolenic acid with values ranging from 3.8 to 8.3% (Ruales and Nair, 1993; Alvarez-Jubete et al., 2009).

6.3.4 MINERALS AND VITAMINS

Amaranth, buckwheat, and quinoa are considered a good source of minerals containing some minerals in a higher amount than cereals and legumes. Becker et al. (1981) reported that amaranth grain is a rich source of minerals having iron in the range of 72 to 174 mg/ kg, calcium 1,300 to 2,850 mg/ kg, sodium 160 to 480 mg/kg, magnesium 2,300 to 3,360 mg/kg and zinc 36.2 to 40 mg/kg. The mineral content of buckwheat grains and their fractions summarized by Li and Zhang (2001) is 2 to 2.5% in whole grains, 1.8 to 2.0% in the kernel, 2.2 to 3.5% in dehulled grains, 3.4 to 4.2% in hulls and approximately 0.9% in flour. Buckwheat contains higher levels of some minerals such as copper, manganese, and zinc than major cereals like wheat, maize, or rice (Steadman et al., 2001). In buckwheat, the bioavailability of zinc, potassium, and copper K is particularly high. It has been found that approximately 13 to 89% of the recommended dietary allowance for copper, manganese, zinc, and magnesium is achieved by 100 g of buckwheat flour intake. Bran contains a major quantity of these minerals followed by endosperm. The level of some minerals including zinc, copper, magnesium, and manganese is higher in buckwheat flour as compared to wheat flour. Quinoa seeds are also rich in mineral nutrients. The contents of potassium (927 mg/100 g), calcium (149 mg/100 g), magnesium (250 mg/100 g), phosphorus (384 mg/100 g), sulfur (150 to 220 mg/100 g), iron (13.2 mg/100 g), and zinc (4.4 mg/100 g) in quinoa seeds are comparatively higher than those of cereals such as wheat and rice. Presence of minerals like calcium, magnesium, and potassium in biologically suitable forms in quinoa and their concentration in seeds is adequate for a balanced diet.

Vitamins are the organic compounds required in very little quantity for the adequate functioning of the human body. Pseudocereals are also a rich source of vitamins containing riboflavin, thiamine, tocopherol, niacin, and niacinamide, ascorbic acid. Amaranth contains riboflavin (0.19 to 0.23 mg/100 g of flour), ascorbic acid (4.5 mg/100 g), niacin (1.17 to 1.45 mg/100 g), and thiamine (0.07 to 0.1 mg/100 g). Comparatively higher level of vitamins (thiamine, riboflavin, tocopherol, and niacin and niacinamide) is present in buckwheat grains than most cereals. Generally, the content of vitamin B is higher while that of vitamin E is lower in the seeds of tartary variety than that of common buckwheat variety. Thiamine (vitamin B1) is adhered tightly to proteins in tartary buckwheat seeds, and this complex enhances the storage stability of vitamin thiamine and also increases its bioavailability (Li and Zhang, 2001). Processing like germination of buckwheat seeds increases the concentration of some vitamins such as vitamin C, vitamin B1, and B6. Schynder et al. (2001) reported that bran of tartary buckwheat offers approximately 6% of daily therapeutic doses of pyridoxine useful in the lowering homocysteine level of blood plasma. An appreciable amount of vitamins such as riboflavin, thiamine, folic acid, and α-tocopherol and γ-tocopherol is present in quinoa seeds. Quinoa contains a higher level of riboflavin, pyridoxine, and folic acid as compared to other grains such as wheat, maize, oat, barley, rye, and rice. Thiamine is present in the pericarp of the quinoa grains, and its content varies from 0.05 to 0.60 mg/100 g dry matter (DM). It is an admirable source of tocopherol with content higher (4.60 to 5.90 mg of vit E/100 g of DM) relative to wheat (Abugoch James, 2009; Alvarez-Jubete et al., 2010). The ascorbic acid content of quinoa varies in a band of 0–63.0 mg/100 g. It is reported that 100 g of quinoa meets the daily requirements 100% for pyridoxine and folic acid in adults, 80%

for riboflavin in children and 40% for riboflavin in adults (Abugoch James, 2009). It is also a good source of niacin; however, its content does not fulfill the daily requirement.

6.4 FUNCTIONALITY OF PSEUDOCEREALS

6.4.1 ANTIOXIDANT ACTIVITY

Natural anti-oxidants play a significant role in hampering free radicals chain reactions and oxidative reactions at the levels of tissue and membrane. All plants species exhibits own antioxidant defense property due to different secondary metabolites, mainly polyphenolic compounds. Pseudocereal grains are a good source of anti-oxidatively important polyphenolic compounds. Increasing the positive reception of the nutritional value and functional properties of pseudocereals has also supported the initiation of studies about their antioxidant properties. Polyphenols present in plants parts add to the antioxidant activities of different plant foods and their extracts. Klimczak et al. (2002) reported that ethanolic extracts of amaranth have an appreciable amount of in beta-carotene and linoleate model. In the study of tartary buckwheat vinegar by Wang et al. (2012), a positive correlation was noticed between the total flavonoid content and antioxidant activity. Antioxidant activity of tartary buckwheat is reported higher than common buckwheat in different studies, and the reason could be the higher quercetin and rutin in tartary buckwheat. Several quinoa extracts (cultivated in Japan and Bolivia) evaluated for antioxidant activity by Nsimba et al. (2008) showed high antioxidant activity in the quinoa seeds which was also higher than the anti-oxidant activity of amaranth grains. The cooking effects on sweet and bitter quinoa seeds were studied, and results showed anti-oxidant activity in bitter quinoa seeds were higher than in the sweet ones (Dini et al., 2010). Presence of phenols and flavonoid in the bitter seeds was base for higher antioxidant activity whereas the occurrence of phenols, flavonoids, and carotenoids was responsible for antioxidant activity of sweets quinoa seeds. Cooking decreased antioxidant capacity in both bitter and sweet quinoa seed. The anti-oxidant effect reported due to the presence of isoflavones is utilized in the development of several functional foods for the reduction of possibility factors for some chronic problems such as coronary and neurodegenerative diseases, bones related diseases and cancer. Present data suggest that pseudocereal grains can act as a potential source of natural antioxidants.

6.4.2 *HYPOCHOLESTEROLEMIC ACTIVITY*

Production of cholesterol in the liver and absorption through the food is required for metabolic processes. Higher cholesterol intake can persuade oxidative stress resulting in raised blood cholesterol level, leading to the initiation or expansion of chronic diseases like atherosclerosis. Various studies have revealed that amaranth and buckwheat have the ability to change the cholesterol level, which leads to the prevention of cardiovascular diseases (CVD). Several investigations have proposed that the sterols present in plants reduce the absorption of cholesterol in the body. The cholesterol-lowering effect of amaranth oil fraction was studied by Pogojeva et al. (2006) and reported a decrease in blood cholesterol level without observing any increment in liver cholesterol. Mendonca et al. (2009) confirmed that the amaranth protein isolate could also decrease the total and LDL cholesterol level in hamsters. Consumption of tartary buckwheat cookies proved positive correlation with the reduction of cholesterol level in the human body (Wieslander et al., 2011, 2012). Zhang et al. (2007) considered the relationship of some diseases like hypertension, dyslipidemia, and hyperglycemia in human population with utilization of buckwheat seed as main food for whole life period and observed a decline in total cholesterol (TC), triglycerides, and LDL. Phytosterol content in quinoa seeds comprising β-sitosterol, campesterol, and stigmasterol are higher than that recorded for other seeds such as barley, maize, and squash. The consumption of protein isolates of quinoa seeds (> 10% grain) considerably decreased the level of TC level of plasma and liver in mice provided fat-rich foods (Takao et al., 2005). Bioactive properties of quinoa were noticed in other studies involving use of quinoa flour, or hydrolyzed quinoa protein resulted in considerably reduced blood pressure in studied animals including mice and rats.

6.4.3 *ANTI-INFLAMMATORY ACTIVITY*

Inflammation is characterized by some common symptoms like redness, edema, high body temperature, pain, and poor functional activity and it is generally a biological process as a reaction to tissue injury, pathogens infection, and chemical irritation. Current investigations proposed that too much inflammation and oxidative harm can lead to numerous health problems such as autoimmune, neurological, and CVD and cancer. Therefore, prevention of chronic or excessive inflammation is the main way to avoid cancer. Use of steroidal as well as non-steroidal anti-inflammatory drugs (NSAIDs) is common for the treatment of chronic inflammatory diseases; however long-lasting use of these drugs can result in unwanted side-effects. Therefore, natural compounds having anti-inflammatory properties used traditionally are preferred now a day's for the prevention and treatment of inflammations. Bavarva and Narasimhacharya (2013) studied the anti-inflammatory activity of petroleum ether and ethanolic extract of whole plant of *Amaranthus spinosus Linn.* at different doses. The considerable inhibition of carrageenan-induced paw edema was noticed, and effect was dose-dependent. The anti-inflammation potential of amaranth plant extract was similar to the drug Ibuprofen. Anti-inflammatory activity of amaranth plant extract may be associated with the steroids, alkaloids, and flavonoids in the extracts. Nuclear factor-kappa B (NF-κB) helps in the regulation of the genes associated with inflammation. Hole et al. (2009) noticed that different phenolic buckwheat extract (30 to 50 mg/ ml), as well as phenolic acids (8,5-difeluric, feluric, and p-coumaric) present in buckwheat, reduced the NF-κB activity in the U937 cell line. Karki et al. (2013) also found that tartary buckwheat extract and rutin hampered the activation of NF-κB in LPS motivated macrophages. Traditionally, quinoa has been utilized by Andean people as a natural therapy for the curing of inflammations occurred due to muscle sprains, twists, and muscular strains. The tri-terpenes in the grains of quinoa produce it as a natural source of new elements in the drugs formulation. Saponins (monodesmosidic saponins) present in quinoa seeds are associated with the anti-inflammatory activity of quinoa (Kuljanabhagavad and Wink, 2009). Oleanolic acid and hederagenin in quinoa grains and leaves respectively are linked with different anti-inflammatory molecular activities.

6.4.4 ANTICANCER ACTIVITY

Major reason of death in developed countries and the second main cause of death in developing countries is cancer. Diet is considered a causative aspect for occurrence of cancer and some other chronic illnesses. Development of functional food to avoid the occurrence of diseases like cancer is one of the century's chief worldwide challenges. Initiation and propagation of cancer is associated with oxidative DNA damage. Therefore, anticancer activity of functional foods is based on their antioxidant activities. Different dietary phyto-chemicals found in various food products have represented chemo-preventive properties in opposition to cancer, in preclinical trials on animals as well as in epidemiological research works on humans (Wang et al., 2012). The ethanolic extract of *A. spinosus* leaves was feed orally to mice for more than two weeks and a reduction in tumor volume and viable cell count was noticed as compared with the mice of control group. Cao et al. (2008) reported that ethanolic buckwheat extract from tartary and common buckwheat prevented DNA damage by hydroxyl radical. In another study, rutin, quercetin, and methanolic extract of buckwheat flours (common and tartary) represented high antioxidant capacity and inhibited DNA damage in the HepG2 cell line induced by tert-butylhydroperoxide (Vogrinčič, 2013). Gawlik-Dziki et al. (2013) assessed the nutraceutical prospective of quinoa leaves by studying the phenolic components along with their collective bioactivity. It was noticed that in quinoa leaves the phenolic compounds exhibit chemo-preventive and anti-cancer effects, with the involvement of intracellular signaling systems based on oxidative stress. The study confirmed the nutraceutical potential of quinoa leaves for cancer as well as other diseases related to oxidative stress.

6.5 FOOD APPLICATIONS OF PSEUDOCEREALS

Good nutritional value, different technological characteristics of grains and grain fractions, nutraceutical importance and lower cost of production of pseudocereals make them suitable for wide range of applications. The proposed utilization of grains of pseudocereals and their specific processing behavior is summarized in Table 6.3. Pseudocereals are used in composite flour for the production of various bakery products. As they are locally grown and give value addition to the products as well as offer the avenue for local taste. The amaranth seeds are popped and bound with syrup or honey. In India, amaranth seeds are used in making laddoos and in few regions eaten with rice after cooking with rice. In Himalayan regions, amaranth, and buckwheat flours are used to make chapattis. Flours of pseudocereals can be used in biscuit formation. Crispiness of pseudocereal flour biscuits was highest in buckwheat followed by quinoa and amaranth, and biscuits prepared using flours of buckwheat and amaranth were scored higher in sensory evaluation. Buckwheat flours can be utilized the formulation of gluten-free cracker without unfavorably changes in the sensory quality of crackers. Pseudocereal flours can be utilized in the production of pasta. Mixture of flours of amaranth and rice (25:75 ratio) produced pasta with improved the nutritional quality and good cooking behavior (Cabrera et al., 2012). As quinoa, buckwheat, and amaranth have non-glutinous proteins and their flours are utilized in the production of gluten-free food products. Starch is the major component of pseudocereal grains. Starch isolated from their grains can be utilized in various food and non-food applications. Isolated starch is applicable in as thickener, stabilizing, and gelling agent in various products like soup, sauce, pie-filling, noodles, etc. Starches can be utilized as base material in the formation of edible films. Whole grains of pseudocereals can used as breakfast meal after processing such as germination, roasting, steaming, popping (Table 6.3).

TABLE 6.3 Possible Utilization of Pseudocereal Seeds

Process	Products
Cooking	Cooked seeds of amaranth, quinoa, and buckwheat
Puffing	Popped or expanded amaranth seeds
Millin	Wholemealal flour of grains all three pseudocereals
Classification with milling	Different fractions of flour, high protein flours, high starch flours of pseudocereals
Cooking and flaking	Flakes of seeds of pseudocereals
Drum drying	Pre-gelatinized flours of amaranth and buckwheat, partially gelatinized flours of quinoa
Extrusion cooking, extrusion with milling or with flaking	Cooked, expanded or non-expanded products, pre-gelatinized flours, lakes. Buckwheat-very good expansion while quinoa and amaranth are processed by blending witlow-fatat materials
Germination	Sprouts, malt of amaranth, quinoa, and buckwheat
Starch hydrolysis	Concentrates of protein and glucose syrup of pseudocereals
Starch films	Isolated starch from amaranth and buckwheat
Starch isolation	Starch extracts of pseudocereals

Adapted from Berghofer and Schoenlenchner (2007).

Whole or partially broken grains are also used in the form of porridge. Flowers of pseudocereals are Colorful and utilized for the ornamental purpose. As amaranth and quinoa flower are greatly Colored and used as colorant in the foods and medicines. Bioactive components in the pseudocereals make them applicable in medicines or drugs developments and functional food formation. Preparation of gluten-free breads based on pseudocereal is reported in literature. Incorporation of pseudocereals produced increased loaf volume, slower staling and more softness in crumb with improved sensory properties. Native and modified starches of amaranth and buckwheat were utilized for the preparation of edible films in the previous studies (Sindhu and Khatkar, 2018a, b, c). The breads formed of pseudocereals were rich in protein, fiber, minerals, and polyphenolic compounds with raised in-vitro antioxidant activity. Modification in the properties of pseudocereal-based bread depends directly on the type and quantity of pseudocereal added as well as other ingredients and processing conditions (Alvarez-Jubete et al., 2009; 2010). Also beers can be formed from sorghum, maize, amaranth, quinoa, and buckwheat. However, beers may vary in stability of foams, color, flavor, and opacity of color.

KEYWORDS

- **non-steroidal anti-inflammatory drugs**
- **nuclear factor-kappa B**
- **pseudocereals**

REFERENCES

Abugoch, L. E., (2009). Quinoa (*Chenopodium quinoa* Willd.): Composition, chemistry, nutritional and functional properties. *Advances in Food and Nutrition Research, 58*, 1–31.

Ahmed, A., Khalid, N., Ahmad, A., Abbasi, N. A., Latif, M. S. Z., & Randhawa, M. A., (2014). Phytochemicals and biofunctional properties of buckwheat: A review. *Journal of Agricultural Science, 152*, 349–369.

Alvarez-Jubete, L., Arendt, E. K., & Gallagher, E., (2009). Nutritive value and chemical composition of pseudocereals as gluten-free ingredients. *International Journal of Food Science and Nutrition, 60*(1), 240–257.

Alvarez-Jubete, L., Wijngaard, H., Arendt, E., & Gallagher, E., (2010). Polyphenol composition and *in vitro* antioxidant activity of amaranth, quinoa buckwheat and wheat as affected by sprouting and baking. *Food Chem., 119*, 770–778.

Bavarva, J. H., & Narasimhacharya, A. V., (2013). Systematic study to evaluate anti-diabetic potential of Amaranthus spinosus on type-1 and type-2 diabetes. *Cell Mol Biol., 2*(59), 1818–1825.

Becker, R., (1994). Amaranth oil: Composition, processing and nutritional qualities. In: Paredes-Lopez, O., (ed.), *Amaranth–Biology, Chemistry and Technology* (pp. 133–142).

Berghofer, E., & Schoenlechner, R., (2002). Grain amaranth. In: Belton, P. S., & Taylor, J. R. N., (eds.), *Pseudocereals and Less Common Cereals: Grain Properties and Utilization Potential* (pp. 219–260).

Berghofer E., & Schoenlechner R., (2007). *Pseudocereals: An Overview*. http://projekt.sik.se/traditionalgrains/review/Oral%20presentation%20PDF%20files/Berghofer%20.pdf.

Bressani, R., (2003). Amaranth. In: Caballero, B., (ed.), *Encyclopedia of Food Sciences and Nutrition* (pp. 166–173). Oxford: Academic Press.

Bruni, R., Medici, A., Guerrini, A., Scalia, S., Poli, F., Muzzoli, M., & Sacchetti, G., (2001). Wild *Amaranthus caudatus* seed oil, a nutraceutical resource from Ecuadorian flora. *Journal of Agricultural and Food Chemistry, 49*, 5455–5460.

Cabrera-Chávez, F., De la Barca, A. M. C., Islas-Rubio, A. R. *A*., Marengo, M., Pagani, M. A., Bonomi, F., & Iametti, S., (2012). Molecular rearrangements in extrusion processes for the production of amaranth-enriched, gluten-free rice pasta. *LWT-Food Science and Technology, 47,* 421–426.

Cao, W., Chen, W. J., Suo, Z. R., & Yao, Y. P., (2008). Protective effects of ethanolic extracts of buckwheat groats on DNA damage caused by hydroxyl radicals. *Food Res., 41,* 924–929.

Dini, I., Tenore, G. C., & Dini, A., (2010). Antioxidant compound contents and antioxidant activity before and after cooking in sweet and bitter Chenopodium quinoa seeds. *LWT-Food Science and Technology, 43*, 447–451.

Djordjevic, T. M., Šiler-Marinkovic, S. S., & Dimitrijevic, B. S. I., (2010). Antioxidant activity and total phenolic content in some cereals and legumes. *International Journal of Food Properties, 14*(1), 175–184.

Drzewiecki, J., Delgado-Licon, E., Haruenkit, R., Pawelzik, E., Martin- Belloso, O., Park, Y. S., et al. (2003). Identification and differences of total proteins and their soluble fractions in some pseudocereals based on electrophoretic patterns. *Journal of Agricultural and Food Chemistry, 51*(26), 7798–7804.

Gawlik-Dziki, U., Świeca, M., Sułkowski, M., Dziki, D., Baraniak, B., & Czyż, J., (2013). Antioxidant and anticancer activities of *Chenopodium quinoa* leaves extracts–*in vitro* study. *Food and Chemical Toxicology, 57,* 154–160.

Grivennikov, S. I., Greten, F. R., & Karin, M., (2010). Immunity, inflammation, and cancer. *Cell, 140*(6), 883–899.

Hole, A., Grimmer, S., Naterstad, K., Jensen, M., Paur, I., Johansen, S., Balstad, T., Blomhoff, R., & Sahlstrom, S., (2009). Activation and inhibition of nuclear factor κB activity by cereal extracts: Role of dietary phenolic acids. *J. Agric. Food Chem., 57,* 9481–9488.

Karki, R., Dong, C., & Kim, W., (2013). Extract of buckwheat sprouts scavenges oxidation and inhibits pro-inflammatory mediators in lipopolysaccharide-stimulated macrophages (RAW264.7). *Journal of Integrative Medicine, 11*(4), 246–252.

Karki, R., Park, C. H., & Kim, D. W., (2013). Extract of buckwheat sprouts scavenges oxidation and inhibits pro-inflammatory mediators in lipopolysaccharide-stimulated macrophages (RAW264.7). *J. Integr. Med., 11,* 246–252.

Kopicova, Z., & Vavreinova, S., (2007). Occurrence of squalene and cholesterol in various species of Czech freshwater fish. *Czech Journal of Food Sciences, 25,* 195–201.

Kuljanabhagavad, T., & Wink, M., (2009). Biological activities and chemistry of saponins from *Chenopodium quinoa* Willd. *Phytochem. Rev., 8*, 473–490.

Lee, C. C., Shen, S. R., Lai, Y. J., & Wu, S. C., (2013). Rutin and quercetin, bioactive compounds from tartary buckwheat, prevent liver inflammatory injury. *Food Funct., 4,* 794–802.

Li, S., & Zhang, Q. H., (2001). Advances in the development of functional foods from buckwheat. *Critical Reviews in Food Science and Nutrition, 41*(6), 451–464.

Marshall, H. G., & Pomeranz, Y. (1982). Buckwheat description, breeding, production and utilization. In Y. Pomeranz (Ed.), *Advances in Cereal Science and Technology* (pp. 157–212). St. Paul, MN: American Association of Cereal Chemistry.

Mazza, G., & Oomah, B. D., (2003). Buckwheat. In: Caballero, B., (ed.), *Encyclopedia of Food Sciences and Nutrition* (pp. 692–699). Oxford: Academic Press.

Mendonça, S., Saldiva, P. H., Cruz, R. J., & Arêas, J. A. G., (2009). Amaranth protein presents cholesterol-lowering effects. *Food Chemistry, 116,* 738–742.

Mitsunaga, T., Iwashima, A., Matsuda, M., & Shimizu, M., (1986). Isolation and properties of thiamine-binding protein from buckwheat seed. *Cereal Chemistry, 63*, 332–335.

Nsimba, R. Y., Kikuzaki, H., & Konishi, Y., (2008). Antioxidant activity of various extracts and fractions of *Chenopodium quinoa* and *Amaranthus* spp. seeds. *Food Chemistry, 106*(2), 760–766.

Pogojeva, A. V., Gonor, K. V., Kulakova, S. N., Miroshnichenko, L. A., & Martirosyan, D. M., (2006). Effect of amaranth oil on lipid profile of patients with cardiovascular diseases. In: Martirosyan, D. M., (ed.), *Book Functional Foods for Chronic Diseases* (pp. 35–45). Dallas, USA.

Ruales, J., & Nair, B., (1993). Saponins, phytic acid, tannins and protease inhibitors in quinoa (*Chenopodium quinoa*, Willd.) seeds. *Food Chem., 48,* 137–143.

Ryan, E., Galvin, K., O'Connor, T., Maguire, A., & O'Brien, N., (2007). Phytosterol, squalene, tocopherol content and fatty acid profile of selected seeds, grains and legumes. *Plant Foods Hum. Nutr., 62,* 85–91.

Schoenlechner, R., Linsberger, G., Kaczyc, L., & Berghofer, E., (2006). Production of short dough biscuits from the pseudocereals amaranth, quinoa and buckwheat with common bean. *Ernahrung, 30,* 101–107.

Schynder, G., Roffy, M., Pin, R., Flammer, Y., Lange, H., & Eberly, F., (2001). Decreased rate of coronary restenosis after lowering of plasma homocysteine levels. *New England Journal of Medicine, 345,* 1593–1600.

Sindhu, R., & Khatkar, B. S., (2016). Thermal, pasting anmicrostructuralal properties of starch and flour of tartary buckwheat (F. Tataricum). *International Journal of Engineering Research & Technology*, *5*(06), 305–308.

Sindhu, R., & Khatkar, B. S., (2018a). Development of native and hydrothermally modified amaranth starch films. *International Advanced Research Journal in Science, Engineering and Technology, 5*(3), 13–16.

Sindhu, R., & Khatkar, B. S., (2018b). Development of edible films from Native and modified starches of Common buckwheat. *International Advanced Research Journal in Science, Engineering and Technology, 5*(3), 9–12.

Sindhu, R., & Khatkar, B. S., (2018c). Amaranth starch isolation, oxidation, heat-moisture treatment and application in edible film formation. *International Journal of Advanced Engineering Research and Sciences, 5*(3), 136–141.

Steadman, K. J., Burgoon, M. S., Lewis, B. A., Edwardson, S. E., & Obendorf, R. L., (2001). Minerals, phytic acid, tannin and rutin in buckwheat seed milling fractions. *Journal of the Science of Food and Agriculture, 81*, 1094–1100.

Taylor, R. N., & Parker, M. L., (2002). Quinoa. In: Belton, P., & Taylor, J., (eds.), *Pseudocereals and Less Common Cereals* (pp. 93–121). Springer, Berlin.

Vogrinčič, M., Kreft, I., Filipič, M., & Zegura, B., (2013). Antigenotoxic effect of Tartary (*Fagopyrum tataricum*) and common (*Fagopyrum esculentum*) buckwheat flour. *J. Med. Food., 16*(10), 944–952.

Wang, H., Khor, T. O., Shu, L., Su, Z. Y., Fuentes, F., Lee, J. H., & Kong, A. N., (2012). Plants vs. Cancer: A review on natural phytochemicals in preventing and treating cancers and their druggability. *Anticancer Agents Med. Chem., 12*(10), 1281–1305.

Wieslander, G., Fabjan, N., Vogrincic, M., Kreft, I., Janson, C., Spetz-Nyström, U., Vombergar, B., Tagesson, C., Leanderson, P., & Norbäck, D., (2011). Eating buckwheat cookies is associated with the reduction in serum levels of myeloperoxidase and cholesterol double-blind crossover study in day-center staffs. *Tohoku J. Exp. Med., 225,* 123–130.

Zhang, Z., Li, Y., Li, C., Yuan, J., & Wang, Z., (2007). Expression of a buckwheat trypsin inhibitor gene in Escherichia coli and its effect on multiple myeloma IM-9 cell proliferation. *Acta Biochim. Biophys. Sin., 39,* 701–707.

CHAPTER 7

Development of Functional Food Products from Food Waste

VASUDHA BANSAL, PANKAJ PREET SANDHU, and
NIDHI BUDHALAKOTI

*Department of Food Engineering and Nutrition, Center of Innovative and Applied Bioprocessing (CIAB), Mohali, Punjab, India,
E-mail: vasu22bansal@gmail.com*

ABSTRACT

The inclination towards attaining health from natural food products has raised the necessity for the development of functional foods. Owing to the burgeoning of clinical disorders in terms of diabetes mellitus, cancer, obesity, hypertension, and neuro-based degenerative diseases, the consumption of functional foods has been incremented tremendously. Functional foods are confined towards plant-based products, thereby; value addition to plant-based food waste will provide optimum utilization with potential benefits. Therefore, the unutilized food products viz. peels, shreds, pulp, seed coats, kernels, can be converted to innovative functional products. The waste products of fruits and vegetables contain vitamin C, pectin, and active phytochemicals. The composition of phytophenolic and their antioxidant capacities can be employed for the development of valuable food products.

7.1 INTRODUCTION

These days' consumers are looking for those foods that are not only possessed with high nutritive value but also with potent functional value. Therefore, the functional value of plant-based foods (in terms of fruits and vegetables) has always been considered. It is not only the edible part of the fruits and vegetables that contain functional components; also, their peels and pulp are of equal value. The non-edible part are usually been regarded as waste and dumped into landfills. The uncontrolled activity of dumping not only adding to environmental pollution, but also, is making the management of food waste a major issue (Zhang et al., 2013). Besides this problem, it is being estimated that the landfills will soon be filled in coming years and thereby, result in soil and water pollution owing to the leaching out of the organic compounds. There are many areas where we can utilize these wastes in the form of production of chemicals, fuels, formation of biodegradable film, etc. Similarly, these food wastes can be potentially utilized in the food processing industry in order to produce value-added edible products.

7.2 SOLID WASTE GENERATION OF FRUITS

Among the fruits, particularly, orange, apple, banana which are consumed in all over parts of the world, create significant amount of solid waste. Till now, technologists have investigated the effect of the addition of peel and pulp of in bakery products, chocolates, ice-cream, beverages, and nutraceutical. However, the percentage of utilization is still not large. The utilization of these solid wastes and their addition to food products also depends on the consumer acceptance. Since, consumers are getting inclined towards functionally enriched food products with the addition of natural additives. Therefore, the optimized addition of these solid wastes in forming the novel products will be advantageous. Banana peels are the major organic wastes, which can be utilized for the synthesis and extraction of various organic materials which might be beneficial in many ways. These waste products are sometimes used as feed for animals or converted into manure for soil. As a feed for cattle's, pigs, etc. these peels provide major nutrients, which might also benefit humans. It is rich in various micro and macronutrients, making it nutritionally valuable asset. Due to the ever increasing world population, there has been high demand for utilizing alternative resources of food products for energy and essential nutrients (Mohammadi, 2006).

Similarly, the major waste of apple juice manufacturing industry is apple pomace (AP), it accounts for a total of 3.0–4.2 × 10^6 million tons (MT) annually worldwide (Djilas et al., 2009). AP is a leftover residue after the extraction of juice containing peel, seeds, and remaining solid parts and constitute about 25–35% of the weight of the fresh apple processed. AP is a rich source of polyphenols, minerals, and dietary fiber. After processing of apples into applesauce and canned apples, the peel is also generated as waste. Apple peels are a good source of antioxidants and can be utilized. Around 267 million pounds of apple processed in New York State generates around 20 million pounds of the peel.

The waste generation from the apple industry can be categorized into two parts. The first part is at the beginning of the apple processing, when partially bruised and spoiled apples are removed on the sorting belt and the second kind of waste is the AP that is generated after the extraction of the juice. AP is wet, and the moisture content is very high. So it is very important to dispose of the waste very safely in order to avoid environmental pollution. Large amount of substantial cost is also involved in disposal of this kind of waste. As the waste is rich in various health beneficiary components, this waste can be utilized for conversion or extraction of various components, which can help in avoiding the disposal cost. In India, approximately only 10,000 tons of AP is being utilized out of total production of 1 million tons per annum (Manimehalai, 2007). So a huge quantum of AP waste is not utilized and is dumped in the fields causing environmental pollution because of its high chemical oxygen demand (COD) and biological oxygen demand (BOD).

Traditionally AP is utilized as a cattle feed, but only a fraction of it is being utilized, as the spoilage rate is quite high because of large moisture content. In India, it is estimated that approximately 50,000 is paid every month for its disposal. In the United States, the disposal fee of AP is estimated to be around 10 million dollars per year.

7.3 DEVELOPMENT OF VALUE ADDED PRODUCTS FROM WASTE

7.3.1 ORANGE PEEL

Orange fruit is one of the major citrus fruit that has been consumed in large quantities throughout the world. It is consumed in all forms, viz. whole fruit and in juice. It is the rich source of vitamin C, A, and B, minerals (potassium, calcium, and phosphorus), dietary fiber, bioactive compounds as hydroxycinnamic acids, flavonoids, carotenoids, and many other phytochemicals (Roussos, 2011). The usage of orange juice creates 8–20 MTs of solid residues globally in a year. Despite of the richness of nutrients present in the waste, it has not produced much economical value. Correspondingly, the usage has been confined to animal feed, fertilizer, production of ethanol, biogas, and extraction of oils and pectin. Thereby, food-based alternative processing methods are needed to be carried out in order to add value to this solid and liquid waste and also, minimizing the environmental threats. The soluble sugars present in the orange peel are glucose, sucrose, and fructose. Orange peel is composed of a cellular wall made of pectin, cellulose, and hemicelluloses (Rezzadori et al., 2012). Usually, after the extraction of juice from the orange fruit, the pulp and seeds are used for animal feed in crushed form. Commercial industries use the extracts of orange peel in cosmetics, perfumes, inks, solvents, animal feed, etc. and food industries utilize it for mostly beverage processing.

Citrus industry is always keen on finding uses for its byproducts. The major portion for adding value to processed food is the dietary fiber. Dietary fiber of citrus peel is spongy and cellulosic tissue which is white in color and consumption this dietary fiber is advantageous in the prevention of obesity, cardiovascular diseases (CVD), rectal cancers, and diabetes. Moreover, dietary fiber reduces the reduction in fat intake and salt (Larrea et al., 2005), thereby, enrichment with dietary fiber will increase the nutritional value of food, particularly bakery products (cakes, muffins, breads, cookies, biscuits, etc.). However, the optimized addition of dietary fiber is the key factor in satisfying the sensory properties of the consumer (Giuntini et al., 2003). Dietary fiber is categories into two classes: Polymers that are soluble in water as pectins and gums (SDF), and others are water-insoluble as hemicelluloses, cellulose, and lignin (IDF) (Nassar et al., 2008).

Food processing waste not only includes dietary fiber as a valuable by-product, essential fatty acids, minerals, antimicrobials, antioxidants are also of great importance (Yağcı and Göğüş, 2008). Since, the citrus peels are having the advantage of water holding capacity; therefore, their usage in bakery products improve the softness and yield of the product (Bilgiçli et al., 2007). The development of various food products using orange peel and pulp are shown in Table 7.1.

7.3.1.1 APPLICATIONS IN BAKING INDUSTRY

The baking industry is one of the major food industries owing to the generation of vivid products that are consumed all over the globe (Butt et al., 2008). In the baking industry, the addition of fruit fiber is most advantageous as fiber leads to the sustained capacity of water binding, which is a conductive characteristic. Romero-Lopez et al. (2011) reported the development of muffin using 10% of the orange bagasse. Similarly, Bilgiçli et al. (2007) developed cookies from lemon peel. In this study, the wheat flour was replaced with dietary fibers in the ratio of 15%, 20%,

TABLE 7.1 Application of Citrus Waste in Development of Functional Foods

SL. No	Source	Extraction	Amount	Product	Processing	References
1	Orange peel	Essential oils, polyphenols, pectin	Essential oil 4.22 ± 0.03% Polyphenols 50.02 mg GA/100 g dm Pectin 24%	–	Microwave, ultrasound 0.956 W/cm^2, and 59.83 C	Boukroufa et al., 2015
2	Orange peel	Cellulose	40.4% and 45.2% for sodium sulfite and sodium metabisulfite	–	Digestions of sodium sulfite and sodium metabisulphite	Bicu et al., 2011
3	Orange peel	Fiber fractions	Soluble and insoluble dietary fibers 500 g/kg	–	Alcohol and water-based extraction	Chau et al., 2003
4	Orange peel	Essential oil	96–98%		Microwave steam; 12–14 min	Farhat et al., 2011
5	Orange peel	Dietary fiber	1.5–3%	Low-fat fermented sausage		Garcıa et al., 2002
6	Orange pulp		19%	Juice	Reverse osmosis	Garcia-Castello et al., 2011
7	Orange peel		3–7%	Extruded snack	Extrusion	Yağcı et al., 2008
8	Orange bagasse	Dietary fiber	10%	Bakery product muffin		Romero-Lopez et al., 2011
9	Lemon	Dietary fiber	1–30%	Cookies		Bilgiçli et al., 2007
10	Orange pulp		5–25%	Cookies/biscuits	Extrusion	Larrea et al., 2005
11	Orange peel and pulp		5–25%	Biscuits		Nassar et al., 2008
12	Orange peel	Dietary fiber		Ice cream		Crizel et al., 2014
	Orange bagasse	Dietary fiber		Ice cream		Crizel et al., 2015
	Orange seed	Dietary fiber		Ice cream		Crizel et al., 2016

and 30%. It was found that the addition of lemon peels reduced the phytic acid content of the cookies. Similarly, *in-vitro* protein digestibility got increased. However, no effects were observed on the total phenolics and antioxidant activity ($p>0.05$). In addition, the orange pulp was used for developing biscuits (Nassar et al., 2008). Biscuits were developed with the concentration of 5, 15, and 25%. It was found that addition of orange pulp increased the fiber content of the biscuits by 2 to 15%. It was observed that till 155 of the pulp can be incorporated for the acceptable biscuits. The developed biscuits showed decreased content of fat and carbohydrate. Thereby, the addition of orange pulp, which usually gets wasted after the juice extraction can be employed as a source of dietary fiber and biochemical compounds in the production of biscuits. These days' extruded snacks are getting attention. Yağcı and Göğüş, (2008), developed extruded rice grit snacks with the addition of orange peel in the ration of 3–7%. The snack was produced well in terms of sensory in combination with partially defatted hazelnut flour and fruit waste.

7.3.1.2 MISCELLANEOUS

It was found that the concentration of 1.5–30% is suitable for incorporating the fruit fibers. However, it further depends on the type of the final product formulation. For example, in development of bakery products, we can add higher amount of fruit fiber in comparison to addition to meat-based products. Garcia et al. (2002) found that 3% of fruit fiber was not acceptable to add to sausage owing to their cohesiveness and hardness. However, the best results were obtained with 1.5% orange fiber while incorporating in pork sausage. By adding fiber to sausage, we can reduce the concentration of fat and can satisfy the consumer acceptance as well.

Consumers are always fond of low-fat products. Particularly, development of ice-cream by replacing their fat with fruit peels is an innovative step. It was found that ice cream developed with orange peels (1% dietary fiber) resulted in 51% reduction of lipid content (Crizel et al., 2014). However, the addition of peels did not put any effect on the color parameters of the ice-cream. Concurringly, the non-significant results were found on the texture of fiber added ice-cream with respect to control. Therefore, without affecting the texture and color of the ice-cream, we can decrease the content of fat in the ice-cream, which is the main criteria of consumption among the consumers.

7.3.2 APPLE WASTE

Apple is a widely used fruit, mainly used in the human diet, because of its whole year availability, great taste, and it is also rich in various nutrients such as vitamins and other bioactive compounds. (Jakobeket al., 2015). The total production of apple all over the world is estimated to be 2.6 MTs. (USFDA report, 2017). It is the most favored fruit, which is widely being grown all over the globe. The United States is the second largest apple producer worldwide, behind China. U.S. production is followed by Poland, Italy, and France. The United States grows approximately 200 unique apple varieties. The total average production of apples in the European Union is about 10 MTs, with Poland, France, Italy, and Germany as the most productive countries (European Union Report, 2014). India contributes about one-third of the total production in the world, making it to the 9th position. The annual production is about 1.42 MTs.

7.3.2.1 INDUSTRIAL USAGE OF APPLE

Apples are processed into a variety of products, but by far the largest volume of processed apple products is in the form of juice. In the U.S approximately 45% of the total U.S raw apple is utilized in the processing industry, this is approximately 13.7% of the total U.S crop of the processed apples 75% is utilized in making applesauce, and other different kinds of spiced apple products. Rest of 7.2% are dried and utilized in various products. 71% of the total apple production in India is consumed as such, about 20% of it is processed into value-added products, out of which 65% is transformed into apple juice concentrate. The remaining portion of the apple produce is utilized into other products like ready-to-serve (RTS) apple (Kaushal et al., 2002).

7.3.2.2 UTILIZATION OF APPLE WASTE IN FOOD PRODUCTS

The main by-product of the apple processing industry is AP, which is obtained mainly after the extraction of juice, or a byproduct in the manufacturing of apple cider vinegar. In different parts of the world, only 25%-30% of AP is used for the development of fertilizer, fuel, feedstuff, ethanol, vinegar or some industrial materials, and rest usually goes as waste. However, in India, it is not at all utilized and merely goes waste because little efforts have been made on its utilization. AP is rich in many vital components that can be utilized AP typically contains 66.4–78.2% (wb) moisture and 9.5–22.0% carbohydrates. AP contains 26.4% dry matter (DM), 4.0% proteins, 3.6% sugars, 6.8% cellulose, 0.38% ash, 0.42% acid and calcium, 8.7 mg/100 g of wet AP. Many different technologies have been standardized to produce various different kind of food products like pectin, ethanol, lactic acid, citric acids, incorporation into cookies, cakes, and muffin, or extraction of polyphenols that are rich in antioxidants. The AP can be utilized after removal of moisture from it that can be done by utilizing different drying techniques. At the moment, the main way of making use of AP is as a feed ingredient, after drying. AP is a rich source of fiber, which is a very important component of our diets.

7.3.2.2.1 Bakery and Confectionary Products

AP, a cheap by-product of apple juice production is rich in pectin, antioxidants, and flavor compounds. It could be used for several applications, such as pectin recovery (Schieber et al., 2003), jam, and jelly production (Royer et al., 2006).

Usage of the dried AP powder has been studied as a replacement of pectin in the manufacturing of jam (Szabó-Nótin et al., 2014). Based on the research conducted, dried AP powder showed suitable properties like strength and stability of jams, making it suitable to replace pectin up to 40% without changing rheological properties of bakery jam. The AP used in this study for dried up to 5% moisture content at 80°C and was then grinded and sieved through 0.2mm sieve producing AP powder. This study revealed that AP powder could be used to replace pectin and also improve storage stability.

The utilization of AP as a source for the development of new functional foods is also being evolved (Reis et al., 2014). Baked product as scones and extruded products were formulated, which resulted in the increase of dietary fiber and antioxidant activity. Extruded snacks and baked scones were formulated with increasing levels (0–30%) of AP. The incorporation of up to 20% of AP in extruded snacks and in baked scones showed an increase in the fiber content, phenolic content and antioxidant capacity (DPPH radical scavenging activity (SA), ferric reducing antioxidant power (FRAP) and b-carotene/linoleic acid system) increased when compared to the products to which no AP was added. Chlorogenic acid and quercetin were the major phenolic compounds found in the products.

Yadav et al. (2015) formulated the noodles using AP and evaluated its phytochemical and antioxidant activity. AP is known to be a good source of bioactive compounds like phytochemicals, antioxidants, and dietary fiber. In the study, AP was incorporated for extruded product (noodles) at three different levels (10, 15 & 20%) and its impact on the cooking properties and sensory characteristics of noodles was studied. It was observed that the total dietary fiber and protein content of the noodles increased from 6.0 to 13.28% and 10.20 to 11.80%, respectively, as compared to the control noodles. The AP enriched noodles exhibited improved antioxidant activity. The study revealed that AP can be used as a potential source for the development of functional foods. GC-MS analysis of AP extract revealed the presence of 4H-Pyran-4-one, 2, 3-dihydro-3, 5-dihydroxy-6-methyl-, 2, 6, 10, 14, 18, 22 Tetracosahexaene, 2, 6, 10, 15, 19, 23-hexamethyl-,(all E)-, γ–Sitosterol, and 2-Furancarboxaldehyde, 5(hydroxymethyl), Vitamin E and fatty acids, all of which have significant therapeutic uses.

Furthermore, Sudha et al. (2007) utilized AP as a source of dietary fiber and polyphenols in cake making. The study showed that cakes prepared from 25% of AP had a dietary fiber content of 14.2% The total phenol content in wheat flour and AP was 1.19 and 7.16 mg/g respectively whereas cakes prepared from 0% and 25% AP blends had 2.07 and 3.15 mg/g indicating that AP can serve as a good source of both polyphenols and dietary fiber. Utilization of AP increased the water absorption capacity of the flour, also effected the elastic and pasting properties. Concordantly, the utilization of apple skin powder that is also generated as by-product of apple processing has also been reported (Rupasinghe et al., 2008). The majority of phenolic of apples are present in the skin, which is an underutilized food-processing by-product. Therefore, the studies revealed the use of apple and its by-products (skin powder, pomace) as an alternative dietary fiber source or specialty food ingredient for muffins, other bakery products, or selected functional foods and nutraceuticals.

7.3.2.2.2 Fiber Concentrates

A research by Fernando et al. (2005) evaluated some functional properties of fiber concentrates from apple, which was obtained by washing, coring, chopping, and separation of juice from pomace by pressing. AP was washed twice with warm water at 30°C, and then it was dried at 60°C during 30 min in an air tunnel drier and ground to a particle size of 500–600 μm and these were used as potential fiber sources in the enrichment of foods. All the fiber concentrates had a high content of dietary fiber (between 44.2 and 89.2 g/100 g DM), with a high proportion of insoluble dietary fiber. Protein and lipid contents ranged between 3.12 and 8.42 and between 0.89 and 4.46 g/100 g DM, respectively. The caloric values of concentrates were low (50.8–175 kcal/100 g or 213–901 kJ/100 g). So, therefore, these fiber concentrates can be used in the development of fiber-enriched foods.

7.3.2.2.3 Pectin

Pectin is usually considered as a complex polysaccharide which consists of α-1,4-linked d-galacturonic acid, which is partly methyl esterified, and the side chain contains various neutral sugars, such as l-rhamnose, l-arabinose, and d-galactose (Mohnen, 2008; Xie, Li and Guo, 2008). The properties of Pectin include gelatinization, thickening, and stabilization, due to which it can be widely used in the food industry (Sato et al., 2011). Gazala et al. (2017) studied the extraction of pectin. In the study, apple pomace from two different sources was reported for pectin extraction. Extraction was done with hot acid extraction method using three different acids Hydrochloric acid (HCl), sulfuric acid (H_2SO_4) and citric acid. The pomace was sieved through the different mesh, and all these meshes were subjected to extraction under pH 1.0 and temperature 70°C. The comparison of the pectin Extracted was made on the basis of yield of pectin and physiochemical features between sources as well as between the acids and mesh sizes. AP from source S1 showed a higher percent yield of pectin (52.6%) whereas between acids and mesh sizes citric acid and meshm1 showed the highest percent yield of 52%. Under the different extraction conditions used in this study, the pectin recovered presented variable characterized features which varied with the type of acid and mesh size. CitRic acid extraction revealed a superior quality of pectin. So this gives an idea about the utilization of citric acid for extraction of pectin. The pectin recovered can be effectively used in forming rigid gels for emulsification and in various food formulations.

Apple peel is also a major waste in apple processing industries, which contains 1.21% pectin. Virk and Shogi (2004) studied the extraction of pectin from apple peels in citric acid. The pectin powder was prepared by triple extraction with a citric acid solution (1%) clarification through sedimentation, concentration (24°B), precipitation using ethyl alcohol, vacuum drying, and grinding. The study shows the Physico-chemical properties of pectin powder having a moisture content of (10.0%), total ash (1.4%), equivalent weight (652.48), methoxyl content (3.7%), anhydro uronic acid (62.82%), degree of esterification (33.44%), acetyl value (0.68), and jelly grade (80). This extracted pectin can be used in the food industry for various food preparations.

7.3.2.2.4 Source of Antioxidant Enrichment

Polyphenolic compounds are found in apples in high amount (Wojdylo et al., 2008). The major polyphenolic groups are polymeric proanthocyanidins composed of several flavan-3-ol molecules, monomeric flavan-3-ols (flavanols), flavonols, anthocyanins, dihydrochalcones, and hydroxycinnamic acids (Khanizadeh et al., 2008). Flavonols and anthocyanins are usually found in the peel, although some red-fleshed apples can have anthocyanins in the flesh as well. Proanthocyanidins/flavan-3-ols, dihydrochalcones, and hydroxycinnamic acids are the major polyphenol groups found in the apple flesh (Jakobek et al., 2013). This reveals that the leftover pomace after the processing is also rich in antioxidants, which can be extracted using different extraction techniques and utilized as a source of antioxidants in various food products. Ultrasound technology can also be utilized for polyphenols extraction from apples. Ultrasound-assisted extraction (UAE) technique is a green technology widely recognized as "green and innovative," which typically involves reduced operating and maintenance costs, moderate energy consumption and small processing time, low quantity of water and solvents (Chemat et al., 2017). Wang et al. (2018) studied the ultrasound-assisted aqueous extraction (UAE) for recovery of Catechin and total phenolic contents (TPC) from flesh and peel of apple tissues.

7.3.2.2.5 Usage in Food Packaging Films

Edible packaging films are emerging these days. Riaz et al. (2018) extracted polyphenols from apple peel and incorporated it into chitosan (CS) to develop a novel functional film. Structure, potential interaction, and thermal stability of the prepared films were studied. Physical properties including moisture content, density, color, opacity, water solubility, swelling ratio, and water vapor permeability were also measured. The results revealed that addition of polyphenols into chitosan improved the physical properties of the film by increasing its thickness, density, solubility, opacity, and swelling ratio whereas moisture content and water vapor permeability were decreased. Antioxidant and antimicrobial activities of the film prepared were higher. This study indicates that it could be developed as bio-composite food packaging material for the food industry and can also contribute to a shelf life extension of the food products.

7.3.3 BANANA WASTE

Banana production is generally grouped into two different categories; the vast majorities being the small scale farmers that produce banana mainly for self-consumption and for the domestic market while the other group involves large plantations and companies that supply to both domestic as well as international markets (Padam et al., 2014). According to FAO (2014), global banana production was around 118 MT the year 2012 and 2013. India is a leading producer followed by China, Philippines, and Brazil sequentially (The Daily Records, 2018). The annual production of bananas in India is around 27,575,000 tonnes/year. Tamil Nadu is the largest banana producing state in India with the production of 5136.2 metric tonnes annually. It significantly contains a high-nutritious value for the human consumption, making it increasingly in high-demand among the various consumers in the state of Tamil Nadu (The Daily Records, 2018). Most of the edible bananas are cultivated for their fruits, and this leads to the generation of a large amount of wastes consisting of some valuable underutilized commodities. Therefore, without proper agricultural waste management practices can cause serious ecological damages (Essien et al., 2005; Shah et al., 2005; Yabaya and Ado, 2008). Banana peels consist of 35% weight of the total banana fruit.

7.3.3.1 UTILIZATION OF BANANAS AND BANANA PEEL BIOMASS FOR THE PRODUCTION OF FOOD PRODUCTS

Consuming unhealthy foods leads to different extremities, lack of nutrition on one hand, and aggravation of chronic diseases and obesity on the other. A possible solution to these problems is the creation of healthy and convenient foods. Extrusion processing can serve the food industry with food products which are convenient as well as nutritious by providing a platform for the enrichment of food products with fiber, protein, antioxidants, and vitamins (Chahal, 2015). Banana peels are a rich source of carbohydrates, proteins, vitamin A it's precursor's, i.e., β-carotene, α-carotene, etc., they show considerably good antioxidant activity and are a rich source of antioxidant compounds including phenols. Besides, they are a very rich source of dietary fibers (soluble and insoluble). Extracting these nutrients and incorporating them into various food products as fortifying agents may enhance the nutritive as well as the product value. Banana peels are also a good source of minerals like calcium and phosphorus. The nutritive value of the peels varies according to the stage of ripening and also the variety. As the banana ripens the nutritive value of the peels decreases. Plantain or green banana peels can be considered to be a better source of nutrients than bananas in different stages of ripening. The application of banana peels to food products is shown in Table 7.2.

Yellow noodles are typically made by adding alkaline salt to the ingredients. The alkaline salt added imparts the unique flavors to noodles with pH 9.0–11.0. Substituting wheat flour with flour of green and ripe banana peel will enhance the dietary fiber content, thus slowing the rate of starch hydrolysis in yellow noodles (Ramli et al., 2009). Green (unripe) banana is a good source of resistant starch, non-starch polysaccharides including dietary fiber, antioxidants, polyphenols, essential minerals such as potassium, and various vitamins, e.g., provitamin A (carotenoid), vitamin B1, B2 and C which are important for human health. Functional gluten-free cakes can be formulated by substituting green banana peel flour (GBPF) with rice flour (5%, 10%, 15%, and 20%) while investigating their physical properties (Turker et al., 2016). GBPF can also be substituted partially (10% or 20%) for wheat flour for the formulation of bread (Gomes et al., 2016). Ripe banana peel flour can be substituted for wheat flour for developing donuts (Futeri et al., 2014).

Biomass can be used for various purposes like extraction of bio-colorants, development of various food products by incorporating safe for consumption extracts from biomass, etc. The main food bio-colorants are carotenoids, flavonoids, anthocyanidins, chlorophyll, betalain, and crocin, which are extracted from several horticultural plants (Rymbai et al., 2011). These natural colorant's or nutrients, especially $^{\beta}$carotene, anthocyanins, etc. can also be used for value and nutrient enhancement of functional foods (Jenshi et al., 2011). Banana inflorescence is helpful towards bronchitis and dysentery and on ulcers; cooked banana flowers are given to diabetics (Kumar et al., 2012).

Novel approach of biotransformation of banana pseudostem extract can be used for converting into a functional juice, containing high-value non-digestible oligosaccharides (Sharma et al., 2017). Tender core of banana pseudo-stem flour (TCBPF) has good functional properties. The TCBPF has high water holding capacity, swelling power, and solubility (Aziz et al., 2011). Presence of some essential proteins also makes it a natural food preservative, as these proteins show antifungal activity which might assist in enhancing the shelf life of the bakery products such as breads (Yasmin

TABLE 7.2 Potential Application of Banana and its by-Products

SL. No.	Banana Parts	Potential Food Applications	Processing	Reference
	A. Banana Pulp			
1	Banana pulp	Extruded products	Freeze dried unripe banana flour used for total 16 extruded products formulation, and the extruded product were cut into small strands with a length of about 8 cm and dried in a tray drier.	Chahal, 2015
		Banana muffins	Gluten-free muffins were prepared by replacing rice flour with green banana pulp flour at the ratio of 50:50.	Kaur et al., 2017
		Banana chips	Banana chips were prepared using fully matured unripe banana fruits. The fruit was steeped in 2% sodium chloride salt solution for 15 min, wiped with a cloth and 1.75 to 2 mm thick uniform slices were chipped using hand slicer and directly put into the coconut oil frying medium at 160°C, keeping fat and material ratio 4:1.	Manikantan et al., 2012
		Banana liqueur	Production of banana fruit wine using various recombinant wine yeast strains	Byarugaba-Bazirake, 2008
2	Banana peels	Noodles	Partial substitution of wheat flour with green Cavendish banana peels formulated noodles having low glycemic indices.	Ramli et al., 2009
		Cakes	Functional gluten-free cake formulation by substituting green banana peel flour with rice flour (5, 10, 15 and 20%).	Türker et al., 2016
		Doughnuts	Formulation of doughnuts having high organoleptic properties by substitution of wheat flour with banana peel flour at the ratio of 1:10.	Futeri et al., 2014
		Breads	Substitution of green banana (with its peel) flour (GBF) as a partial substitute (10% or 20%) for wheat flour for the production of bread.	Gomes et al., 2016

TABLE 7.2 *(Continued)*

SL. No.	Banana Parts	Potential Food Applications	Processing	Reference
3	Banana inflorescence	Food colorant and source of anthocyanins in functional food products	Extraction of anthocyanins that are slightly higher in amounts, ranging from 14–32 mg anthocyanin/100 g bracts than red cabbage.	Jenshi et al., 2011
		Medicinal functional food	Natural bio-colorant such as the anthocyanins not only have high health-promoting properties but also the increase in demand on natural or functional foods.	Rymbai et al., 2011
		Cooked inflorescence	Cooked flowers benefit towards diabetes.	Kumar et al., 2012
4	Banana leaves	Covering agent for the production of fermented fish food product	Dried fish covered with dried banana leaves before subjecting to fermentation for 2–3 months. Contributing as an energy-efficient technique for bioconversion into useful edible products. Imparts subtle sweet flavor and aroma to the food product, protects food from burning and keeps the juices within.	Narzaraya et al., 2016
5	Banana pseudostem	Processed banana pseudo stem juice	The bioprocess involves the employment of membrane separation techniques, and the biocatalysts executing glucosyltransferase and D-fructose epimerization activities.	Sharma et al., 2017
6	Banana pith or tender core of banana pseudostem	Source of total dietary fiber, lignin, hemicellulose, and cellulose	Potential functional food ingredient for products containing high dietary fiber	Aziz et al., 2011

and Saleem, 2014). Besides, bakery products like buns, breads, etc. can be prepared utilizing the dietary fibers extracted from the banana peels. This will not only enhance the fiber content of the products but will also enrich it with vital nutrients. Proteins extracted from banana peels can be used as protein powders or can be incorporated into various food products for protein enrichment. The quality of protein extracted from the banana peels depend upon the method used for its extraction and purification. Banana peels (*Musa* spp.) are a good example of a plant tissue where protein extraction is challenging due to the abundance of interfering metabolites. Sample preparation is a critical step in proteomic research and is critical for good results (Zhang et al., 2012). Banana peels are high in $^{\beta}$carotene content and protein content. They also show potential antioxidant activity. Thereby, it can be considered as a good secondary food by-product for the extraction of these compounds and further purification and utilization commercially. Insoluble dietary fibers extracted from banana peels is not only a good source of $^{\beta}$carotene and protein, but it also shows some antioxidant activity; therefore, substituting it in edible food products would not only enhance the fiber content but also enhance its value as a food fortificant and as a pro-vitamin A precursor.

7.4 CONCLUSION

The fruit peels of orange, apple, and banana can be effectively utilized in the various food products. The optimization of addition of these fibers needs to be standardized in order to scale up of food-based products (like noodles, pasta, cakes, muffins, jams, ice-creams, chocolates) at the industrial level. The effective utilization of these wastes will reduce their dumping activities, which thereby, reduce the environmental pollution. In addition, novel applications in terms of biodegradable and edible films need to be developed in order to increase the efficacy of the waste.

KEYWORDS

- **bioactive components**
- **functional food products**
- **plant-based waste**
- **value addition**

REFERENCES

Aziz, N. A. A., Ho, L. H., Azahari, B., Bhat, R., Cheng, L. H., & Ibrahim, M. N. M., (2011). Chemical and functional properties of the native banana (*Musa acuminata* × *balbisiana Colla cv.* Awak) pseudo-stem and pseudo-stem tender core flours. *Food Chemistry, 3*(1), 748–753.

Bicu, I., & Mustata, F., (2011). Cellulose extraction from orange peel using sulfite digestion reagents. *Bioresource Technology, 102*(21), 10013–10019.

Bilgiçli, N., İbanoglu, Ş., & Herken, E. N., (2007). Effect of dietary fiber addition on the selected nutritional properties of cookies. *Journal of Food Engineering, 78*(1), 86–89.

Boukroufa, M., Boutekedjiret, C., Petigny, L., Rakotomanomana, N., & Chemat, F., (2015). Bio-refinery of orange peels waste: A new concept based on integrated green and solvent-free extraction processes using ultrasound and microwave techniques to obtain essential oil, polyphenols and pectin. *Ultrasonics Sonochemistry, 24*, 72–79.

Butt, M. S., Tahir-Nadeem, M., Ahmad, Z., & Sultan, M. T. (2008). Xylanases and their applications in baking industry. *Food Technology and Biotechnology, 46*(1), 22–31.

Byarugaba-Bazirake, G. W. (2008). *The Effect of Enzymatic Processing on Banana Juice and Wine* (Doctoral dissertation, Stellenbosch: Stellenbosch University), pp. 1–177

Chahal, B. K., (2015). *Development of Fruit Based Products Using Extrusion-Drying Process for Need-Based Applications.* Master thesis, Department of Food Science and Agricultural Chemistry Macdonald Campus, McGill University Montreal.

Chau, C. F., & Huang, Y. L., (2003). Comparison of the chemical composition and physicochemical properties of different fibers prepared from the peel of *Citrus sinensis* L. Cv. Liucheng. *Journal of Agricultural and Food Chemistry, 51*(9), 2615–2618.

Chemat, F., Rombaut, N., Sicaire, A. G., Meullemiestre, A., Fabiano-Tixier, A. S., & Abert-Vian, M., (2017). Ultrasound-assisted extraction of food and natural products. Mechanisms, techniques, combinations, protocols and applications. A review. *Ultrasonics Sonochemistry, 34*, 540–560.

Crizel, T. D. M., Araujo, R. R. D., Rios, A. D. O., Rech, R., & Flôres, S. H., (2014). Orange fiber as a novel fat replacer in lemon ice cream. *Food Science and Technology, 34*(2), 332–340.

Djilas, S., Canadanovic-Brunet, J., & Cetkovic, G., (2009). By-products of fruits processing as a source of phytochemicals. *Chem. Ind. Chem. Eng. Q., 15*, 191–202.

Essien, J. P., Akpan, E. J., & Essien, E. P., (2005). Studies on mold growth and biomass production using waste banana peel. *Biore Technol., 96*, 1451–1456.

Farhat, A., Fabiano-Tixier, A. S., El Maataoui, M., Maingonnat, J. F., Romdhane, M., & Chemat, F., (2011). Microwave steam diffusion for extraction of essential oil from orange peel: Kinetic data, extract's global yield and mechanism. *Food Chemistry, 125*(1), 255–261.

Fernando, F., Marıa, L. H. B., Ana, M. E. B., Italo, C. B., & Fernando, A., (2005). Fiber concentrates from apple pomace and citrus peel as potential fiber sources for food enrichment. *Food Chemistry, 91*, 395–401.

Fresh Deciduous Fruit: World Markets and Trade (Apples, Grapes, & Pears), (2017). United States Department of Agriculture. Foreign Agriculture Service Report.

Futeri, R., & Pharmayeni, (2014). Substituting wheat flour with banana skin flour from mixture various skin types of banana on making donuts. *International Journal on Advanced Science Engineering Information Technology, 4*(2), 2088–5334.

Garcıa, M. L., Dominguez, R., Galvez, M. D., Casas, C., & Selgas, M. D., (2002). Utilization of cereal and fruit fibers in low fat dry fermented sausages. *Meat Science, 60*(3), 227–236.

Garcia-Castello, E. M., Mayor, L., Chorques, S., Argüelles, A., Vidal-Brotons, D., & Gras, M. L., (2011). Reverse osmosis concentration of press liquid from orange juice solid wastes: Flux decline mechanisms. *Journal of Food Engineering, 106*(3), 199–205.

Gazala, K., Masoodi, F. A., Masarat, H. D., Rayees, B., & Shoib, M. W. (2017). Extraction and characterisation of pectin from two apple juice concentrate processing plants. *International Food Research Journal, 24*(2), 594.

Giuntini, E. B., & Lajolo, F. M., (2003). Dietary fiber potential in Iberian-American countries: Food, products and residues. *Latin American Nutrition Archives, 53*(1), 14–20.

Jakobek, L., Boc, M., & Barron, A. R., (2015). Optimization of ultrasonic-assisted extraction of phenolic compounds from apples. *Food Analytical Methods, 8*(10), 2612–2625.

Jakobek, L., García-Villalba, R., & Tomás-Barberán, F. A., (2013). Polyphenolic characterization of old apple varieties from Southeastern European region. *J. Food Compos. Anal., 31*, 199–211.

Jenshi, R. J., Saravanakumar, M., Aravinthan, K. M., & Suganya, D. P., (2011). Antioxidant analysis of anthocyanin extracted from *Musa acuminata* bract. *J. Pharm. Res., 4*(5), 1488–1492.

Kaur, K., Singh, G., & Singh, N., (2017). Development and evaluation of gluten-free muffins utilizing green banana flour. *Bioved., 28*(2), 359–365.

Kaushal, N. K., Joshi, V. K., & Sharma, R. C., (2002). Effect of stage of apple pomace collection and the treatment on the physical-chemical and sensory qualities of pomace papad (fruit cloth). *J. Food Sci. Technol., 39*, 388–393.

Khanizadeh, S., Tsao, R., Rekika, D., Yang, R., Charles, M. T., & Rupasinghe, H. P. V., (2008). Polyphenol composition and total antioxidant capacity of selected apple genotypes for processing. *J. Food Compos. Anal., 21*, 396–401.

Kumar, K. P. S., Bhowmik, D., Duraivel, S., & Umadev, M., (2012). Traditional and medicinal uses of banana. *Journal of Pharmacognosy and Phytochemistry, 1*(3), 51–63.

Larrea, M. A., Chang, Y. K., & Martinez-Bustos, F., (2005). Some functional properties of extruded orange pulp and its effect on the quality of cookies. *LWT-Food Science and Technology, 38*(3), 213–220.

Manikantan, M. R., Sharma, R., Kasturi, R., & Varadharaju, N., (2012). Storage stability of banana chips in polypropylene based nanocomposite packaging films. *J. Food Sci. Technol., 51*(11), 2990–3001.

Manimehalai, N., (2007). Fruit and waste utilization. *Bev. Food World, 34*(11), 53–54, 56.

Mohammadi, I. M., (2006). Agricultural waste management extension education (AWMEE). The ultimate need for intellectual productivity. *Am. J. Environ. Sci., 2*(1), 10–14.

Mohnen, D., (2008). *Mohnen Pectin Structure and Biosynthesis Current Opinion in Plant Biology, 11*(3), 266–277.

Narzarya, Y., Brahmab, J., Brahmac, C., & Das, S., (2016). A study on indigenous fermented foods and beverages of Kokrajhar, Assam, India. *Journal of Ethnic Foods, 3*(4), 284–291.

Nassar, A. G., AbdEl-Hamied, A. A., & El-Naggar, E. A., (2008). Effect of citrus by-products flour incorporation on chemical, rheological and organoleptic characteristics of biscuits. *World J. Agric. Sci., 4*(5), 612–616.

Özbay, N., Apaydın-Varol, E., Uzun, B. B., & Pütün, A. E., (2008). Characterization of bio-oil obtained from fruit pulp pyrolysis. *Energy, 33*(8), 1233–1240.

Padam, B. S., Tin, H. S., Chye, F. Y., & Abdullah, M. I., (2014). Banana by-products: An under-utilized renewable food biomass with great potential. *J. Food Sci. Technol., 51*(12), 3527–3545.

Ramli, S., Alkarkhi, A. F. M., Yong, Y. S., & Easa, A. M., (2009). Utilization of banana peels as a functional ingredient in yellow noodles. *Asian Journal of Food and Agro-Industry.*, *2*(03), 321–329.

Rasidek, N. A. M., Nordin, M. F. M., & Shameli, K., (2016). Formulation and evaluation of semisolid jelly produced by *Musa acuminata* Colla (AAA Group) peels. *Asia Pacific Journal of Tropical Biomedicine, 6*(1), 55–59.

Reis, S. F., Rai, D. K., & Abu-Ghannam, N., (2014). Apple pomace as a potential ingredient for the development of new functional foods. *International Journal of Food Science & Technology, 49*(7), 1743–1750.

Rezzadori, K., Benedetti, S., & Amante, E. R., (2012). Proposals for the residues recovery: Orange waste as raw material for new products. *Food and Bioproducts Processing, 90*(4), 606–614.

Riaz, A., Lei, S., Akhtar, H. M. S., Wan, P., Chen, D., Jabbar, S., Abid, M., Hashim, M. M., & Zeng, X., (2018). Preparation and characterization of chitosan-based antimicrobial active food packaging film incorporated with apple peel polyphenols. *International Journal of Biological Macromolecules, 114*, 547–555.

Romero-Lopez, M. R., Osorio-Diaz, P., Bello-Perez, L. A., Tovar, J., & Bernardino-Nicanor, A., (2011). Fiber concentrate from orange (*Citrus sinensis* L.) bagase: Characterization and application as bakery product ingredient. *International Journal of Molecular Sciences, 12*(4), 2174–2186.

Roussos, P. A., (2011). Phytochemicals and antioxidant capacity of orange (*Citrus sinensis* (l.) Osbeck cv. Salustiana) juice produced under organic and integrated farming system in Greece. *Scientia Horticulturae, 129*(2), 253–258.

Royer, G., Madieta, E., Symoneaux, R., & Jourjon, F., (2006). Preliminary study of the production of apple pomace and quince jelly. *LWT - Food Sci. Technol., 39*, 1022–1025.

Rupasinghe, H. V., Wang, L., Huber, G. M., & Pitts, N. L., (2008). Effect of baking on dietary fiber and phenolics of muffins incorporated with apple skin powder. *Food Chemistry, 107*(3), 1217–1224.

Rymbai, H., Sharma, R. R., & Srivastav, M., (2011). Biocolorants and its implications in health and food industry. *Int. J. Pharmtech. Res., 3*(4), 2228–2244.

Sato, M. D. F., Rigoni, D. C., Canteri, M. H. G., Petkowicz, C. L. D. O., Nogueira, A., & Wosiacki, G., (2011). Chemical and instrumental characterization of pectin from dried pomace of eleven apple cultivars. *Acta Scientiarum Agronomy, 33*(3), 383–389.

Schieber, A., Hilt, P., Streker, P., Endress, H. U., Rentschler, C., & Carle, R., (2003). A new process for the combined recovery of pectin and phenolic compounds from apple pomace. *Innov. Food Sci. Emerg., 4*(1), 99–107.

Shah, M. P., Reddy, G. P., Banerjee, R., Ravinra, B. P., & Kothari, I. L., (2005). Microbial degradation of banana waste under solid state bioprocessing using two lignocellulosic fungi (Phyloticta spp. MPS-001 and *Aspergillus* spp. MPS-002). *Process Biochem., 40*, 445–451.

Sharma, M., Patel, S. N., Sangwan, R. S., & Singh, S. P., (2017). Biotransformation of banana pseudostem extract into a functional juice containing value-added biomolecules of potential health benefits. Center of Innovative and Applied Bioprocessing (CIAB), Department of Biotechnology (DBT). *Indian Journal of Experimental Biology, 55*, 453–462.

Sudha, M. L., Baskaran, V., & Leelavathi, K., (2007). Apple pomace as a source of dietary fiber and polyphenols and its effect on the rheological characteristics and cake making. *Food Chemistry, 104*(2), 686–692.

Szabó-Nótin, B., Juhász, R., Barta, J., & Stéger-Máté, M., (2014). Apple pomace powder as natural food ingredient in bakery jams. *Acta Alimentaria, 43*(1), 140–147.

The Daily Records, (2018). Top 10 largest banana producing States in India. http://www.thedailyrecords.com/2018–2019–2020–2021/world-famous-top-10 list/india/largest-banana-producing-states-india-varieties/18383/ (Accessed on 12 June 2019).

Turker, B., Savlak, N., & Kasikci, M. B., (2016). *Effect of Green Banana Peel Flour Substitution on Physical Characteristics of Gluten Free Cakes*, 4 (Special Issue Conference).

Wang, L., Boussetta, N., Lebovka, N., & Vorobiev, E. (2018). Selectivity of ultrasound-assisted aqueous extraction of valuable compounds from flesh and peel of apple tissues. *LWT, 93*, 511–516.

Wojdylo, A., Oszmiański, J., & Laskowski, P., (2008). Polyphenolic compounds and antioxidant activity of new and old apple varieties. *J. Agric. Food Chem., 56*, 6520–6530.

Xie, L., Li, X., & Guo, Y., (2008). Ultrafiltration behaviors of pectin-containing solution extracted from citrus peel on a ZrO_2 ceramic membrane pilot unit. *Korean Journal of Chemical Engineering, 25*(1), 149–153.

Yabaya, A., & Ado, S. A., (2008). Mycelia protein production by *Aspergillus niger* using banana peels. *Sci. World J., 3*(4), 9–12.

Yadav, S., & Gupta, R. K., (2015). Formulation of noodles using apple pomace and evaluation of its phytochemicals and antioxidant activity. *Journal of Pharmacognosy and Phytochemistry, 4*(1).

Yağcı, S., & Göğüş, F., (2008). Response surface methodology for evaluation of physical and functional properties of extruded snack foods developed from food-by-products. *Journal of Food Engineering, 86*(1), 122–132.

Yasmin, and Saleem, (2014). Biochemical characterization of fruit-specific pathogenesis-related antifungal protein from Basra banana. *Microbial Research, 169*, 369–377.

Zhang, A. Y. Z., Sun, Z., Leung, C. C. J., Han, W., Lau, K. Y., Li, M., & Lin, C. S. K., (2013). Valorization of bakery waste for succinic acid production. *Green Chemistry, 15*(3), 690–695.

Zhang, L. L., Feng, R. J., & Zhang, Y. D., (2012). Evaluation of different methods of protein extraction and identification of differentially expressed proteins upon ethylene-induced early-ripening in banana peels. *J. Sci. Food Agric.,* 1–10.

CHAPTER 8

Roselle (*Hibiscus sabdariffa* L.) Calyces: A Potential Source of Natural Color and Its Health Benefits

V. H. SHRUTHI and C. T. RAMACHANDRA

Department of Agricultural Engineering, University of Agricultural Sciences, Bengaluru, GKVK, Bengaluru–560 065, Karnataka, India, E-mail: shruthihgowda@gmail.com, ramachandract@gmail.com

ABSTRACT

Roselle (*Hibiscus sabdariffa* L.) belongs to the family *Malvaceae* and is a popular vegetable in Indonesia, India, West Africa, and many tropical regions. The calyces of roselle are rich in anthocyanin, ascorbic acid, and other phenolic compounds. It is water soluble with brilliant and attractive red color and with sour and agreeable acidic taste, which aid digestion. Roselle has been used by people for preparing soft drinks and in traditional medicine. It has been observed that its components, such as vitamins (C and E), polyphenols acids and flavonoids, mainly anthocyanins, have functional properties. They contribute benefit to health as a good source of antioxidants as well as a natural food colorant. The other health benefits of this plant include diuretic and choloratic properties, intestinal antiseptic, and mild laxative actions. It also used in treating heart and nerve disorder, high blood pressure, and calcified arteries. Due to perceived safety and physiological advantage of the natural colorants over synthetic ones, interest is being geared into the search of new natural colorants and the verification of the safety of existing ones. In this respect, roselle calyces appear to be good and promising sources of water-soluble red colorants that could be utilized as natural food colorants. Anthocyanins present in Roselle are delphinidin 3-sambubioside, cyanidin 3-sambubioside, delphinidin 3-glucoside, and cyanidin 3-glucoside.

8.1 INTRODUCTION

Roselle (*Hibiscus sabdariffa* Linn.) is an annual herbaceous shrub belonging to the family–*Malvaceae*. It is thought to be a native to Asia (India to Malaysia) or Tropical Africa. The plant is widely grown in tropics like Caribbean, Central America, India, Africa, Brazil, Australia, Hawaii, Florida, and the Philippines as a home garden crop. In addition to Roselle, in English-speaking regions, it is called as Rozelle, Sorrel, Red sorrel, Jamanica sorrel, Indian sorrel, Guinea sorrel, Sour-sour, Queensland jelly plant, Jelly okra, Lemon bush, and Florida cranberry. In Indian languages, it is called as Lal-ambari, Patwa (Hindi), Lal-ambadi (Marathi), Gongura, Yerra gogu (Telugu), Pulichchai kerai (Tamil), Pulachakiri, Pundibija (Kannada), Polechi, Pulichchai (Malayalam), and Chukiar (Assam) (Mahadevan et al., 2009). The species *H. sabdariffa* comprises a large number of cultivated types which, on the basis of their growth habit and end use, two botanical types of Roselle are recognized, *Hibiscus sabdariffa* var *subdariffa* and *Hibiscus sabdariffa* var *altissima* (Eltayeib and Hamade, 2014). Former is generally bushy and pigmented and cultivated for the edible calyces; the latter includes tall-growing, unbranched types bearing inedible calyces and mainly cultivated for the stem fiber, Roselle (Gautam, 2004).

Roselle is more than just a pretty fruit: it's been used in dishes, beverages, and traditional medicine for centuries. Julia Morton attests in her book, "Fruits of Warm Climates" that Roselle originated in the regions between and including India and Malaysia; shortly thereafter, she states it was brought to Africa. According to the Food and Agriculture Organization of the United Nations, the world's largest producers of Roselle are China and Thailand. Interestingly, the FAO ranks Roselle from Sudan as of the best quality, but poor packaging and distribution inhibit fruits from this part of the world from gaining global popularity. China and Thailand are the largest producers and control much of the world supply. Thailand invested heavily in Roselle production, and their product is of superior quality (Anon., 2013). In Sudan, it is a major crop of export, especially in the western part, where it occupies second place area wise after pearl millet followed by Sesamum (Gautam, 2004).

The Roselle calyces production, in developing countries, is of great importance since its production represents a very important income for people from rural communities. Roselle is a robust and thriving industry in India: In fact, the country is the fourth largest exporter of Roselle plant parts to Germany and the fifth largest exporter of plant parts to the US for the nation's thriving tea companies. Roselle tolerates a remarkable variety of climates and soil conditions. These factors make them a suitable crop found in many states including West Bengal, Assam, Bihar, Orissa, Uttar Pradesh, Tripura, Andhra Pradesh, Karnataka, Tamil Nadu and Meghalaya (Anon., 2013). Figures 8.1–8.3 represent fresh Roselle, Roselle calyces, and dried Roselle calyces, respectively.

FIGURE 8.1 (See color insert.) Fresh Roselle.

FIGURE 8.2 (See color insert.) Roselle calyces.

FIGURE 8.3 **(See color insert.)** Dried Roselle calyces.

8.2 PLANT DESCRIPTION AND ECOLOGY

The plant has been found to thrive on a wide range of soil conditions. It can perform satisfactorily on relatively infertile soils, but for economic purposes, a soil well supplied with organic materials and essential nutrients is essential. It can tolerate relatively high temperature throughout the growing and fruiting periods. The plant requires an optimum rainfall of approximately 45–50 cm distributed over a 90–120-day growing period (Adanlawo and Ajibade, 2006). The plant cultivated mainly for its flowers. Seeds and leaves have some uses in traditional medicine. The plant is about 3.5 m tall and has a deep penetrating taproot. It has a smooth or nearly smooth, cylindrical, typically dark green to red stems. Leaves are alternate, 7.5–12.5 cm long, green with reddish veins and long or short petioles. Leaves of young seedlings and upper leaves of older plants are simple; lower leaves are deeply 3 to 5 or even 7-lobed, and the margins are toothed. Flowers, borne singly in the leaf axils are up to 12.5 cm wide, yellow or buff with a rose or maroon eye and turn pink as they wither at the end of the day. The typically red calyx, consists of 5 large sepals with a collar (epicalyx) of 8–12 slim, pointed bracts (or bracteole) around the base, they begin to enlarge at the end of the day, 3.2–5.7 cm long and fully enclose the fruit. The fruit is a velvety capsule, 1.25–2 cm long, which is green when immature, 5-valved, with each valve containing 3–4 seeds. The capsule turns brown and splits open when mature and dry. Seeds are kidney-shaped, light-brown, 3–5 mm long and covered with minute, stout, and stellate hairs (Mahadevan et al., 2009). The plant takes about 3–4 months to reach the commercial stage of maturity before the flowers are harvested. Roselle plants are suitable for tropical climates with well-distributed rainfall of 1500–2000 mm/year, from sea-level to about 600 m in altitude. The plant tolerates a warmer and more humid climate with nighttime temperature, not below 21°C and is most susceptible to damage from frost and fog. In addition, it requires 13 hours of sunlight during the first months of growth to prevent premature flowering (Ismail et al., 2008).

8.3 NUTRITIONAL AND MEDICINAL IMPORTANCE OF ROSELLE CALYCES

8.3.1 NUTRITIONAL IMPORTANCE OF ROSELLE CALYCES

Roselle is rich in anthocyanins and protocatechuic acid. The dried calyces contain the flavonoids gossypetine, hibiscetine, and sabdaretine. The major pigment, formerly reported as hibiscine has been identified as daphniphylline. Small amounts of myrtillin (delphinidin 3-monoglucoside), chrysanthenin (cyanidin 3-monoglucoside) and delphinidin are also present. Roselle seeds are a good source of lipid-soluble antioxidants, particularly γ-tocopherol (Mohamed et al., 2012). The anthocyanin content of *H. sabdariffa* in five strains of the plant reportedly reached 1.7% to 2.5% of the dry weight during calyx growth (Khafaga and Koch, 1980b). *H. sabdariffa* calyces contain high amounts of organic acids, namely: citric acid, malic acid, tartaric acid and hibiscus protocatechuic acid (Kerharo, 1971; Khafaga and Koch, 1980a; Tseng et al., 1996). The acid content of the calyces increases during growth but decreases when it reaches maturity or ripens. The aqueous extract of *H. sabdariffa* calyces has a very rich red pigmentation due to the presence of anthocyanins, and the color properties have been the subject of intense scientific investigations (Ali et al., 2005; Salazer et al., 2012; Aishah et al., 2013). *Hibiscus sabdariffa* calyces were found to contain a higher amount of iron content (164.78 mg/kg) (Maregesi et al., 2013). The plant is also found to be rich in minerals, especially potassium and magnesium. Vitamins (ascorbic acid, niacin, and pyridoxine) were also present in appreciable amounts (Puro et al., 2014). The chemical composition of dried Roselle calyces is shown in Table 8.1.

TABLE 8.1 Chemical Composition of Roselle Calyces

Element	Calyx types		
	Green	**Red**	**Dark red**
Crude protein (%)	17.9	17.4	8.6
Ether extract (%)	3.2	2.1	2.9
Crude fiber (%)	11.2	8.5	9.8
Ash (%)	6.6	6.5	6.8
Ascorbic acid (mg/100 g)	86.5	63.5	54.8
Moisture (FW)%	88.3	86.5	85.3
Calcium (mg/100 g)	1209	1583	1602
Magnesium (mg/100 g)	235	316	340
Potassium (mg/100 g)	1850	2060	2320
Sodium (mg/100 g)	9.5	5.5	6.5
Iron (mg/100 g)	32.8	37.8	34.6
Zinc (mg/100 g)	5.8	6.5	6.3

Source: Reprinted from Babalola et al. (2001). Open access.

The chemical components contained in the flowers of *Hibiscus sabdariffa* include anthocyanins, flavonoids, and polyphenols (Lin et al., 2007). The petals are potentially a good source of antioxidant agents as anthocyanins and ascorbic acid (Prenesti et al., 2007). Roselle calyx contains a rich source of dietary fiber, vitamins, minerals, and bioactive compounds such as organic acids, phytosterols, and polyphenols, some of them with antioxidant properties. The phenolic content in the plant consists mainly of anthocyanins like delphinidin-3-glucoside, sambubioside, and cyanidin-3- sambubioside mainly contributing to their antioxidant properties (Aurelio et al., 2007). Recently, the biological activities of anthocyanins, such as antioxidant activity and anticarcinogenic activity, have been investigated (Tsai et al., 2002). The flowers of *Hibiscus sabdariffa* are rich in anthocyanins (Cisse et al., 2009). The anthocyanins (Figure 8.4) are responsible for the red color, while the acid taste is due to the presence of some organic acids. Sepal's acidity may also contribute to their color variation (Abou-Arab et al., 2011).

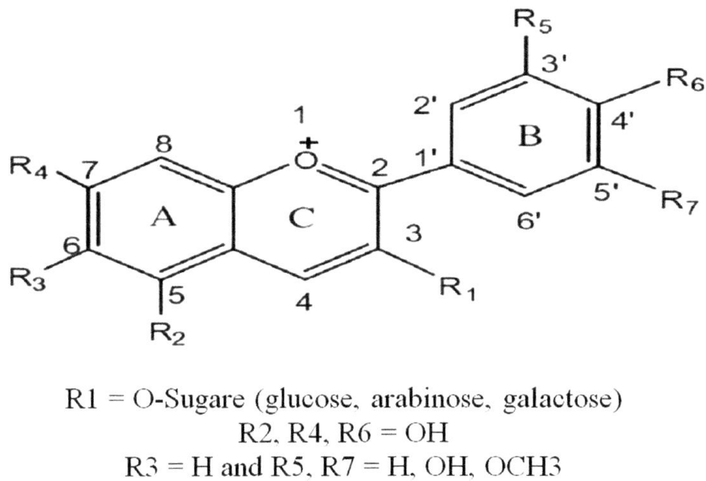

FIGURE 8.4 Structure of anthocyanins. Source: Reprinted from Abou-Arab et al. (2011). Open access.

8.3.2 MEDICINAL IMPORTANCE OF ROSELLE CALYCES

Roselle, also known as sorrel, Jamaica flower, and karkade, has been used by people for preparing soft drinks and in traditional medicine. It has been observed that its components, such as vitamins (C and E), polyphenols acids and flavonoids, mainly anthocyanins, have functional properties. Today, several studies have shown that compounds found in aqueous and ethanol Roselle calyces extracts may have antioxidant properties. These compounds could work in several ways in humans; for instance, they could have anti-cancer characteristics. They may also reduce chronic diseases such as diabetes mellitus, dyslipidemias, hypertension, and cardiovascular diseases (CVD). Some of these compounds (flavonoids and anthocyanins) are natural, which have no toxic or mutagenic effects (Cid and Guerrero, 2014). It is reported to be antihypertensive, antiseptic, sedative, diuretic, digestive, purgative, emollient, demulcent, and astringent (Odigie et al., 2003).

The calyces are used to treat heart ailments, hypertension, and leukemia. They are also reported to have diuretic, aphrodisiac, antiseptic, astringent, cholagogue, sedative, laxative, and antimicrobial activity. They are also used as a remedy for pyrexia and abscesses. The flowers and fruits are used for the treatment of cough and bronchitis (Maregesi et al., 2013). Anthocyanins present in Roselle are delphinidin 3-sambubioside, cyanidin 3-sambubioside, delphinidin 3-glucoside, and cyanidin 3-glucoside. They contribute benefit for health as a good source of antioxidants as well as a natural food colorant. The blending of Roselle juice with tropical fruit juices is anticipated to give products with high nutritional value and functional activity (Kilima et al., 2014). The calyx of *Hibiscus sabdariffa* is widely used by humans like food, jams, jellies, juice drinks, wine, and as medicinal syrups (Akanya et al., 1997). It is used effectively in folk medicines for the treatment of hypertension, inflammatory diseases, and cancer (Lin et al., 2007). The calyces are used to decrease blood viscosity and reduce hypertension (Christian et al., 2006). Hibiscus pigments reduce the incidence of liver lesions, including inflammation, leucocyte infiltration, and necrosis (Kong et al., 2003). Also potent effects in the reduction of urinary concentrations of creatinine, uric acid, citrate, tartrate, calcium, sodium, potassium, and phosphate have been demonstrated (Kuo et al., 2012; Laikangbam and Damayanti Devi, 2012). Moreover, it's anti-inflammatory, antioxidative, hepatoprotective, and anti-tumoral effects have been highlighted (Fernandez et al., 2011; Lin et al., 2012; Wang et al., 2000).

The aqueous methanolic extract of Roselle exhibits antimicrobial activity against many bacterial spp. Roselle - Hibiscus anthocyanins (HAs), which are a group of natural pigments existing in the dried calyx exhibited antioxidant activity and liver protection (Puro et al., 2014). The red color of Roselle petals, essentially the anthocyanins, is an attractive source of natural food colorants (Goda et al., 1997). Anthocyanins possess a high thermostability and contribute towards antioxidative, anti-inflammatory, cardioprotective, and hepatoprotective activities (Azevedo et al., 2010). According to many authors, anthocyanins inhibit the growth of human cancer cells and low-density lipoprotein (LDL) oxidation. Therefore, the addition of natural anthocyanins as food colorants would not only enhance the decorative value of the food but also improve its beneficial properties (Abeda et al., 2014). Also, Roselle extracts are reported to have an antimicrobial effect on different pathogenic and food spoilage microorganisms due to its metabolites of phenolic compounds such as anthocyanins. They are associated with the prevention of illnesses generated by oxidative stress (Heba et al., 2014). Pharmacological studies have demonstrated the antihypertensive effect produced by *Hibiscus sabdariffa* extracts (Onyenekwe et al., 1999; Odigie et al., 2003; McKay et al., 2010; Alonso et al., 2012). *H. sabdariffa* calyx extracts have been studied extensively as a food colorant (Duangmal et al., 2008; Alobo and Offonry, 2009) and found to be a suitable replacement for various artificial colorants.

Extracts of hibiscus (*Hibiscus sabdariffa*) contain a high amount of polyphenolic compounds, such as anthocyanins, protocatechuic acid, and other polyphenols exhibiting, significantly reducing potential (Sayago et al., 2007). These polyphenolic compounds can trap reactive oxygen species (ROS) in the interstitial fluid of the arterial wall or plasma, and they probably inhibit LDL oxidation, therefore, suppressing an atherosclerosis onset (Hou et al., 2004). Hibiscus extract is supposed to have a number of positively relevant effects on lipid metabolism, antihypertensive activity and apoptosis (Kumar et al., 2008; Sayago et al., 2007) leading to its use in pharmacological mixtures and teas. Moreover, a free radical scavenging activity (SA) against hydroxyl, superoxide, and hydrogen peroxide radicals, and reduced nitric oxide formation was observed in foods (Kumar et al., 2008). Flowers of H. sabdariffa are widely used in Latin America as beverages, and 66% of the total extractable polyphenolic compounds present in the hibiscus flower are known to have a high antioxidant capacity (Sayago et al., 2007). Both the inhibition of the formation of heterocyclic amines as an additive in marinades (Gibis and Weiss, 2010) and an antimutagenic effect was described for colon carcinogens such as heterocyclic amines and azoxymethane in the Salmonella mutation assay (Chewonarin et al., 1999; Gibis et al., 2014).

8.4 USES OF ROSELLE CALYCES

Different food products, fermented foods, and beverages of *Hibiscus sabdariffa* are widely used in different countries. From the fresh flowers of *Hibiscus sabdariffa,* cold and hot beverages are produced. The juice from the calyces of Sorrel (Roselle) is called "zoborodo" (soborodo), a non-alcoholic drink in Nigeria, which involves the production process of solid-liquid extraction leaving the calyx pulp as raffinate, which is a heavy organic material that could be converted as glucose (Ajayi et al., 2012). In Mexico, this beverage is called "flor de Jamaica." In Senegal (a state of West Africa) this *Hibiscus sabdariffa* known as bissap and drink is prepared through aqueous extraction from a solid-to-solvent ratio. This drink consumption is widespread in Asia and Africa. In Senegal, it is more popular, and more consumption is observed during the month of Ramadan (Padmaja et al., 2014). In Egypt, this beverage is called a "drink of the Pharaohs." In Sudan, it is called as "tea Karkade." In Mali (Africa) it is called as "da Bilenni." Tea from the *Hibiscus sabdariffa* also called Sudan tea, sour tea which is prepared from ground dried calyces also have medicinal properties and is considered as herbal tea, and Roselle is also used as the main ingredient in many other tisanes (herbal tea) (Diane et al., 2009).

Fresh calyx (the outer whorl of the flower) is eaten raw in salads, or cooked and used as a flavoring in cakes and is also used in making jellies, soups, sauces, pickles, puddings, etc. The calyx is rich in citric acid and pectin and is useful for making jams, jellies (Pacome et al., 2014). The seeds contain 17 to 20% edible fixed oil, which is similar in its properties to cottonseed oil. On the other hand, the color extract from the dry calyces is rich in anthocyanin, amino acids, organic compounds, mineral salts, and source of vitamin C (Al-Ansary et al., 2016). Calyces extract is also a potential source of natural colorant to replace red synthetic coloring agents for carbonated soft drinks, jams, juices, jellies, sauces, chutneys, wines, preserves, and other acidic foods (Delgado and Parcedes, 2003). Anthocyanins are approved as food colorants in the USA under the category of fruit (21 CFR 73.250) or vegetable (21 CFR 73.260) juice color. The EU classifies anthocyanins as 'natural colorants' under classification number E163. The use of anthocyanins may show benefits over that of synthetic colors. A potential use for Roselle extract may include the production of fruit juice, drinks, and jam (Duangmal et al., 2008).

Fleshy calyces (sepals) are commercially important for the production of beverages, juices, jams, and syrup in the food industry. Furthermore, these calyces are a good source of natural food colorants because of their high pigment content (Bridle and Timberlake, 1997). Moreover, the dried calyces are consumed worldwide in hot infusions and in cold drinks (Hervert and Goni, 2012). Besides its extended consumption as a beverage and its uses in the food industry, Roselle is also used in animal feed, nutraceuticals, cosmetics, and pharmaceuticals (Wang et al., 2012; Borras et al., 2015).

8.5 EXTRACTION OF NATURAL COLOR FROM ROSELLE CALYCES

Food colorants are either natural or synthetic depending on the source. Natural colorants are extracted from renewable sources such as plant materials, insects, algae, etc., while the synthetic colorants are manufactured chemically and are the most commonly used dyes in the food, pharmaceutical, and cosmetic industries. Due to this limitation and worldwide tendency towards the consumption of natural products, the interest in natural colorants has increased significantly. Of special interest to the food industry is the limited availability of red pigments; therefore, research into natural sources of red pigments has increased recently (Abou-Arab et al., 2011).

Color is one of the most important quality attributes affecting the consumer's acceptance of food since it gives the first impression of food quality. Many convenience foods such as confectionery products, gelatin desserts, snacks, cakes, puddings, ice creams, and beverages would be colorless and would thus appear undesirable without the inclusion of colorants (Hirunpanich et al., 2006). In this respect, Roselle calyces appear to be good and promising sources of water-soluble red colorants that could be utilized as natural food colorants (Abou-Arab et al., 2011).

Anthocyanins are one of the most important groups of water-soluble pigments visible to the human eye. They are responsible for many of the attractive colors, from scarlet to blue, of flowers, fruits, leaves, and storage organs. HAs were identified as having delphinidin-3-sambubioside (Dp-3-sam) (70% of the anthocyanins) and cyanidin-3-sambubioside (Cyn-3-sam) as the major pigments, with delphinidin-3-glucoside (Dp-3-glu) and cyanidin-3-glucoside (Cyn-3-glu) as the minor ones (Amor and Allaf, 2009). Cahlikova et al. (2015) explained that the active constituents of the extracts of Roselle, which widely used in folk medicine to combat many illnesses and have been shown on several occasions to be anthocyanins and they ascertained through UHPLC-ESI MS/MS that delphinidin-3-sambubioside is the major one but the predominant one is cyaniding-3-sambubioside.

Plant materials generally contain a small amount of high added value active solute. Extraction and purification of bioactive compounds from natural sources have become very important for the utilization of phytochemicals in the preparation of dietary supplements or nutraceuticals, functional food ingredients, food additives, pharmaceutical, and cosmetic products. The polar character of anthocyanins makes them soluble in several types of polar solvents such as methanol, ethanol, acetone, and water. Solvent extraction of anthocyanins is the initial step in the determination of total and individual anthocyanins prior to quantification, purification, separation, and characterization (Rivas, 2003). The extraction of anthocyanins is commonly carried out under cold conditions using methanol or ethanol containing a small amount of acid in order to obtain the flavylium cation (FC) form, which is red and stable in a highly acid medium (Du and Francis, 1973; Amor and Allaf, 2009).

Extraction of anthocyanins is commonly carried out under cold conditions with methanol or ethanol containing a small amount of acid with the objective of obtaining the FC form, which is red and stable in a highly acid medium. However, the acid may cause partial hydrolysis of the acyl moieties in acylated anthocyanins, especially in anthocyanins acylated with dicarboxylic acids such as malonic acid (Bronnum and Flink, 1985). Abou-Arab, et al. (2011) used ethanol acidified with 1.5 N/L HCl (85:15), ethanol acidified with 1% citric acid, 2% citric acid solution and distilled water in extracting the pigments from Roselle calyces. They concluded that the addition of acids to water or ethanol increased the efficiency of anthocyanins extraction compared with distilled water alone. In general, HCl was more effective than citric acid. Ethanol acidified with HCl, showed the strongest influence on the amount of anthocyanins extracted, followed by 2% citric acid solution and ethanol acidified with 1% citric acid. Frimpong et al. (2014) investigated the potential of aqueous extract of *Hibiscus sabdariffa* calyces as a coloring agent in three pediatric oral pharmaceutical formulations and they found that Aqueous extract of *H. sabdariffa* calyces at a concentration of 33% w/v solution was successfully used to color three pediatric oral formulations. The formulations colored with the extract were susceptible to deterioration on exposure to light, pH, and high temperatures. They concluded that the extract of *H. sabdariffa* calyces could be a good substitute for amaranth as a coloring agent for pediatric oral pharmaceutical formulations when buffered at pH 5.0 and protected against high temperatures and light.

8.6 STABILITY OF ROSELLE ANTHOCYANIN

Anthocyanins are highly unstable molecules in the food matrix. The color stability of Anthocyanins is strongly affected by pH, solvents, temperature, anthocyanin concentration and structure, oxygen, light, enzymes, and other accompanying substances (Arueya and Akomolafe, 2014). Color stability of anthocyanins depends on a combination of various factors, including structure of anthocyanins, pH, temperature, oxygen, light, and water activity. Enzymatic degradation and interactions with food components such as ascorbic acid, sugars, metal ions, sulfur dioxide, and copigments are no less important (Abou-Arab, et al., 2011). Among these factors, pH is one of the major factors that significantly influenced the pigment color variations and stability. In general, anthocyanins are more stable in acidic media at low pH values than in alkaline solutions (Aishah et al., 2013). Anthocyanin is relatively unstable and because of their high reactivity it may be easily degraded and form colorless or undesirable brown-colored compounds during extraction processing and storage (Durst and Wrolstad, 2001) Indeed, temperature, pH, light oxygen, metals, organic acids, sugars, ascorbic acid, enzymes, sulfur dioxide, co-pigmentation, and interactions with food components may affect both the structure and stability of anthocyanins (Zuhaili et al., 2012).

In general, anthocyanins are more stable in acidic media at low pH values than in alkaline solutions. Cavalcanti et al. (2011) explained the anthocyanins' color shift based on their chemical structures. In aqueous solutions, anthocyanins exist basically in the form of four species in equilibrium depending on the pH: quinonoidal base (QB), FC, carbinol or pseudobase (PB) and chalcone (CH). Under acidic conditions (pH < 2), the anthocyanins exist primarily in the form of deep-red FC. Increasing pH values causes a rapid loss of the proton producing blue or violet QB forms. At the same time hydration of FC occurs, generating the carbinol or PB, which reaches equilibrium slowly with the colorless CH. The relative amounts of FC, QB, PB, and CH forms at the equilibrium condition vary according to pH (Cavalcanti et al., 2011). In other words, the anthocyanins ionic nature enables reversible changes of the molecule structure according to the prevailing pH, resulting in different colors and hues at different pH values. It is believed that through pH adjustment, the natural stabilization process for anthocyanins can be achieved; thus, the knowledge may become of great value to food manufacturers (Aishah et al., 2013).

Bronnum and Flink (1985) reported that the efficiency of extracting solvent increased with increasing the concentration of citric acid and concluded that pH of extracting medium was the determining factor for anthocyanins extractability. Anthocyanins are easily susceptible to pH changes due to the ionic nature of anthocyanin. Anthocyanins exist as four equilibrium forms, namely the QB, the FC, the carbinol (PB), and the CH. Under low pH, the anthocyanin exists primarily in the form of FC in red. As the pH is raised (>5), a rapid loss of proton occurred to form a quinoidal base that tends to become blue or violet. In addition, the increasing of pH causes the hydration of the FC to form a carbinol (PB) or CH, which are colorless (Amelia et al., 2013). Light is another factor, which affects the stability of anthocyanin. Palamidis and Markakis (1975) have studied the role of light on the stability of anthocyanin in grape juice and showed that exposure of the pigments to light accelerates their destruction. Their experiments showed that after placing the juice samples containing anthocyanin in the dark for 135 days at 20°C, almost 30% of the pigments were destroyed, but placing the same samples in the same temperature and same period of time in the presence of light destroyed more than 50% of total pigments. Palamidis and Markakis (1975) have studied the effect of temperature on the stability of anthocyanin in soft drinks and have shown that an increase in the storage temperature greatly accelerates the destruction of pigments in soft drinks. Maccarone et al. (1985) reported the anthocyanin in red orange juice in 15°C, 25°C, and 35°C during a 15 day period and found that the increase in temperature accelerates the destruction of anthocyanins.

8.7 ENCAPSULATION OF ROSELLE ANTHOCYANIN

Encapsulation facilitates light and heat-labile molecules to maintain their stability and improve their shelf lives and activity. It is a rapidly expanding technology, highly specialized, with affordable costs. Among the diverse encapsulation techniques available, only a few were evaluated for anthocyanin encapsulation, and these include: spray drying using different coating materials, gelation using polymers as sodium alginate, pectin, curdlan, and glucan and lyophilization (Santos et al., 2013). Spray drying technique has been widely used in the microencapsulation of food ingredients susceptible to deterioration by external agents and consists of entrapping an active agent (solid particles, liquid droplets or gaseous compounds) in a polymeric matrix, in order to protect it from adverse conditions. The immediate drying of the mixture leads to the formation of a matrix system in which the polymer forms a tridimensional network, which contains the encapsulated material (Tonon et al., 2010).

Anthocyanins are susceptible to degradation through factors such as the presence of light, pH (mainly pH higher than 7), temperatures higher than 60–80°C depending the anthocyanin group, the presence of sulfite, ascorbic acid, enzymes (such as glycosidases, galactoside, peroxidases, and phenolases) among other factors (Giusti and Wallace, 2009). One possible way to effectively protect the anthocyanin compound, from product processing to consumption, could be the use of encapsulation techniques. Encapsulation facilitates light- and heat-labile molecules to maintain their stability and improve their shelf lives and activity. It is a rapidly expanding technology, highly specialized, with affordable costs (Santos and Meireles, 2010; Cavalcanti et al., 2011). Among the diverse encapsulation techniques available, only a few were evaluated for anthocyanin encapsulation, these include: spray drying using different coating materials (Cai and Corke, 2000), gelation using polymers as sodium alginate, pectin, curdlan (Ferreira et al., 2009), glucan (Xiong et al., 2006) and lyophilization (Gradinaru et al., 2003; Santos et al., 2013).

Microencapsulation has been used by the food industry in order to protect sensitive food ingredients during storage, to mask or preserve aromas and flavors, to protect food against nutritional losses or even to add nutritive materials to food after processing (Re, 1998). The spray dry technique minimizes the loss of nutrients content during processing and storage, the powder obtained is soluble and convenient to carry anywhere, it requires less storage space, and it is one of the most commonly used methods to transform a wide range of liquid food products into powder. Some of the most used materials include pectin, whey protein, carrageenan, carboxymethyl cellulose (CMC), gelatin, and xanthan gum amongst others, trying to protect the core of the particles against adverse environmental conditions, providing at the same time, an easy handling and release rate control (Diaz et al., 2015). Protecting polyphenols with carriers instead of administer them as free compounds can overcome the disadvantages of their instability (Chen et al., 2006), improve the bioavailability and enhance their half-life in vivo and in vitro, as well as may counteract the need of administering high concentration of molecules of interest required to obtain beneficial results. Among the vegetables rich in polyphenols, Roselle is known for its high anthocyanin content (Sayago et al., 2007).

Serrano et al. (2013) obtained films of Roselle's polyphenols in binary and ternary mixtures of CMC, whey protein, and pectin concluding that CMC avoids the release of Roselle's polyphenols while whey protein and to a lesser extent pectin, causes the opposite effect. Bandera et al. (2013) reported the kinetic release of Roselle's polyphenols encapsulated in gelatin beads coated with sodium alginate showing that the release can be well controlled manipulating the number of alginate coats and the immersion time in a $CaCl_2$. Idham et al. (2012) encapsulated Roselle's polyphenols, via spray dry, using pullulan, fruit fiber, and maltodextrin-gum Arabic-soluble starch as carrier materials and reported the thermal and color stability at storage conditions, but the release of polyphenols from its matrix was not evaluated. The spray dry technique minimizes the loss of nutrients' content during processing and storage, the powder obtained is soluble and convenient to carry anywhere, it requires less storage space, and it is one of the most commonly used methods to transform a wide range of liquid food products into powder. Some of the most used materials include pectin, whey protein, carrageenan, CMC, gelatin, and xanthan gum amongst others, trying to protect the core of the particles against adverse environmental conditions, providing at the same time, an easy handling and release rate control (Bandera et al., 2015).

Supercritical fluids (SFCs) processes have become an attractive alternative to extract and encapsulate natural pigments due to the use of environmentally friendly solvents. In recent years, novel particle formation techniques using SFCs have been developed in order to overcome some of the disadvantages of conventional techniques. Some of the drawbacks of conventional techniques are: (a) poor control of particle size and morphology; (b) degradation and loss of biological activity of thermosensitive compounds; (c) low encapsulation efficiency; and (d) low precipitation yield. Additionally, the use of SFCs as phase separating agents has been intensively studied to minimize the amount of potentially harmful residues in the capsules (Jacobson et al., 2010; Santos and Meireles, 2010; Santos et al., 2013).

8.8 CONCLUSION

Roselle has been reported to use as a refrigerant in the form of tea, flavoring for sauces, jellies, marmalades, and soft drinks or to use as a colorant for foods which Roselle appear to be good and promising sources of water-soluble natural red pigments. Roselle has been used as a therapeutic plant for centuries. Traditionally, extracts treat toothaches, urinary tract infections, colds, and even hangovers. Several pharmacological and clinical studies suggest Roselle can be a good source of natural color in the form of antioxidant, antimicrobial, antihypertension, and many other properties like anti-inflammatory, hepatoprotective, and antitumoral effects have been highlighted. There is a huge demand for commercial cultivation of Roselle and quantification and purification of anthocyanins found in the calyces of Roselle *viz.*, delphinidin-3-sambubioside, cyanidin-3-sambubioside, delphinidin-3-monoglucoside, and cyanidin-3-monoglucoside. This will further help in generating entrepreneurship opportunities and upliftment of the socio-economic condition of the farming community.

KEYWORDS

- **anthocyanin**
- **flavonoids**
- **polyphenols**
- **roselle calyces**

REFERENCES

Abeda, H. Z., Kouassi, M. K., Yapo, K. D., Koffi, E., Sie, R. S., Kone, M., & Kouakou, H. T., (2014). Production and enhancement of anthocyanin in callus line of Roselle (*Hibiscus sabdariffa* L.). *International Journal of Recent Biotechnology*, *2*(1), 45–56.

Abou-Arab, A. A., Abu- Salem, F. M., & Abou-Arab, E. A., (2011). Physicochemical properties of natural pigments (anthocyanin) extracted from Roselle calyces (*Hibiscus subdariffa*). *Journal of American Science*, *7*(7), 445–456.

Adanlawo, I. G., & Ajibade, V. A., (2006). Nutritive value of the two varieties of Roselle (*Hibiscus sabdariffa*) calyces soaked with wood ash. *Pakistan Journal of Nutrition*, *5*(6), 555–557.

Aishah, B., Nursabrina, M., Noriham, A., Norizzah, A. R., & Shahrimi, H. M., (2013). Anthocyanins from *Hibiscus sabdariffa*, *Melastoma malabathricum* and *Ipomoea batatas* and its color properties. *International Food Research Journal*, *20*(2), 827–834.

Ajayi, O. A., Olawale, A. S., & Adefila, S. S., (2012). Conversion of sorrel (*Hibiscus sabdariffa*) calyces to glucose. *International Journal of Scientific & Technology Research*, *1*(8), 130–138.

Akanya, H. O., Oyeleke, S. B., Jigam, A. A., & Lawal, F. F., (1997). Analysis of some drink. *Nigerian Journal of Biochemistry and Molecular Biology*, *12*, 77–81.

Al-Ansary, A. M. F., Nagwa, R. A. H., Ottai, M. E. S., & El-Mergawi, R. A., (2016). Gamma irradiation effect on some morphological and chemical characters of Sudani and Masri Roselle varieties. *International Journal of Chem. Tech. Research*, *9*(3), 83–96.

Ali, B. H., Wabel, N. A., & Blunden, G., (2005). Phytochemical, pharmacological and toxicological aspects of *Hibiscus sabdariffa* L.: Review. *Phytotherapy Research*, *19*(5), 369–375.

Alobo, A. P., & Offonry, S. U., (2009). Characteristics of colored wine produced from Roselle (*Hibiscus sabdariffa*) calyx extract. *Journal of the Institute of Brewing*, *115*(2), 91–94.

Alonso, A. J., Zamilpa, A., Aguilar, F. A., Herrera, R. M., Tortoriello, J., & Jimenez, J. E., (2012). Pharmacological characterization of the diuretic effect of *Hibiscus sabdariffa* Linn (Malvaceae) extract. *Journal of Ethnopharmacology*, *139*(3), 751–756.

Amalia, F., & Afnani, G. N., (2013). Extraction and stability test of anthocyanin from buni fruits (*Antidesma Bunius* L) as an alternative natural and safe food colorants. *Journal of Food and Pharmaceutical Sciences*, *1*(2), 49–53.

Amor, B. B., &. Allaf, K., (2009). Impact of texturing using instant pressure drop treatment prior to solvent extraction of anthocyanins from Malaysian Roselle (*Hibiscus sabdariffa*). *Food Chemistry*, *115*(3), 820–825.

Anonymous, (2013). *All About Roselle, The Earth of India*, http://theindianvegan.blogspot.in/2013/02/all-about-roselle.html (Accessed on 12 June 2019).

Arueya, G. L., & Akomolafe, B. O., (2014). Stability studies of microencapsulated anthocyanins of Roselle (*Hibiscus Sabdariffa* L) in native starch and its potential application in jam production. *IOSR Journal of Environmental Science, Toxicology and Food Technology (IOSR-JESTFT)*, *8*(7), 112–122.

Aurelio, D., Edgardo, R. G., & Navarro, G. S., (2007). Thermal kinetic degradation of anthocyanins in a Roselle (*Hibiscus sabdariffa* L. cv. ′Criollo′) infusion (Online). Available http://www.blackwell-synergy.com/doi/pdf/10.1111 /j.1365–2621.2006.01439.x (Accessed on 12 June 2019).

Azevedo, J., Fernandes, I., Faria, A., Oliveira, J., Fernandes, A., Freitas, V., & Mateus, N., (2010). Antioxidant properties of anthocyanidins, anthocyanidin-3-glucosides and respective portisins. *Food Chemistry*, *119*(2), 518–523.

Babalola, S. O., Babalola, A. O., & Aworth, O. C., (2001). Compositional attributes of the calyces of Roselle. *Journal of Food Technology* in *Africa*, *6*(4), 133–134.

Bandera, D. D., Villanueva, C. A., Garcia, D. O., Quintero, S. B., & Dominguez, L. A., (2013). Release kinetics of antioxidant compounds from *Hibiscus sabdariffa* L. encapsulated in gelatin beads and coated with sodium alginate. *International Journal of Food Science & Technology*, *48*, 2150–2158.

Bandera, D. D., Villanueva, C. A., Garcia, D. O., Quintero, S. B., & Dominguez, L. A., (2015). Assessing release kinetics and dissolution of spray-dried Roselle (*Hibiscus sabdariffa* L.) extract encapsulated with different carrier agents. *LWT-Food Science and Technology, 64*(2), 693–698.

Borras, L. I., Fernandez, A. S., Arraez, R. D., Palmeros, S. P. A., Díaz, D. R., Andrade, G. I., Fernandez, G. A., Gomez, L. J. F., & Segura, C. A., (2015). Characterization of phenolic compounds, anthocyanidin, antioxidant and antimicrobial activity of 25 varieties of Mexican Roselle (*Hibiscus sabdariffa*). *Industrial Crops and Products, 69*, 385–394.

Bridle, P., & Timberlake, C. F., (1997). Anthocyanins as natural food colors–selected aspects. *Food Chemistry, 58,* 103–109.

Bronnum, H. K., & Flink, M. J., (1985). Anthocyanin colorants from elderberry (*Sambucus nigra* L.). 2. Process considerations for production of freeze-dried product. *Journal of Food Technology, 20,* 714–723.

Cahlikova, L., Ali, B. H., Havlikova, L., Locarek, M., Siatka, T., Opletal, L., & Blunden, G., (2015). Anthocyanins of *Hibiscus sabdariffa* calyxes from Sudan. *Natural Product Communications, 10*(1), 77–79.

Cai, Y. Z., & Corke, H., (2000). Production and properties of spray-dried Amaranthus betacyanin pigments. *Journal of Food Science, 65,* 1248–1252.

Cavalcanti, R. N., Diego, T. S., & Maria, A. A. M., (2011). Non-thermal stabilization mechanisms of anthocyanins in model and food systems - An overview. *Food Research International, 44,* 499–509.

Chen, L., Remondetto, G., & Subirade, M., (2006). Food protein-based materials as nutraceutical delivery systems. *Trends in Food Science & Technology, 17,* 272–283.

Chewonarin, T., Kinouchi, T., Kataoka, K., Arimochi, H., Kuwahara, T., & Vinitketkumnuen, U., (1999). Effects of roselle (*Hibiscus sabdariffa* Linn.), a Thai medicinal plant, on the mutagenicity of various known mutagens in *Salmonella typhimurium* and on formation of aberrant crypt foci induced by the colon carcinogens azoxymethane and 2-amino-1-methyl-6-phenylimidazo[4,5-b]pyridine in F344 rats. *Food and Chemical Toxicology, 37,* 591–601.

Christian, K. R., Nair, M. G., & Jackson, J. C., (2006). Antioxidant and cyclooxygenase inhibitory activity of sorrel (*Hibiscus sabdariffa*). *Journal of Food Composition & Analysis, 19*(8), 778–783.

Cid, O. S., & Guerrero, B. J. A., (2014). Roselle calyces particle size effect on the physicochemical and phytochemicals characteristics. *Journal of Food Research, 3*(5), 83–94.

Cisse, M., Dornier, M., Sakho, M., Ndiaye, A., Reynes, M., & Sock, O., (2009). Le bissap (*Hibiscus sabdariffa L.*): Composition et principles utilizations. *Fruits, 64,* 179–193.

Delgado, V. F., & Parcedes, L. O., (2003). *Natural Colorants for Food and Nutraceutical Uses* (p. 327). CRC Press, LLC: Boca Raton, FL.

Diane, L. M., Chen, O., Saltzman, E., & Blumberg, J. B., (2009). *Hibiscus sabdariffa* L. tea (tisane) lowers blood pressure in prehypertensive and mildly hypertensive adults. *The Journal of Nutrition, 140*(2), 298–303.

Diaz, B. D., Villanueva, C. A., Dublan, G. O., Quintero, S. B., & Dominguez, L. A., (2015). Assessing release kinetics and dissolution of spray-dried Roselle (*Hibiscus sabdariffa* L.) extract encapsulated with different carrier agents. *LWT-Food Science & Technology, 64*(2), 693–698.

Du, C. T., & Francis, F. J., (1973). Anthocyanins of Roselle (*Hibiscus sabdariffa* L.). *Journal of Food Science, 38*(5), 810–812.

Duangmal, K., Saicheua, B., & Sueeprasan, S., (2008). Color evaluation of freeze-dried Roselle extract as a natural food colorant in a model system of a drink. *LWT- Food Science & Technology*, *41*(8), 1437–1445.

Durst, R., & Wrosltad, R., (2001). *Separation and Characterization of Anthocyanins by HPLC*. Food analytical chemistry John Wiley and Sons, Inc., USA.P.F1.3.1.

Eltayeib, A. A., & Hamade, H., (2014). Phytochemical and chemical composition of water extract of *Hibiscus Sabdariffa* (red karkade calyces) in North Kordofan state-Sudan. *International Journal of Advanced Research in Chemical Science (IJARCS), 1*(6), 10–13.

Fernandez, A. S., Rodriguez, M. I. C., Beltran, D. R., Pasini, F., Joven, J., Micol, V., Segura, C. A., & Fernandez, G. A., (2011). Quantification of the polyphenolic fraction and *in vitro* antioxidant and *in vivo* anti-hyperlipemic activities of *Hibiscus sabdariffa* aqueous extract. *Food Research International, 44,* 1490–1495.

Ferreira, D. S., Faria, A. F., Grosso, C. R. F., & Mercadante, A. Z., (2009). Encapsulation of blackberry anthocyanins by thermal gelation of curdlan. *Journal of Brazilian Chemical Society, 20,* 1908–1915.

Frimpong, G., Adotey, J., Kwakye, K. O., Kipo, S. L., & Dwomo, F. Y., (2014). Potential of aqueous extract of *Hibiscus sabdariffa* calyces as coloring agent in three pediatric oral pharmaceutical formulations. *Journal of Applied Pharmaceutical Science*, *4*(12), 001–007.

Gautam, R. D., (2004). Sorrel - a lesser-known source of medicinal soft drink and food in India. *Natural Product Radiance*, *3*(5), 338–342.

Gibis, M., & Weiss, J., (2010). Inhibitory effect of marinades with hibiscus extract on formation of heterocyclic aromatic amines and sensory quality of fried beef patties. *Meat Science, 85*(4), 735–742.

Gibis, M., Zeeb, B., & Weiss, J., (2014). Formation, characterization, and stability of encapsulated hibiscus extract in multilayered liposomes. *Food Hydrocolloids*, *38*, 28–39.

Giusti, M. M., & Wallace, T. C., (2009). Flavonoids as natural pigments. In: Bechtold, T., & Mussak, R., (eds.), *Handbook of Natural Colorants* (pp. 261–268). Chichester: John Wiley & Sons.

Goda, Y., Shimizu, T., Kato, Y., Nakamura, M., Maitani, T., Yamada, T., Terahara, N., & Yamaguchi, M., (1997). Two acylated anthocyanins from purple sweet potato. *Phytochemistry*, *44*(1), 183–186.

Gradinaru, G., Biliaderis, C. G., Kallithraka, S., Kefalas, P., & Garcia, V. C., (2003). Thermal stability of *Hibiscus sabdariffa* L. anthocyanins in solution and in solid state: Effects of co-pigmentation and glass transition. *Food Chemistry, 83,* 423–436.

Heba, A., Lisa, J. M., & Michael, R. A. M., (2014). Comparative chemical and biochemical analysis of extracts of *Hibiscus sabdariffa*. *Journal of Food Chemistry, 164,* 23–29.

Hervert, H. D., & Goni, I., (2012). Contribution of beverages to the intake of polyphenols and antioxidant capacity in obese women from rural Mexico. *Public Health Nutrition, 15,* 6–12.

Hirunpanich, V., Utaipat, A., Morales, N. P., Bunyapraphatsara, N., Sato, H., Herunsale, A., & Suthisisang, C., (2006). Hypocholesterolemic and antioxidant effects of aqueous extracts from the dried calyx of *Hibiscus sabdariffa* L. in hypercholesterolemic rats. *Journal of Ethnopharmacology*, *103*(2), 252–260.

Hou, L., Zhou, B., Yang, L., & Liu, Z. L., (2004). Inhibition of human low-density lipoprotein oxidation by flavonols and their glycosides. *Chemistry and Physics of Lipids, 129,* 209–219.

Idham, Z., Muhamad, I. I., & Sarmidi, M. R., (2012). Degradation kinetics and color stability of spray-dried encapsulated anthocyanins from *Hibiscus sabdariffa* L. *Journal of Food Process Engineering, 35,* 522–542.

Ismail, A., Ikram, E. H. K., & Nazri, H. S. M., (2008). Roselle (*Hibiscus sabdariffa* L.) seeds–nutritional composition, protein quality and health benefits. *Food, 2*(1), 1–16.

Jacobson, G. B., Shinde, R., McCullough, R. L., Cheng, N. J., Creasman, A., & Beyene, A., (2010). Nanoparticle formation of organic compounds with retained biological activity. *Journal of Pharmaceutical Sciences, 99,* 2750–2755.

Kerharo, J., (1971). Senegal bisap (*Hibiscus sabdariffa)* or Guinea sorrel or red sorrel. *Plant Medicine & Phytotherapy*, *5*(4), 277–281.

Khafaga, E. R., & Koch, H., (1980a). Stage of maturity and quality of Roselle (*Hibiscus sabdariffa* L. var. *sabdariffa*). 1. Organic acids. *Angewandte Botanik, 54,* 287–293.

Khafaga, E. R., & Koch, H., (1980b). Stage of maturity and quality of Roselle (*Hibiscus subdariffa* L. var. *sabdariffa*). 2. Anthocyanins. *Angewandte Botanik, 54,* 295–300.

Kilima, B. M., Remberg, S. F., Chove, B. E., & Wicklund, T., (2014). Physio-chemical, mineral composition and antioxidant properties of Roselle (*Hibiscus sabdariffa* L.) extract blended with tropical fruit juices. *African Journal of Food, Agriculture, Nutrition & Development, 14*(3), 8963–8978.

Kong, J. M., Chia, L. S., Goh, N. K., Chia, T. F., & Brouillard, R., (2003). Analysis and biological activities of anthocyanins. *Phytochemistry*, *64*(5), 923–933.

Kumar, S., Kumar, D., & Prakash, O., (2008). Evaluation of antioxidant potential, phenolic and flavonoid contents of *Hibiscus tiliaceus* flowers. *Electronic Journal of Environmental, Agricultural and Food Chemistry, 7*(4), 2863–2871.

Kuo, C., Kao, E., Chan, K., Lee, H., Huang, T., & Wang, C., (2012). *Hibiscus sabdariffa* L. extracts reduce serum uric acid levels in oxonate-induced rats *Hibiscus sabdariffa* L. extracts reduce serum uric acid levels in oxonate-induced rats. *Journal of Functional Foods, 4,* 375–381.

Laikangbam, R., & Damayanti, D. M., (2012). Inhibition of calcium oxalate crystal deposition on kidneys of urolithiatic rats by *Hibiscus sabdariffa* L. extract. *Urological Research, 40,* 211–218.

Lin, H., Chan, K., Sheu, J., Hsuan, S., Wang, C., & Chen, J., (2012). *Hibiscus sabdariffa* leaf induces apoptosis of human prostate cancer cells *in vitro* and *in vivo*. *Food Chemistry, 132,* 880–891.

Lin, T., Lin, H. H., Chen, C. C., Lin, M. C., Chou, M. C., & Wang, C. J., (2007). *Hibiscus Sabdariffa* extract reduces serum cholesterol in men and women. *Nutrition Research, 27*(3), 140–145.

Maccarone, E. A., & Rapisared, P., (1985). Stabilization of anthocyanins of blood orange fruit juice. *Journal of Food Science*, *50*(4), 901–904.

Mahadevan, N., Shivali, & Kamboj, P., (2009). *Hibiscus sabdariffa* Linn. - An overview. *Natural Product Radiance*, *8*(1), 77–83.

Margesi, S., Kagashe, G., & Dhokia, D., (2013). Determination of iron contents in *Hibiscus sabdariffa* calyces and *Kigelia Africana* fruit. *Scholars Academic Journal of Biosciences (SAJB)*, *1*(4), 108–111.

McKay, D. L., Chen, C. Y., Saltzman, E., & Blumberg, J. B., (2010). *Hibiscus sabdariffa* L. tea (tisane) lowers blood pressure in prehypertensive and mildly hypertensive adults. *Journal of Nutrition, 140,* 298–303.

Mohamed, B. B., Sulaiman, A. A., & Dahab, A. A., (2012). Roselle (*Hibiscus sabdariffa* L.) in Sudan, cultivation and their uses. *Bulletin of Environment, Pharmacology & Life Sciences*, *1*(6), 48–54.

Odigie, I. P., Ettarh, R. R., & Adigun, S. A., (2003). Chronic administration of aqueous extract of *Hibiscus sabdariffa* attenuates hypertension and reverses cardiac hypertrophy in 2K-1 C hypertensive rats. *Journal of Ethnopharmacology*, *86,* 181–185.

Onyenekwe, P. C., Ajani, E. O., Ameh, D. A., & Gamaniel, K. S., (1999). Antihypertensive effect of roselle (*Hibiscus sabdariffa*) calyx infusion in spontaneously hypertensive rats and a comparison of its toxicity with that in Wistar rats. *Cell Biochemistry & Function, 17,* 199–206.

Pacome, O. A., Bernard, D. N., Sekou, D., Joseph, D. A., David, N. D., Mongomake, K., & Hilaire, K. T., (2014). Phytochemical and antioxidant activity of Roselle (*Hibiscus Sabdariffa* L.) petal extracts. *Research Journal of Pharmaceutical, Biological & Chemical Sciences*, *5*(2), 1453–1465.

Padmaja, H., Sruthi, S., & Vangalapati, M., (2014). Review on *Hibiscus sabdariffa*–A valuable herb. *International Journal of Pharmacy & Life Sciences*, *5*(8), 3747–3752.

Palamidis, N., & Markakis, T., (1975). Structure of anthocyanin. *Journal of Food Science*, *40*(5), 104–106.

Prenesti, E., Berto, S., Daniele, P. G., & Toso, S., (2007). Antioxidant power quantification of decoction and cold infusions of *Hibiscus sabdariffa* flowers. *Food Chemistry, 100,* 433–438.

Puro, K., Sunjukta, R., Samir, S., Ghatak, S., Shakuntala, I., & Sen, A., (2014). Medicinal uses of Roselle plant (*Hibiscus sabdariffa* L.): A mini review. *Indian Journal of Hill Farming*, *27*(1), 81–90.

Re, M. I., (1998). Microencapsulation by spray drying. *Drying Technology*, *16*(6), 1195–1236.

Rivas, G. J., (2003). Analysis of polyphenols. In: Santos-Buelga, C., & Williamson, G., (eds.), *Methods in Polyphenols Analysis* (pp. 95–98, 338–358). Royal Society of Chemistry (Athenaeum Press, Ltd.): Cambridge, U.K.

Salazar, G. C., Vergara, B. F. T., Ortega, R. A. E., & Guerrero, B. J. A., (2012). Antioxidant properties and color of *Hibiscus sabdariffa* extracts. *Ciencia e Investigacion Agraria*, *39*(1), 79–90.

Santos, D. T., & Meireles, M. A. A., (2010). Carotenoid pigments encapsulation: Fundamentals, techniques and recent trends. *The Open Chemical Engineering Journal, 4,* 42–50.

Santos, D. T., Albarelli, J. Q., Beppu, M. M., & Meireles, M. A. A., (2013). Stabilization of anthocyanin extract from jabuticaba skins by encapsulation using supercritical CO_2 as solvent. *Food Research International, 50*(2), 617–624.

Sayago, A. S. G., Arranz, S., Serrano, J., & Gontili, I., (2007). Dietary fiber content and associated antioxidant compounds in roselle flower (*Hibiscus sabdariffa* L.) beverage. *Journal of Agricultural and Food Chemistry, 55*(19), 7886–7890.

Serrano, C. M. R., Villanueva, C. A., Rosales, E. J. M., Davila, J. F. R., & Dominguez, L. A., (2013). Controlled release and antioxidant activity of Roselle (*Hibiscus sabdariffa* L.) extract encapsulated in mixtures of carboxymethyl cellulose, whey protein, and pectin. *LWT-Food Science and Technology*, *50*(2), 554–561.

Tonon, R. V., Brabet, C., & Hubinger, M. D., (2010). Anthocyanin stability and antioxidant activity of spray-dried acai (*Euterpe oleracea* Mart.) juice produced with different carrier agents. *Food Research International*, *43*(3), 907–914.

Tsai, P. J., McIntosh, J., Pearce, P., Camden, B., & Jordan, R. B., (2002). Anthocyanin and antioxidant capacity in Roselle (*Hibiscus sabdariifa* L.) extract. *Journal of Food Research International, 35,* 351–356.

Tseng, T. H., Wang, C. J., Kao, E. S., & Chu, C. Y., (1996). Hibiscus protocatechuic acid protects against oxidative damage induced by tertbutyl hydroperoxide in rat primary hepatocytes. *Chemico- Biological Interactions, 101*(2), 137–148.

Wang, C., Wang, J., Lin, W., Chu, C., Chou, F., & Tseng, T., (2000). Protective effect of hibiscus anthocyanins against tert-butyl hydroperoxide-induced hepatic toxicity in rats. *Food & Chemical Toxicology, 38,* 411–416.

Wang, M. L., Morris, B., Tonnis, B., Davis, J., & Pederson, G. A., (2012). Assessment of oil content and fatty acid composition variability in two economically important Hibiscus species. *Journal of Agricultural & Food Chemistry, 60,* 6620–6626.

Xiong, S., Melton, L. D., Easteal, A. J., & Siew, D., (2006). Stability and antioxidant activity of black currant anthocyanins in solution and encapsulated in glucan gel. *Journal of Agricultural & Food Chemistry, 54,* 6201–6208.

Zuhaili, I., Muhamad, I. I., Setapar, S. H. M., & Sarmidi, M. R., (2012). Effect of thermal processes on Roselle anthocyanins encapsulated in different polymer matrices. *Journal of Food Processing & Preservation, 36*(2), 176–184.

CHAPTER 9

Phytochemical, Pharmacological, and Food Applications of Asparagus (*A. racemosus*)

ANAMIKA RANJAN[1] and PRAMOD K. PRABHAKAR[2]

[1]2413 Via Palermo APT 1623, Fort Worth, Texas, 76109, USA

*[2]Department of Food Science and Technology, National Institute of Food Technology Entrepreneurship and Management,
Kundli, Sonepat, Haryana, India
E-mails: pramodkp@niftem.ac.in, pkprabhakariitkgp@gmail.com*

ABSTRACT

Asparagus, a popular Indian herb, belongs to the family Asparagaceae (formerly *Liliaceae*) is an herbaceous perennial plant. It has about 300 species distributed throughout the world. It includes commercially important species of vegetable as well as some ornamentally or medicinally important species. It represents a highly valuable plant species containing therapeutic as well as nutraceutical importance. *Asparagus officinalis* is cultivated for its edible shoots. *A. racemosus* Willd, usually called as Satawar, Shatavari, or Shatamuli is a woody climber, 1–2 m tall and commonly occurring throughout India and the Himalayas. The roots of *A. racemosus* are finger-shaped and bitter in taste and are sweet oleaginous, indigestible, cooling, and an appetizer. It is predominant in the Ayurveda system of medicine because of its ability to play an important role in preventing and curing multiple diseases. The leaves and the tuberous roots are used in the treatment of several diseases. They are used in different medicines and found effective in the treatment of gastric and duodenal ulcers, diarrhea, and dysentery associated with bleeding and treating many diseases such as nervous disorders, dyspepsia, liver diseases, inflammation, throat infections, tuberculosis, cough bronchitis, jaundice, etc. It has also galactagogue effect and thereby increases milk secretions in nursing mothers. The plant has numerous therapeutic application *viz.* antioxidant, anti-ADH, anticancerous, antiulcerogenic, anti-inflammatory, antibacterial, antidepressant, antitussive, immunomodulatory, cardioprotective, anti-diarrheal activity, antihepatotoxic activity. The phytochemical extract of Shatavari roots has many bioactive compounds indicated the presence of carbohydrates, phenolic compounds, steroidal glycosides, and tannins in ethanolic and aqueous extract whereas steroids, terpenes, and saponins in ethanolic extract. The presence of essential oils, polyphenols, flavonoids (kaempferol, quercetin, rutin), alkaloids (racemosol), vitamins, resins and some amino acids viz. asparagine, arginine, tyrosine, and others make it more widespread. The roots of *A. racemosus* contain saponins (shatavarin I–IV), which are also known as the defense system as it is antimicrobial in nature that inhibits mold and hence prevents plants from insects. This chapter comprises the comprehensive outlook of medicinal and phytochemistry properties of Shatavari extracts and their application in some extensively used dairy and bakery products like bread, biscuits, milk, and milk products, etc.

9.1 INTRODUCTION

The genus *Asparagus* is an herbaceous perennial plant consists of about 300 species around the world, out of which 22 species are recorded in India (Singla and Jaitak, 2014). It is most commonly used in indigenous medicine and mainly originated from Asia, Africa, and Europe (Velvan et al., 2007; Prohens et al., 2008). This genus comprising of three subgenera, *Asparagus*, Proto*asparagus*, and Myrsiphyllum which is distributed from sub-arid to arid regions and grown throughout the world. Total 14 species of Asparagus has been deducted in Pakistan (Dahlgren et al., 1985; Clifford and Conran, 1987; Chen and Tamanian, 2000; Ali and Khan, 2009). Among all, the most economically important *Asparagus* species is garden *Asparagus* (commonly known as *Asparagus officinalis* L.), which is ranked very high in spring vegetables, folk name sparrow grass (Stajner et al., 2002). Garden asparagus has an edible organ, known as spears, which are tender and unexpanded shoots (Rubatzky and Yamaguchi, 1997). Garden Asparagus has very good economic value. A fungus known as *Fusarium* spp causes a disease in Asparagus crop, called "crown and root rot" has reduced significantly the production and use of this crop (Farr et al., 1989).

Different species of *Asparagus* has different values for the mankind like *A. plumosus, A. pyramidalis, A. cooperi, and A. densiflorus* is widely used in horticultural and ornamental purpose whereas *Asparagus officinalis* has a significant role as a vegetable crop. *Asparagus racemosus* Willd. is well known as Shatavari, Satawar or Satmuli in Hindi. *A. racemosus* Willd. is a well-known Ayurvedic Rasayana. *A. racemosus* 'Rasayana' literally means the path that 'Rasa' takes (Rasa means plasma; Ayana means path). 'Rasayana' is a specialized section of Ayurveda, which deals with the preservation and promotion of health by revitalizing the metabolism and enhancing immunity. Our ancestors also had the knowledge of *A. racemosus;* we can find its use written in Ayurveda (Charaka Samhita) thousands of year back (Chawala et al., 2011; Sharma and Dash, 2003).

Asparagus racemosus is thought to be medicinally important due to the occurrence of steroid saponins, and sapogenins. *A. racemosus* is widely used for various medicinal, therapeutic, nutraceutical, and food consumption purposes (Sabnis, 1968; Shasnayet al., 2003). *Asparagus* possess saponins as well as fructans. Saponins have anti-tumor activity, and fructans involve in reducing the risk of disorder such as diarrhea, constipation as well as osteoporosis, obesity, cardiovascular disease (CVD), rheumatism, and diabetes (Shao et al., 1997). It has many different pharmacological properties like antiulcer, anticancer, antibacterial, antidysenteric antioxidant, antifungal, anti-inflammatory, anti-abortifacient, anticoagulant effects and also helpful in female reproductive problems, etc. (Sharma et al., 2000).

Morphologically, *A. racemosus* plant is a woody creeper, whitish grey or brown Colored, and 100–200 cm in height (Figure 9.1). Its leaves are small, uniform, and pine like needles in shape; flowers are tiny white Color and have spikes while the plant itself is bittersweet in taste (Chawala et al., 2011). The roots have a finger-like structure and are clustered in nature. The roots of *A. racemosus have* significant medicinal properties. In powdered form, root tuber is morphologically pale yellow in Color, odorless, and slightly sweetish in taste (Shaha and Bellankimath, 2017).

The species of *Asparagus* has high heterozygosity. For the betterment of crop and conservation of gene pool, assessment of genetic diversity between different species is highly recommended. There is a wide variation between the types and numbers of different chromosomes of these species when their chromosomes characteristic is analyzed. At the very early stage of evolution phenomenon of duplication or translocation between the chromosomes occurs, causing an increase or decrease in a number of secondary chromosomes (Kumari and Mukhopadhyay, 2015). The present investigation has clearly indicated that minute structural changes in chromosomes have been responsible for the evolution of these species (Kumari et al., 2014).

FIGURE 9.1 (See color insert.) Photographs of *Asparagus racemosus* showing (A) shoot, leaves, and berries, (B) tuberous roots, (C) pine like needle shaped leaves, and (D) a bundle of cultivated garden asparagus (spears).

9.2 PHYTOCHEMICAL APPLICATION

The major bioactive components present in the plant are steroidal saponins. Shatavari mainly contains a group of steroidal saponins (Shatavarin I-VI), which are the glycosides of sarsasapogenin. The phytochemical extract of Shatavari roots has many bioactive compounds indicated the presence of carbohydrates, phenolic compounds, steroidal glycosides, and tannins in ethanolic and aqueous extract whereas steroids, terpenes, and saponins in ethanolic extract. The presence of polyphenols, flavonoids (kaempferol, quercetin, rutin), alkaloids (racemosol), and vitamins make it more widespread. Shatavari contains essential oils, resin, tannins, and some amino acids viz. asparagine, arginine, tyrosine, (Mehta, 2013; Venkatesan et al., 2005).

1. **Steroids:** In Shatavari, the major bioactive constituents called steroidal saponins are present (Mishra et al., 2005). A group of steroidal saponins is presently known as Shatavarins, which are the glycoside of sarsasapogenin, which is generally occurring in two types of skeletons furostanols and spirostanols rhamnose (Choudhary and Sharma, 2014). It mainly contains six components, (Shatavarin I-VI). Shatavarin IV, a major glycoside being present in the roots of the plants. The structure of shatavarin IV comprises of two molecules of rhamnose along with a single molecule of glucose and is mainly found in the leaf, fruits, and roots of the plant (Chawla et al., 2011). Rhamnose and 3-glucose moieties attached to sarsapogenin in Satavarin I (a glycoside) (Singh et al., 2014a). Twenty-nine steroidal saponins (1–29) were reported from *A. racemosus* (Singla and jaitak, 2014). From the root powder of *A. racemosus* (shatavari), ten shatavarins (I-X)has been extracted in which shatavarins I-V are steroidal saponins, and VI-X are phytoestrogen compounds). Shatavarin IV is a glycoside compound of sarsapogenin (Hayes et al., 2008). Some other compounds have also been reported recently viz. asparinins, shatavarin V, ascarosides, oligospirostanoside (Immunoside) curillins, and curillosides. From the roots of *A. racemosus,* five new steroidal saponins have been identified which areshatavarin I (or asparoside B), shatavarin IV (or asparinin B), shatavarin V, immunoside, and schidigera saponin D5 (or asparanin A) including shatavarins VI-X (Hayes et al., 2006). Satavarins are very beneficial for treating multiple diseases like kidney problems, chronic fevers, epilepsy, stomach ulcers, and problem in milk secretion in nursing mothers, irregular sexual behavior, liver cancer, etc. This depicts the plant with tremendous potential in both healthcare and trade. *A. racemosus*is are famous for its phytoestrogenic properties so used for the purpose of hormone modulation (Mayo, 1998).
2. **Alkaloids and Flavonoids:** A cage like a pyrrolizidine and a polycyclic alkaloid called Aspargamine A. It has the presence of isoflavones majorly racemosol and asparagamine A, which gives it antioxidant properties. Racemofuran, a furan compound, and racemosol both are the new antioxidant compound which is isolated from leaves and roots of *A. racemosus*, respectively (Wiboonpun, 2004). Racemofuran contains rutin and flavonoid glycosides called quercetin-3-glucuronide. The ripe fruit of *A. racemosus* contains cyanidin-3-glucorhamnoside and cyanidin-3-galactoside, whereas normal fruits have rutin, steroidal saponins, glycosides of

quercetin, and hyperoside (Joy et al., 1998). There are three classes of flavonoids: prenylated, isoflavones, and coumestans flavonoids, among the phytoestrogens, are the one which has the most potent oestrogenic action, and these phytoestrogen classes can bind the estrogen receptor (ER) even though their binding affinity is lower than that of endogenous estradiol. Sarsapogenin and Kaepfrol are the two components, which are extracted from the woody portion of the tuberous root of *A. racemosus.* Roots have undecanyl, cetanoate sterols viz. sitosterol, 4, 6-dihydryxy2-O (-2-hydroxy isobutyl), and benzaldehyde (Chatterjee and Pakrashi, 2009).

3. **Vitamins and Minerals:** The plant of *A. racemosus* contains vitamins (viz. vitamin A, vitamin B1, vitamin B2, vitamin C, vitamin E, and folic acid) as well as minerals. As reported by Sahrawat et al., the different parts of *A. racemosus* like root, stem seed, flowers, and leaves contain the presence of different minerals like Mg, Ca, Co, Cu, Fe, K, and Zn (Sahrawat et al., 2014). Some minerals are reported in trace amount viz. Zn, Mn, Cu, Co whereas some in high concentration like Manganese (Mn). Some of them have significant use is potential herbal medicine like Mn for inflammation, infertility, and strain; Iron (Fe) for patients who are suffering from anemia; Co for a constituent of vitamin B12 which in turn maintains normal bone marrow for producing erythrocytes (Singh et al., 2018). Some other trace element like Zn is used for potential herbal medicine viz. mental disorder, growth, and hair loss (Negi et al., 2010).
4. **Miscellaneous Compounds:** In roots of *A. racemosus,* some other compounds like eugenol thymol and essential fatty acids; for example, gamma linolenic acids, vitamin A, quercetin 3-glucuronides, diosgenin, etc., are also present. N-containing compounds like borneol, myrtenol, perillaldehyde, etc. are present in high concentration. Isolation of a 5-membered sulfur-containing heterocyclic compound, which is also known as asparagusic acid (1, 2-dithiolane-4carboxylic acid), shows a distinctive feature to Asparagus by imparting an abnormal odor to urine following *asparagus* ingestion (Mitchell and Waring, 2014).

9.3 PHARMACOLOGICAL APPLICATION

Shatavari is known as the "Queen of herbs" in Ayurveda, which is the most important herb for health benefits in female (Sharma et al., *2013). Asparagus* is used in almost 67 ayurvedic preparations and commonly mentioned as 'Rasayana' in Ayurveda due to its medicinal uses (Choudhary and Sharma, 2014). The plant has numerous therapeutic application *viz.* antioxidant, diuretic, antidepressant, antiepileptic, antitussive, anti-HIV, immunostimulant, hepato-protective, cardioprotective, antibacterial, anti-ulcerative, neurodegenerative (Singla and Jaitak, 2014).

1. **Galactogogue Effect:** *A. racemosus* is very useful in case of threatened abortion or premature birth, galactagogue, and in addition, remedial activity as it is helpful in women's complaints. Ayurveda recommends the root extracts of *A. racemosus* to increase milk secretion after pregnancy (Gyawali and Kim, 2011). Ricalax tablet (*A. racemosus* mixed with other herbal products) enhances milk production in females who has problems with less milk secretion (Mishra et al., 2013). It helps in increasing the mass of mammary gland, lobuloalveolar tissue, which in turns increases the production of milk of prolactin and estrogen secretion. The galactagogue impact has likewise been studied in buffalo as depicted by Patel and Kanitkar (1969). It was seen that *A. racemosus* significantly increased plasma prolactin levels in buffaloes, in this manner, increasing milk production (Sharma and Sharma, 2017).
2. **Antisecretory and Antiulcer Activity:** Ulcer is a very scorching problem in developing and up to certain extents in developed countries. It is induced because of imbalance between the body's aggressive and protecting factors like gastric acid, pepsin, gastric mucosa, prostaglandin, and bicarbonate. Traditional systems of medicine have been using *A. racemosus* for the treatment of acid peptic disease and ulcers since many centuries due to following reasons (Singh et al., 2018; Sharma and Sharma, 2017).

    - It is an equally effective antiulcer-genic agent as ranitidine hydrochloride.
    - Neutralizes the gastric acid secretion. It has an inhibitory consequence on the discharge of gastric hydrochloric acid (HCl) and protection of gastric mucosal destruction by binding to the ulcer bed as coating agents.
    - *A. racemosus* significantly heals the peptic ulcers and increased mucosal protective aspects viz. cellular mucus, mucus secretion, and life period of cells.
    - Cytoprotective effect by preventing and healing gastric ulcers and erosions.

- It provides substantial defense against acute ulcers caused by CRS (cold restraint stress), duodenal ulcers induced by cysteamine and aspirin plus pyloric ligation.

Hence, the roots of the Shatavari plant in the form of powder can be administered to chronic ulcer patients, along with other patients (Mangal et al., 2004).

3. **Gastrointestinal Effects and Anti-Dyspepsia Effects:** In Ayurveda, the powdered form of the root (dried) of *A. racemosus* is used for the treatment of Dyspepsia. Dyspepsia is the condition characterized by the incapability to digest, commonly stated as impaired digestion which is often related with the gastritis. It can also be the first sign for Gastric ulcer and cancer. *A. racemosus* is also used for the cure of ulcerative disorder of the stomach, as mentioned in Ayurveda. It has been reported in *Asparagus racemosus* that dried powdered roots stimulate gastric, and its action seems comparable with synthetic dopamine antagonists' metoclopramide (Dalvi et al., 1990). This has a similar effect in reducing gastric emptying time during dyspepsia as we see for the new era allopathic drug metoclopramide which acts as a dopamine antagonist. It's a mild dopamine agonist. Patients suffering from duodenal ulcers also get restorative relief from the juice of its fresh root (Dalvi et al., 1990; Kanwar and Bhutani, 2010).
4. **Antitussive Effect:** *A. racemosus is* widely used in the treatment of cough as well as in minor upper respiratory tract infection, exhibiting the antitussive properties. The methanolic extract of *A. racemosus* roots is reported to possess important antitussive action on cough in mice which is induced by Sulfur dioxide (Shashi et al., 2013). The methanolic root extract administered at the concentration of 200, 400 mg/kg, and codeine phosphate (a drug obtained from opium) was taken as a standard antitussive reference drug. The 58.5% and 40% of cough inhabitation are comparable to that of 10–20 mg/kg of codeine phosphate where observed inhabitation is 55.4% and 36% respectively (Saxena and Chourasia, 2000). Hence this extract can be used against the opium-based drugs, since there are no side effects like nausea, sweating, tiredness, which can be observed, by use of codeine phosphate associated drugs.
5. **Antibacterial and Antiprotozoal Activity:** The root extracts of *A. racemosus have* been studied for antibacterial activity employing standard cylinder method. Methanolic root extracts showed significant in vitro antibacterial efficiency against various pathogens

such as *Salmonella typhimurium, Bacillus subtilis, Staphylococcus aureus, Escherichia coli, Shigella sonnei, Salmonella typhi, Pseudomonas pectida, Shigella dysenteriae, Shigella flexneri,* and *Vibriocholerae*. Methanol extract has a similar effect as chloramphenicol (Thakur et al., 2009). These findings were further supported by other studies where various extracts of root and leaf of the plant exhibited antibacterial activity (Aggarwal et al., 2013; Patel and Patel, 2013). *A. racemosus* alcoholic root extracts showed an inhibitory effect on the growth of *Entamoeba histolytica* was observed in vitro (Agrawal et al., 2008).

6. **Effect on Uterus:** Various extracts of *A. racemosus have* been found to inhibit the contraction of the uterus (Sharma et al., 2017; Suwannachat et al., 2012). The ethanolic extracts of *A. racemosus* roots exhibited significant uterine muscle relaxation, more so in pregnant rats than non-pregnant ones. Mechanisms involved are both calcium-dependent and -independent pathways (Suwannachat et al., 2012). A polyherbal formulation containing *A. racemosus*, similarly exhibited decreased spasmogen-induced contraction, and also increased uterine weight (Mitra et al., 2012). The effects of this plant on the uterus have been shown to mimic that of estrogen (Pandey et al., 2005; Bhatnagar and Sisodia, 2005).
7. **Antihepatotoxic Activity:** The alcoholic root extract of *A. racemosus* was found to significantly decline the increased levels of hepatic marker enzymes, namely alkaline phosphatase (ALP), aspartate aminotransferase (AST) and alanine aminotransferase (ALT) in CCl_4-induced hepatic damage in rats, demonstrating antihepatotoxic potential of *A. racemosus* (Muruganadanet al., 2000). The increased level of aspartate aminotransferase and ALT indicated increased permeability and hepatic cell damage. The ethanolic root extract of the plant was also found to produce hepatoprotective action in isoniazid-induced hepatotoxicity in rats (Palanisamy Manian, 2011). Similarly, the aqueous root extract of *A. racemosus* was found to attenuate the hepatotoxic effects caused because of lead in Swiss albino mice. In paracetamol-induced liver injury in rats, there is increased levels of SGOT, SGPT, serum bilirubin, and serum ALP, upon treatment with the ethanolic roots extract and reversal in their levels demonstrating the hepatoprotective activity (Sharma et al., 2012).
8. **Cardiovascular Effects:** *Asparagus racemosus* root powder supplements causes reduction of lipid peroxidation as well as dose-dependent in lipids profiles. It has been found in *A. racemosus* that its

alcoholic root extract exhibits positive ionotropic and chronotropic impact with a lower dose and cardiac arrest with the higher dose on frog's heart. It shows a cholinergic mechanism of action as it can generate hypertension in cats, which was initially blocked by atropine medication. In plasma and liver, the reduction of total lipids and cholesterol, as well as triglycerides, has been reported including a decrease in low-density lipoprotein (LDL) and very high-density lipoprotein (VLDL). In rabbits, on the administration of the extract of *A. racemosus,* the bleeding time was increased significantly and on clotting time no impact was observed (Dhingra and Kumar, 2007; Shankar et al., 2012).

9. **Immunomodulatory Activity:** The dried powdered root of *Asparagus racemosus* act as 'Immunomodulatory agents' that helps in increasing or modulating the action of immune systems and thereby reducing the inflammatory response. It stimulates the immune system to fight against infections, cancer, and immunodeficiency disease like AIDS. It boosts the antibody production against different vaccinations containing more effective cell-mediated immune response and thereby protecting against different viral, bacterial, and many other diseases. It has been studied that the effect of root extract of *Asparagus racemosus* was very helpful for providing better protection against infections by amplifying the humoral and cell-mediated immune system (Singh and Sinha, 2014b).

   Immunodeficiency disorders are the group of disorders in which the body's defense system is compromised, making it to be less effective against foreign invaders. As a result, the person with an immunodeficiency disorder will have frequent infections that are generally more severe and remain longer than usual. Immunostimulant produces leukocytosis and major neutrophilia along with increased phagocytic activity in traditional therapy, the immunoadjuvant nature of *Asparagus racemosus* is also well recognized so it can be also used to avoid side effects of synthetic chemotherapeutic drugs (Sharma et al., 2009; Thakur et al., 2012).

10. **Antioxidant Effects:** Antioxidant is a substance that may prevent or delay some type of cell damage by inhibiting oxidation of other molecules. The methanolic extract of the root possesses significant anti-oxidant properties when directed through the oral method by Aarati K., (2015). With the significant reduction in lipid peroxidation the level of enzymes viz. superoxidase, catalase, dismutase, and ascorbic acid increases. The antioxidant properties were mainly

exhibited due to the presence of Isoflavones (Wiboonpun et al., 2004). Crude and aqueous extract of *A. racemosus* has been proven for its antioxidant effect (Kamat et al., 2000). The activity was verified in rat liver cell mitochondrial membrane damage induced by generated free radicals (Takeungwongtrakul et al., 2012). The antioxidant property was studied on the basis of scavenging activity (SA) of the stable DPPH (1, 1diphenyl-2-picrylhydrazyl) free radical. The antioxidant property detected was due to their redox property of the phenolic compounds present in the ethanolic root extract (Karmakar et al., 2012b).

11. **Antidepressant Activity:** In mice, two different experiments TST (tail suspension test) and FST (forced swim test) were done to evaluate antidepressant activity. The methanolic seed extract of *Asparagus racemosus* decreases immobility periods significantly in TST, FST, which was analogous to imipramine, and increased the levels of dopamine in vitro, which indicated significant antidepressant activity (Sravani and Krishna, 2013; Dhingra and Kumar, 2007). Moreover, it was found to significantly reduce brain monoamine oxidase (MAO-A and MAO-B) levels as compared to control. It has been found that the methanolic extract possesses antidepressant activity probably by inhibiting MAO-A and MAO-B; and through interaction with adrenergic, dopaminergic, serotonergic, and GABAergic systems (Gamma-aminobutyric acid) (Dhingra and Kumar, 2007).
12. **Anti-Inflammatory Effects:** Ethanolic extract of *A. racemosus* showed the significant decreases in weight of tissue and thickness of skin and helped in inflammatory cytokine production. It has been studied that the ethanolic root extracts of *A. racemosus* showed significantly anti-inflammatory action in carrageenan-induced rat paw edema. It also exhibited antiarthritic activity in rats by significantly decreasing the paw volume as well as arthritic score in Freund's adjuvant-induced arthritis. In another study, liposomes prepared from *A. racemosus* root extracts displayed significant anti-inflammatory activity in vitro, which was comparable to the action of dexamethasone (Plangsombat et al., 2016). Ethanolic leaf extract of the plant also showed significant anti-inflammatory activity and inhibitory effect on prostaglandin release (Lee et al., 2009; Battu and Kumar, 2010)
13. **Diuretic Activity:** The diuretic property has been confirmed by a suitable experimental model was highlighted in Ayurveda. In the study for its diuretic activity, an aqueous roots extract of *A. racemosus*

was taken using three dose vials 800 mg/kg, 1600 mg/kg and 3200 mg/kg in comparison with furosemide (a standard drug) and normal saline as control rats after performing acute toxicity tests (show no fatalness even with the highest dose). A significant diuretic activity was seen with at the dose of 3200 mg/kg, and it was comparable to the standard drug, furosemide (Kumar et al., 2010).

14. **Anti-Diarrheal Activities:** From a long time Diarrhea has been a major cause of infant deaths and childhood mortality in developing countries. An estimated death rate of 2.2 million people globally, predominantly in the developing countries. The percentage of death caused by this disease was almost 75% of all deaths in infant and children below the age of 5 years. The scenario has improved after the introduction of oral dehydration therapy. But even today the chronic diarrhea is a serious problem in regions where there is malnutrition. It still causes child death in those regions. The fatal cases were found to be in children of less than 5 years of age (Bopana and Saxena, 2007). It has been described in Ayurveda texts that Shatavari plant is very helpful in diarrhea (Atisar), dysentery (Pravahika) and gastritis (Pittaj shool) treatments (Nanal et al., 1974). The antidiarrheal activity showed in castor oil induced diarrhea in rats by using the ethanolic and aqueous root extract of *A. racemosus.*

    Castor oil releases ricinoleic acid that causes irritation and inflammation of intestinal mucosa results in the secretion of 'prostaglandin E,' which causes diarrhea in experimental individuals. Thus, this extracts prevent the biosynthesis of prostaglandin, in turn, inhibits diarrheal effect, i.e., gastrointestinal motility and secretion (Venkatesan et al., 2005).

15. **Anti-Stress Activity :** Anti-stress activity showed by the methanolic and aqueous extract of roots of *A. racemosus.* It shows an inhibitory effect on proinflammatory cytokines viz. interleukin-1β, nitric acid secretion, and tumor necrosis factor (α) (Shankar et al., 2010).

9.4 CYTOTOXICITY, ANALGESIC, AND POTENTIAL TO PREVENT HEPATOCARCINOGENESIS

For biological action, the ethanolic extracts of *A. racemosus* were examined and was found to be very helpful in the cytotoxicity, analgesic, and antidiarrheal activity (Karmakar et al., 2012a). The isolated shatavarin IV, along with AR-2B comprising 5.05% shatavarin IV exhibited potent cytotoxicity as described by Shankar et al. (2012). In a study for analgesic activity on albino mice, exhibited a significant action in Eddy's plate as well as heat conduction models (Rahiman et al., 2011). In another study for analgesic activity, using mice induced with acetic acid, the ethanolic extract inhibited writhing reflex 67.47% significantally to a dose of 500 mg/kg body weight (Karmakar et al., 2012a). In albino mice, in Eddy's plate as well as heat conduction models showed significant analgesic action by using alcoholic and aqueous extracts of roots of *A. racemosus* (Rahiman et al., 2011). By using writhing model of mice induced by acetic acid, significant inhibition of writhing reflex 67.47% to a dose of 500 mg/kg body weight produced by crude ethanolic extract of *A. racemosus* (Karmakar et al., 2012b). Natural products have long been used for treatment against cancer. There are at least 10000 species of plants, known to have anti-cancerous properties. Hepatocarcinogenesis occurs in hepatic tissues can be prevented by using aqueous root extract of *A. racemosus* in rats pretreatment. The aqueous extract of *A. racemosus* with the help of diethylnitrosamine (DEN) exhibited the amelioration of oxidative stress and hepatotoxicity and hence prevents hepatocarcinogenesis (Agrawal et al., 2008).

9.5 FOOD APPLICATION

A popular herb, *Asparagus racemosus* (Shatavari) belongs to the family Asparagaceae (formerly *Liliaceae*) which is used as the conventional food items and plays a significant role in making them more nutritive rich products. It is used in functional food viz. food and beverages that provide nutritional health benefits, which is non-inherent. *Asparagus*, a popular Indian herb, belongs to the family Asparagaceae (formerly *Liliaceae*) is an herbaceous perennial plant.

1. **Milk and Milk Products:** Milk is one of the most widely liquid foods in the world, which provides the necessary nutrients for growth and maintenance of the body. The use of satavari extract in milk, milk beverages, and other milk product is a very recent development. For many popular beverages, milk act as a key ingredient. It has been studied in different studies that by using aqueous extract (freeze-dried) of *A. racemosus* helps in milk fortifying has an immunomodulatory and antioxidative ability. This was investigated on mice. On adding aqueous extract (freeze-dried) of *A. racemosus* at 1% concentrate (1 g/100 ml of milk) showed a significant increase in phagocytic

activity of macrophages, increase in proliferative responses of lymphocytes, decrease in pH and decreased glutathione content and lipid peroxidation. These results showed that the enhancement in the immune system, as well as antioxidative property, could be related to the phytochemicals (glycosides, flavonoids, saponins, antioxidants, polyphenols, vitamins, etc.) present in satavari extract. Shatavari 'ghrita' is given to lactating ladies to enhance lactation (Veena et al., 2005; Singh et al., 2018).

2. **Biscuits:** In current years, the most commonly used bakery items are biscuits. Nowadays, due to the fast life quick everyday schedule, people want ready to eat food items. In such cases, biscuits are more preferable as it also contains nutritional and medicinal supplements. The nutritive and medicinal value increased by the use of ayurvedic powder of satavari, ashwagndha, and yastimadhu in plain refined flour (Mehta, 2013). With the increasing storage time, the moisture content was found increasing, but shatavari powder biscuits were found to be the maximum water-soluble minerals. Cookies with different value were made by changing the percentage of refined flour with different proportions with shatavari, ashwagandha, and ginger root powder. On the basis of overall acceptability, the best suitable formulations for cookies were found to be a 90% refined flour, 5.0% (Shatavari), 2.5% (ashwagandha) and 2.5% (ginger) root powder (Kumari and Gupta, 2016; Singh et al., 2018).

3. **Bread:** In present days, the constant intake of bakery items viz. biscuit and bread is almost 80% (Indian Biscuits Manufacturers Association, 2017). In the daily diet, a majority of humans consume bakery products like bread as it is convenient to use and functionally it could be a great choice. The incorporation of Shatavari (medicinal plant) in breads can provide more nutritional values like carbohydrate, fat, protein, and fiber and hence adds value to the product, thus beneficial for malnutrition patient. A CCRD (Central composite rotatable design) experiment was used, in which a different proportion of yeast and *A. racemosus* powder was taken. It was found that for making bread, the most suitable formulation is 3.5% for shatavari and 4.96% for yeast. Bread enriched with the powdered root of satavari has a significant amount of Steroidal saponins and polycyclic alkaloid. In shatavari-enriched bread, shatavari root powder increases texture by its hardness, adhesiveness, chewiness, cohesiveness as it contains steroidal saponins

and polycyclic alkaloid (Singh and Sinha, 2014a; Indian Biscuits Manufacturers Association, 2017).

9.6 FOOD SUPPLEMENTARY FOR DAIRY ANIMALS

A. racemosus is purely an effective herbal supplement for improving milk production and works naturally to give strength to the reproduction capacity of dairy animals. Ayurveda mainly prefers the powdered form of herbs because after in taking the herb starts the digestive system and sends signals to the body to initiate its own supportive mechanisms. Shatavari can be used in the feeding of the dairy animals, helps in the boosting of the immune system, which results in preventing the infection of the reproductive and udder organs of cows. It is very helpful in improving their productivity and effectively reduces the dairy animal's stress so, therefore, producing clean and healthy milk. Shatavari contains a very high quality of nutrients and thus considered to be a very good supplement of feed for animals. When we take Shatavari powder in our diet it enhances the milk production, colostrum's, total immunoglobulin and decreases the total cholesterol (TC) in milk It has been warned by the researchers that the use of oxytocin for enhancing the production of dairy cows and buffaloes causes health hazards (Kumar et al., 2008; Divya et al., 2015).

9.7 ECONOMIC IMPORTANCE

Among all, the most economically important *Asparagus* species is garden *Asparagus* (commonly known as *Asparagus officinalis* L.), which is a highly prized vegetable. Garden asparagus has an edible shoot, which are tender and unexpanded, known as spears (Rubatzky and Yamaguchi, 1997; Stajner, 2002). This is ranked very high in spring vegetables, folk name sparrow grass (Stajner, 2002; Rubatzky, and Yamaguchi, 1997).

9.8 CONCLUSIONS AND FUTURE PROSPECTS

The therapeutic efficacy of *A. racemosus* extensively used in Indian System of Medicine has been established through modern testing and evaluation in different disease conditions. The phytochemicals and minerals of these plants will enable to exploit its therapeutic use. The dynamic development and improvement of the pharmacological, phytochemical, and technological work demonstrate that *Asparagus* sp. (specially *Asparagus racemosus*) has received a great popularity because of its wide spectrum of significant therapeutic properties including antioxidant, antibacterial, antidiarrheal, strengthening, and cytotoxic, etc. Significant work has been completed to explore the biological and therapeutic uses of the plant; still, there are numerous possibilities of pharmacological applications, which need to be explored. The major studies were reported using extracts of the plant; still, the active principle involved behind these activities needs to be explored. The conservation of this plant is very essential, because of its several uses the demand of *Asparagus racemosus* is increasing gradually through the supply is relatively insufficient. Consistency can be achieved by using the biotechnological approaches like micro-propagation and callus culture.

KEYWORDS

- **food applications**
- ***Asparagus racemosus***
- **phytochemical application**
- **phytochemistry**
- **Shatavari**

REFERENCES

Aggarwal, H., Gyanprakash, R. A., & Chhokar, V., (2013). Evaluation of root extracts of *Asparagus racemosus* for antibacterial activity. *Am. J. Drug Discov. Dev.* doi: 10.3923/ajdd.

Agrawal, A., Sharma, M., Rai, S. K., Singh, B., Tiwari, M., & Chandra, R., (2008). The effect of the aqueous extract of the roots of *Asparagus racemosus* on hepatocarcinogenesis initiated by diethylnitrosamine. *Phytother. Res.*, *22*(9), 1175–1182.

Ali, S. I., & Khan, S. W., (2009). Asparagaceae. In: Ali, S. I., & Qaiser, M., (eds.), *Flora of Pakistan* (Vol. 217, pp. 1–24). Inst. Plant Conser., Univ. Karachi, Karachi and Missouri Bot. Press, Missouri Bot. Garden, St. Louis, Missouri, USA.

Battu, G. R., & Kumar, B. M., (2010). Anti-inflammatory activities of leaf extract of *Asparagus racemosus* Willd. *Int. J. Chem. Sci.*, *8*(2), 1329–1338.

Bhatnagar, M., & Sisodia, S. S., (2005). Antiulcer and antioxidant activity of *Asparagus racemosus* Willd and Withania somnifera Dunal in rats. *Ann. N.Y. Acad. Sci.*, *1056*, 261–278.

Bopana, N., & Saxena, S., (2007). *Asparagus racemosus*: Ethnopharmacological evaluation and conservation needs. *Journal of Ethnopharmacology*, *110*, 1–15.

Chatterjee, A., & Pakrashi, S. C., (2009). *The Treatise on Indian Medicinal Plants* (Vol. 6). Reprint 2001, Publication and Information Directorate C. S. I. R., New Delhi.

Chawla, A., Chawla, P., & Mangalesh, R. R. C., (2011). *Asparagus racemosus* (Wild): Biological activities & its active principles. *Indo-Global Journal of Pharmaceutical Sciences, 1*(2), 113–120.

Chen, X., & Tamanian, K. G., (2000). Asparagus. In: Wu, Z., & Raven, P. H., (eds.), *"Flora of China"* (Vol. 24, pp. 139–146). (Flagellariaceae through Marantaceae). Science Press, Beijing, and Missouri Botanical Garden Press, St. Louis.

Choudhary, D., & Sharma, D., (2014). A phytopharmacological review on *Asparagus racemosus*. *International Journal of Science and Research (IJSR), 3*(7), 2319, 7064.

Clifford, H. T., & Conran, J. G., (1987). Asparagaceae. In: George, A. S., (ed.), *"Flora of Australia"* (pp. 159–164). Australian Government Publishing Service, Canberra.

Dahlgren, R. M. T., Clifford, H. T., & Yeo, P. F., (1985). Asparagaceae. In: Dahlgren, R. M. T., Clifford, H. T., & Yeo, P. F., (eds.), *"The Families of the Monocotyledons"* (pp. 140–142). Springer-Verlag Berlin, Heidelberg.

Dalvi, S. S., Nadkarni, P M., & Gupta, K C., (1990). Effect of *Asparagus racemosus* (Shatavari) on gastric emptying time in normal healthy volunteer. *Journal of Postgraduate Medicine, 36*(2), 91–94.

Dhingra, D., & Kumar, V., (2007). Pharmacological evaluation for antidepressant-like activity of *Asparagus racemosus* wild: In Mice. *Pharmaco., 3*, 133–152.

Divya, K. K., Choudhary, P. L., Khare, A., Choudhary, K. K., Saxena, R. R., Shukla, N., Bharati, K. A., & Kumar, A., (2015). Effect of feeds supplemented with *Asparagus racemosus* on milk production of indigenous cows. *Mintage J. Pharma. Med. Sci., 1*, 1–6.

Farr, D. F., Bills, G. F., Chamuris, G. P., & Rossman, A. Y., (1989). *Fungi on Plants and Plant Products in the United States* (p. 1252). APS Press–The American Phytopathological Society, St. Paul, Minnesota.

Gyawali, R., & Kim, K. S., (2011). Bioactive volatile compounds of three medicinal plants from Nepal. *Kathmandu Uni. J. Sci. Eng. Technol., 8*, 51–62.

Hayes, P. Y., Jahidin, A. H., Lehmann, R., Penman, K., Kitching, W., & De Voss, J. J., (2006). Structural revision of shatavarins I and IV, the major components from the roots of *Asparagus racemosus. Tetrahedron Lett., 47*, 6965–6969.

Hayes, P. Y., Jahidin, A. H., Lehmann, R., Penman, K., Kitching, W., & De Voss, J. J., (2008). Steroidal saponins from the roots of *Asparagus racemosus. Phytochemistry, 69*, 796–804.

Indian Biscuits Manufacturers Association. Biscuit industry in India. Available at: www.ibmabiscuit.in/industry-statistics.htm (Accessed on 12 June 2019).

Joy, P. P., Thomas, J., Mathew, S., & Skaria, B. P., (1998). *Medicinal Plants.* Kerala Agricultural University, Aromatic and Medicinal Plants Research Station Odakkali, Kerala, India.

Kamat, J. P., Boloor, K. K., Devasagayam, T., & Venkatachalam, S., (2000). Antioxidant properties of *Asparagus racemosus* against damage induced by γ-radiation in rat liver mitochondria. *J. Ethnopharmacol., 71*, 425–435.

Kanwar, A. S., & Bhutani, K. K., (2010). Effects of *Chlorophytum arundinaceum, Asparagus adscendens* and *Asparagus racemosus* on proinflammatory cytokine and corticosterone levels produced by stress. *Phytother. Res., 24*, 1562–1566.

Karmakar, U. K., Sadhu, S. K., Biswas, S. K., Chowdhury, A., Shill, M. C., & Das, J., (2012a). Cytotoxicity, analgesic and antidiarrheal activities of *Asparagus racemosus. J. Appl. Sci., 12*, 581–586.

Karmakar, U., Biswas, S., Chowdhury, A., Raihan, S., Akbar, M., Muhit, M., & Mowla, R., (2012b). Phytochemical investigation and evaluation of antibacterial and antioxidant potentials of *Asparagus racemosus*. *Int. J. Pharmacol.*, *8*, 53–57.

Kumar, M. C. S., Udupa, A., Sammodavardhana, K., Rathnakar, U., Shvetha, U., & Kodancha, G., (2010). Acute toxicity and diuretic studies of the roots of *Asparagus racemosus* wild in rats. *West Indian Med. J.*, *59*, 3–5.

Kumar, S., Mehla, R. K., & Singh, M., (2014). Effect of Shatavari (*Asparagus racemosus*) on milk production and Immune-modulation in Karan Fries crossbred cows. *Ind. J. Tradit. Know.*, *13*, 404–408.

Kumari, A., & Mukhopadhyay, S., (2015). Karyotype analysis of three different species of *Calathea*. *J. Botan. Soc. Bengal*, *69*(1), 47–51, ISSN: 0971–2976.

Kumari, A., Lahiri, K., Mukhopadhyay, M. J., & Mukhopadhyay, S., (2014). Genome analysis of species of *Calathea* utilizing chromosomal and nuclear DNA parameters. *Nucleus*, *57*, 203–220.

Kumari, S., & Gupta, A., (2016). Utilization of Ashwagandha, ginger and Shatavari root powder for the preparation of value added "cookies." *Int. J. Home Sci.*, *2*, 27–28.

Lee, D. Y., Choo, B. K., Yoon, T., Cheon, M. S., Lee, H. W., Lee, A. Y., & Kim, H. K., (2009). Anti-inflammatory effects of *Asparagus cochinchinensis* extract in acute and chronic cutaneous inflammation. *J. Ethnopharmacol.*, *121*, 28–34.

Mangal, A., Panda, D., & Sharma, M. C., (2004). Peptic ulcer healing properties of Shatavari (*Asparagus racemosus* Willd). *International Journal of Traditional Knowledge*, *5*(2), 227–228.

Mayo, J. L., (1998). Black cohosh and chastberry: Herbs valued by woman for centuries. *Clinical Nutrition Insight*, *6*, 1–4.

Mehta, M., (2013). Development of low cost nutritive biscuits with ayurvedic formulation. *Int. J. Ayurv. Herb. Med.*, *3*, 1183–1190.

Mishra, A., Niranjan, A., Tiwari, S. K., Prakash, D., & Pushpangadan, S., (2005). Nutraceutical composition of *Asparagus racemosus* (Shatavari) grown on partially reclaimed sodic soil. *J. Med. Aroma. Plant Sci.*, *27*, 240–248.

Mishra, P. K., Singh, P., Prakash, B., Kedia, A., Dubey, N. K., & Chanotiya, C. S., (2013). Assessing essential oil components as plant-based preservatives against fungi that deteriorate herbal raw materials. *Int. Biodeterior. Biodegrad.*, *80*, 16–21.

Mitchell, S. C., & Waring, R. H., (2014). Asparagusic acid. *Phytochemistry*, *97*, 5–10.

Mitra, S. K., Prakash, N. S., & Sundaram, R., (2012). Shatavarins (containing Shatavarin IV) with anticancer activity from the roots of *Asparagus racemosus*. *Indian Journal of Pharmacology*, *44*(6), 732–736.

Muruganadan, S., Garg, H., Lal, J., Chamdra, S., & Kumar, D. I., (2000). Studies on the immunostumulant and antihepatotoxic activities of *Asparagus racemosus* root extract. *Med. Arom. Plant Sci.*, *22*, 49–52.

Nanal, B. P., Sharma, B. N., Ranade S. S., & Nande, C. V., (1974). Clinical study of Shatavari (*Asparagus racemosus*). *Journal of Research in Indian Medicine*, *9*, 23–29.

Negi, J. S., Singh, P., Joshi, G. P., Rawat, M. S., & Pandey, H. K., (2010). Variation of trace elements contents in *Asparagus racemosus* (Willd). *Biol. Trace. Elem. Res.*, *135*, 275–282.

Palanisamy, N., & Manian, S., (2011). Protective effects of *Asparagus racemosus* on oxidative damage in isoniazid-induced hepatotoxic rats: An *in vivo* study. *Toxicol. Ind. Health*, *28*(3), 238–244.

Pandey, S. K., Sahay, A., Pandey, R. S., & Tripathi, Y. B., (2005). Effect of *Asparagus racemosus* rhizome (Shatavari) on mammary gland and genital organs of pregnant rat. *Phytother. Res., 19*(8), 721–424.

Patel, A. B., & Kanitkar, U. K., (1969). *Asparagus racemosus* wild form bordi, as a galactogogue in buffaloes. *Indian Veterinary Journal, 46*, 718–721.

Patel, L. S., & Patel, R. S., (2013). Antimicrobial activity of *Asparagus racemosus* wild from leaf extracts–a medicinal plant. *International Journal of Scientific and Research Publications, 3*(3).

Plangsombat, N., Rungsardthong, K., Kongkaneramit, L., Waranuch, N., & Sarisuta, N., (2016). Antiinflammatory activity of liposomes of *Asparagus racemosus* root extracts prepared by various methods. *Exp. Ther. Med., 12*, 2790–2796.

Prohens, J., Nuez, F., & Carena, M. J., (2008). *Handbook of Plant Breeding* (364). Springer.

Rahiman, F. O. M., Srilatha, B., Kumar, R. M., Niyas, M. K., Kumar, S. B., & Phaneendra, P., (2011). Analgesic activity of aqueous and alcohol root extracts of *Asparagus racemosus* wild. *Pharmacologyonline, 2*, 558–562.

Rubatzky, V. E., & Yamaguchi, M., (1997). World vegetables (principles, production and nutritive values). In: *International Thomson* (2nd edn., p. 843). New York, USA.

Sabnis, P. B., Gaitonde, B. B., & Jetmalani, M., (1968). Effect of alcoholic extracts of *Asparagus racemosus* on mammary glands of rats. *Ind. J. Exp. Biol., 6*, 55–57.

Sahrawat, A. K., Sharma, P. K., & Sahrawat, A., (2014). *Asparagus racemosus*- wonder plant. *Int. J. Adv. Res., 2*, 1039–1045.

Saxena, V. K., & Chourasia, S., (2000). 5-Hydroxy-3,6,4'-trimethoxyflavone7-O-β-D-glucopyranosyl[1→4]-O-α-D-xylopyranoside from leaves of *Asparagus racemosus. J. Inst. Chem. (India), 6*, 211–213.

Shaha, P., & Bellankimath, A., (2017). Pharmacological profile of *Asparagus racemosus*: A review. *Int. J. Curr. Microbiol. App. Sci., 6*(11), 1215–1223. ISSN: 2319–7706.

Shao, Y., Poobrasert, O., Kennelly, E. J., Chin, C. K., Ho, C. T., & Huang, M. T., (1997). Steroidal saponins from *Asparagus officinalis* and their cytotoxic activity. *Planta Med., 63*, 258–262.

Sharma, A., & Sharma, D. N., (2017). A comprehensive review of the pharmacological actions of *Asparagus racemosus. Am. J. Pharm. Tech. Res., 7*(1), ISSN: 2249–3387.

Sharma, A., & Sharma, V., (2013). A brief review of medicinal properties of *Asparagus racemosus* (Shatawari). *Int. J. Pure Appl. Bio. Sci., 1*, 48–52.

Sharma, P. C., Yelne, M. B., & Dennis, T. J., (2000). *Data Base on Medicinal Plants Used in Ayurveda* (Vol. 1, pp. 418–430). Delhi: Documentation & publication division: Central Council for Research in Ayurveda and Siddha.

Sharma, R. K., & Dash, B., (2003). *Charaka Samhita-Text with English Translation and Critical Exposition Based on Chakrapani Datta's Ayurveda Dipika*. India: Chowkhamba Varanasi.

Sharma, U., Kumar, N., Singh, B., Munshi, R. K., & Bhalerao, S., (2009). Immunomodulatory active steroidal saponins from *Asparagus racemosus. Med. Chem. Res., 121*, 1–7.

Sharma, V., Verma, R. B., & Sharma, S., (2012). Preliminary evaluation of the hepatic protection by pharmacological properties of the aqueous extract of *Asparagus racemosus* in lead loaded Swiss albino mice. *Int. J. Pharm. Pharm. Sci., 4*(1), 55–62.

Shashi, A., Jain, S. K., Verma, A., Kumar, M., Mahor, A., & Sabharwal, M., (2013). Plant profile, phytochemistry and pharmacology of *Asparagus racemosus* (Shatavari): A review. *Asian Pacific Journal of Tropical Disease, 3*, 242–251.

Shasnay, A. K., Darokar, M. P., Sakia, D., Rajkumar, S., Sundaresan, V., & Khanuja, S. P. S., (2003). Genetic diversity and species relationship in *Asparagus* spp. using RAPD analysis. *J. Med. Aromat. Plant Sci.*, *25*, 698–704.

Singh, A. K., Srivastava, A., Kumar, A., & Singh, K., (2018). Phytochemicals, medicinal and food applications of Shatavari (*Asparagus racemosus*): An updated review. *The Natural Products Journal*, *8*, 32–44.

Singh, A., & Sinha, B., (2014a). *Asparagus racemosus* and its phytoconstituents, an updated review. *Asi. J. Biochem. Pharma. Res.*, *4*, 230–240.

Singh, A., & Sinha, B., (2014b). Pharmacological significance of Satavari: Queen of herbs. *International Journal of Phytomedicine*, *6*, 477–488.

Singh, N., Jha, A., Chaudhary, A., & Upadhyay, A., (2014). Enhancement of the functionality of bread by incorporation of Shatavari (*Asparagus racemosus*). *J. Food Sci. Technol.*, *51*, 20–38.

Singla, R., & Jaitak, V., (2014). Shatavari (*Asparagus racemosus* wild): A review on its cultivation, morphology, phytochemistry and pharmacological importance. *IJPSR*, *5*(3), 742–757.

Sravani, K., & Krishna, K. S., (2013). Antidepressant and antioxidant activity of methanolic extract of *Asparagus racemosus* seeds. *Asian J. Pharm. Clin. Res.*, *6*(5), 102–107.

Štajner N. (2012) Micropropagation of Asparagus by In Vitro Shoot Culture. In: Lambardi M., Ozudogru E., Jain S. (eds) Protocols for Micropropagation of Selected Economically-Important Horticultural Plants. *Methods in Molecular Biology (Methods and Protocols)*, *994*, 341–351. Humana Press, Totowa, NJ.

Suwannachat, A., Kupittayanant, P., & Kupittayanant, S., (2012). Shatavari (*Asparagus racemosus*) extract can relax the myometrium. *Proceedings of the Physiological Society Physiology (Edinburgh) Proc. Physiol. Soc.*, *27*, PC365.

Takeungwongtrakul, S., Benjakul, S., & H-Kittikun, A., (2012). Lipids from cephalothorax and hepatopancreas of Pacific white shrimp (*Litopenaeus vannamei*): Compositions and deterioration as affected by iced storage. *Food Chem.*, *134*(4), 2066–2074.

Thakur, M., Chauhan, N. S., Bhargava, S., & Dixit, V. K., (2009). A comparative study on aphrodisiac activity of some ayurvedic herbs in male albino rats. *Arch. Sex. Behav.*, *38*, 1009–1015.

Thakur, M., Connellan, P., Deseo, M. A., Morris, C., Praznik, W., Loeppert, R., & Dixit, V. K., (2012). Characterization and *in vitro* immunomodulatory screening of fructo-oligosaccharides of *Asparagus racemosus* Willd. *Int. J. Biol. Macromol.*, *50*, 77–81.

Veena, N., Arora, S., Kapila, S., Singh, R. R. B., Katara, A., Pandey, M. M., Rastogi, S., & Rawat, A. K. S., (2014). Immunomodulatory and antioxidative potential of milk fortified with *Asparagus racemosus* (Shatavari). *J. Med. Plant Stud.*, *2*(6), 13–19.

Veena, N., Arora, S., Singh, R. R. B., Katara, A., Rastogi, S., & Rawat, A. K. S., (2015). Effect of *Asparagus racemosus* (Shatavari) extract on physicochemical and functional properties of milk and it's interaction with milk proteins. *J. Food Sci. Technol.*, *52*, 1176–1181.

Velvan, S., Nagulendran, K. R., Mahesh, R., & Hazeena, B. V. V., (2007). The chemistry, pharmacology and therapeutical applications of *Asparagus racemosus*. *A Rev. Pharm. Rev.*, *1*(2), 350–360.

Venkatesa, N. N., Thiyagarajan, V., Narayanan, S., Arul, A., Raja, S., Kumar, S. G. V., Rajarajan, T., & Perianayagam, J. B., (2005). Anti-diarrheal potential of *Asparagus racemosus* wild root extracts in laboratory animals. *Journal of Pharmacology and Pharmaceutical Sciences*, *8*, 39–45.

Wiboonpun, N., (2004). Identification of antioxidant compound from *Asparagus racemosus*. *Phytotherapy Research*, *18*, 771–773.

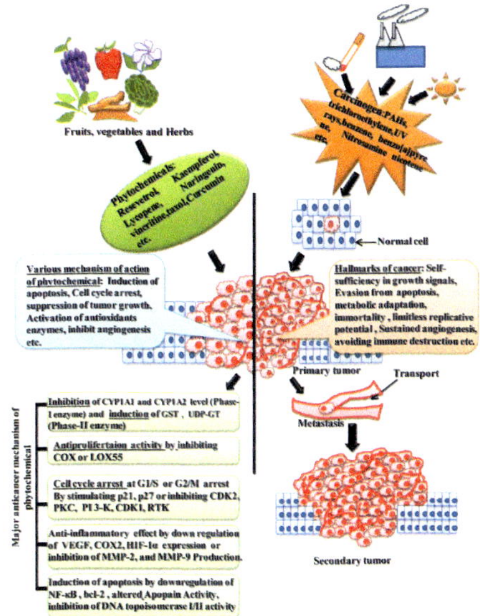

FIGURE 3.1 Various environmental cues generating carcinogenic compound leads to the cancer incidence; however, in assistance with various phytochemicals exhibiting anti-cancer activity through several mechanisms among which some are shown.

FIGURE 5.1 (a) Tall plant of phalsa; (b) Tall variety phalsa fruit.

FIGURE 5.2 (a) Dwarf plant; (b) Dwarf variety phalsa fruit.

FIGURE 6.1 Images of the inflorescence of amaranth plants.

FIGURE 6.2 Images of flowers of buckwheat plants.

FIGURE 6.3 Images of grains of pseudocereals.

FIGURE 8.1 Fresh Roselle.

FIGURE 8.2 Roselle calyces.

FIGURE 8.3 Dried Roselle calyces.

FIGURE 9.1 Photographs of *Asparagus racemosus* showing (A) shoot, leaves, and berries, (B) tuberous roots, (C) pine like needle shaped leaves, and (D) a bundle of cultivated garden asparagus (spears).

CHAPTER 10

Functional and Technological Properties of Probiotics and Prebiotics

BHANJA AMRITA and R. JAYABALAN

Food Microbiology and Bioprocess Laboratory, Department of Life Science, National Institute of Technology, Rourkela, Odisha–769008, India, E-mail: jayabalanr@nitrkl.ac.in

ABSTRACT

The demand for probiotics is increasing very rapidly all over the world. Many researchers are showing keen interest towards probiotics due to their health benefits, as now probiotics have been found for maintaining a healthy gut and a healthy digestive system due to their different physiological functions. Many investigations have been done showing the role of probiotics in controlling obesity, cardiovascular diseases, cholesterol-lowering activity, etc. Many food industries have come forward for producing a combined product containing probiotics and prebiotics for the consumers. Variety of functional foods, probiotics tablets, capsules, and different health supplements are being produced for satisfying consumers need. The probiotic market is estimated to grow at 7.0% CAGR by 2022 due to their increasing benefits and awareness among consumers. This chapter basically deals with the functional and technological properties of probiotics and prebiotics.

10.1 INTRODUCTION

The term probiotics has been explained in many ways, but the most accepted definition of probiotics is given by Food and Agriculture Organization of the United Nations and the World Health Organization FAO/WHO (2006), i.e., these are live microorganisms which when administered in adequate amounts confer a health benefit on the host (Sánchez et al., 2012). AFRC

(1989) proposed the first definition of probiotics in which the term was used for food products containing live microorganisms that benefit consumer's health by maintaining the balance of the microbial population of the gastro-intestinal tract (GIT).

Probiotics carry out their activities by interacting with the local gastro-intestinal environment. Studies in the early 1900s showed the presence of lactic acid bacteria (LAB) in the intestinal ecosystem (Sip and Grajek, 2010). Probiotics are known to colonize the host's health temporarily. Therefore it is required to ingest probiotics on a daily basis to improve the host's health. They are nontoxigenic and nonpathogenic. Every probiotic strain must be screened and considered as GRAS (generally recognized as safe) before its use.

The use of microorganisms for improving human health has been continuously growing over the last two decades. Not only bacteria, but yeasts are also included in the category of probiotics. Most of the *Lactobacillus* and *Bifidobacterium* strains along with the yeast *Saccharomyces boulardii* are well accepted for their probiotic effects (Holzapfel and Schillinger, 2002; Shah, 2007).

The characteristics of a probiotic strain that help them to contribute towards the improvement of human health are:

i. Human origin;
ii. Adherence to human intestinal cells;
iii. Cell surface hydrophobicity;
iv. Acid and bile resistant;
v. Viable in the human intestinal tract;
vi. Resistant towards carcinogenic and pathogenic bacteria;
vii. Production of antimicrobial substances;
viii. Should be efficiently produced in large scale.

Probiotics are delicate and need to survive the intestinal condition to show their effects, due to which the term prebiotics came into existence. Prebiotics can be defined as indigestible dietary plant fibers that work as food for probiotics. It selectively nourishes the good bacteria already present in the colon. Probiotics along with the prebiotics formulate the synbiotic products. Several synbiotic products are available in the market claiming as a functional food.

The normal microflora especially of the gut plays a vital role in the field of probiotics. Most of the existing probiotics that are being used today for formulating synbiotic products are isolated from the human gut microbiota

(Round and Mazmanian, 2009). An important aspect to increase our knowledge about the functioning of gut microflora is the selection of probiotics. It has also been found that microbes isolated from niches other than the gut are found to have improved functional properties since the human body has a high level of microbial diversity (Sánchez et al., 2012).

For a selection of probiotics, it is very much essential for the producers to focus on different functional and technological properties that are required for the successful production of different probiotic products. This chapter completely describes the health-promoting effects of probiotics, their stability, viability, functional characterization along with its large-scale production, the activity of prebiotics as a growth promoter of probiotic and ultimately the formulation of synbiotic products.

10.2 GASTROINTESTINAL PHYSIOLOGY

Before studying the details about the probiotics and its large-scale production, it is important to know about its passage and survival through the unfavorable condition of the GIT. As per Gomes et al. (2014), the digestive system is divided into accessory digestive organs and tubular GIT anatomically. The length of GIT is approximately 6 m starting from the mouth to anus. The GIT consists of four layers and mucosa also known as the mucous membrane is the innermost layer. The mucous membrane is composed of epithelium that performs several functions. One of the functions of the mucosa is that it functions as a barrier to prevent the passage of different strange and toxic components like food components and bacteria (O'Hara et al., 2006).

10.2.1 GASTROINTESTINAL pH

There are several factors that are used to determine the activity of probiotic strains among which pH is considered one of the most important factors for the selection of probiotics. The pH of our body varies from organs to organs. In short, probiotics have to deal with various pH levels during their passage to show their beneficial effects in the colon. The pH of the upper part of the stomach ranges from 4.0–6.5 and decreases to 1.5–4.0 in the lower part of the stomach as it produces hydrochloric acid (HCl) and pepsin. In the duodenum, the pH again increases to 7.0–8.5 and showing a pH range of 7.5 in the ileocolonic region (Maurer et al., 2015).

10.2.2 TRANSIT TIME

To show beneficial effects, probiotics have to reach successfully to the colon in a large number surviving the GIT conditions. Transit time also known as residence time can be defined as the time the probiotics will reside in a target site. It mainly depends on the adhesion to the intestinal surface. The more the percentage of adhesion on the intestinal surface the more will be the transit time of the probiotics. Therefore it is very important for the probiotic to reside at the desired target site and a sufficient concentration to show their effects and also it is very much necessary to have a thorough knowledge about the dosages (Salminen and Gueimonde, 2004).

10.2.3 STOMACH

The stomach is known to be a J- shaped organ of the GIT. It is mostly responsible for carrying out the second stage of digestion after chewing. Being placed in between the esophagus and the small intestine, it acts as a reservoir for food. Here major part of the digestion takes place as it produces a digestive enzyme (pepsin) and HCl that kills most of the bacteria. Therefore it becomes very difficult for probiotic bacteria to resist the condition of the stomach during their passage (Tennant et al., 2008).

10.2.4 SMALL INTESTINE

The small intestine is the part connecting the stomach and the large intestine. It mainly includes three parts, the duodenum, jejunum, and ileum. The partly digested food known as chyme finally moves to the small intestine starting from the stomach. Here the best part of the food comes in contact with bicarbonate, mucus, pancreatic juice, and bile. Due to the antimicrobial activity of the bile, it becomes difficult for the microorganisms, including probiotics to survive in the GIT for a longer period (Hassanzadazar et al., 2012).

10.2.5 LARGE INTESTINE

The large intestine (colon) is extended to the anus. Here a wide range of microorganisms approximately 10^{12} bacteria per gram of the intestinal contents resulting in several metabolic reactions can be found. In the large intestine, prebiotics interacts with the probiotics. The strains mainly identified in the GIT microflora are *Lactobacillus* and *Bifidobacterium*.

10.3 FUNCTIONAL PROPERTIES OF PROBIOTICS

Many investigations have been carried out on *in vitro* and using artificial alimentary tract to discover more about the probiotic properties, may it be functional or technological. Listed below are some of the functional properties:

10.3.1 ADHESION PROPERTIES

Adhesion to the intestinal surface is one of the main criteria for selecting probiotic strains. Adhesion to the GIT increases the probiotic efficacy by increasing the residence time. This, in turn, influences the colonization of human GIT enhancing the probiotic action and thus improving the human health. According to von Ossowski et al. (2010), adherent strains of probiotic bacteria are more effective than non-adherent strains in persisting longer in the intestinal tract. As a result, only adherent probiotic strains are known to stabilize the intestinal mucosal barrier. It also helps in eliminating pathogenic bacteria from the intestinal epithelium. Mucus, the innermost layer of the intestinal surface plays a key role in adhesion. Mucus mainly composed of mucin glycoproteins that are capable of forming a gel to which most of the probiotic strains, including *Lactobacillus* adhere (Van Tassell and Miller, 2011). Initially, probiotics adhere to the mucus to interact with the host cells. Each *Lactobacillus* strains have their adhesive characteristics. The unique property of adhesion might be due to the mucus binding protein (MUB), which has been recently isolated and studied. It has been stated that MUB protein's adhesive property is might be due to the presence of repeated functional domains also referred to as MUB domains. MUB domain is considered as a member of the family MucBP domain.

Adhesion function also depends on the surface composition of bacteria. According to Buck et al. (2005), *L. acidophilus* has been found to show direct adhesion to the GI surface due to the presence of Slp A, an S layer protein. Since some of the *Lactobacillus* species are found associated with S layer along with its contribution towards pathogen exclusion, it has become necessary to conduct more research regarding S layer proteins.

10.3.2 PRODUCTION OF ANTIMICROBIAL SUBSTANCES

To be considered as probiotic, organisms should have the ability to show beneficial effects on the host and production of antimicrobial substances is one among those properties. Scientifically, it can also be termed as bacterial antagonism, which has been taken into consideration for over a century. Particularly this term is getting more attention in describing various strains of LAB (Prabhurajeshwar and Chandrakanth, 2017). Due to the production of antimicrobial substances by the LAB, these organisms are being provided with a competitive advantage over other pathogenic organisms. These antimicrobial substances can enhance the viability of the organism. As per the studied literature, bacteriocins are considered to be special and important antimicrobial compounds.

10.3.2.1 BACTERIOCINS

Bacteriocins are defined as the proteinaceous substances that are secreted by bacteria that work to restrict the growth of other bacterial strains. Andre Gratia in the year 1925 first discovered bacteriocins. According to Saranraj (2013), LAB are in great demand in the field of research due to their antimicrobial activity against foodborne organism mainly, *Bacillus cereus, Listeria monocytogenes, Clostridium botulinum, Staphylococcus aureus,* etc. Bacteriocins are also known to act as great natural food preservatives. *Lactococcus lactis* of dairy origin produce nisin, which is accepted as the only bacteriocin as a food preservative. Most of the bacteriocins that are produced by LAB have a bactericidal effect. However, according to Anand Hoover (2003), and Parada et al. (2007), lactocin 27 produced by *Lactobacillus helveticus* and bacteriocin produced by *Lactobacillus sake* 148 shows bacteriostatic effect.

Apart from these bacteriocins, organic acid, carbon dioxide, and hydrogen peroxide are other antimicrobial substances that are produced by the LAB. It produces lactic acid as one of the major metabolites according to the substrates and microorganisms. LAB also produces acetic acid, which shows its antimicrobial activity up to pH 4.5. Hydrogen peroxide produced by LAB inhibits both Gram-negative as well as Gram-positive such as *Neisseria gonorrhoeae* and *Staphylococcus aureus*, respectively (Pericone et al., 2000).

10.3.3 TOLERANCE TO ACID AND BILE

The literature states that most of the bacteria find difficulties in surviving at low pH values. Thus, according to Shehata et al. (2016), it is very much necessary to screen the microbial cultures for their resistance to gastric acidity before using the culture. After reaching the intestine, their survivability depends upon their ability to show resistance to bile (Ruiz et al., 2013). This is also an important factor used for selecting efficient probiotic strains. Probiotics have to overcome two major challenges while passing through the GIT. They are exposed to acidic pH (=2) in the stomach followed by the bile in the small intestine. It takes approximately 90 minutes for the probiotics to enter and release from the stomach (Hassanzadazar et al., 2012). According to studied literature, the reason for the reduction of survivability of the bacteria is due to the entering of the bile into the duodenal part of the small intestine. Bile which is made up of bile acids such as weak organic acids and bile salts are responsible for causing intracellular acidification. The investigation of the molecular basis of probiotic adaptation and response to bile exposer and acid pH was done by applying proteomics which states that similar strategies are adopted by different probiotic species to overcome adverse environmental conditions. Bile salt hydrolases (BSHs) are the proteins responsible for deconjugating bile salts (Shehata et al., 2016).

To get health benefits, it is necessary to have proper knowledge about the dosage of probiotics as some of these are going to be destroyed during their movement through the GI tract. According to Shah (2000), probiotic bacteria should be maintained at a level of 10^7 CFU of live organisms per milliliter of the food product during the time of consumption. To overcome this problem scientists are showing more interest towards microencapsulation. As per the paper published by Adhikari et al. (2000), and Picot and Lacroix (2004), the survivability of the probiotic bacteria in the food product as well as during their movement through the stomach can be enhanced by microencapsulation. According to the previous report, it is known that in acidic food products the viability of probiotic bacteria is increased by microcapsules made up of alginate polymers.

10.3.4 REDUCTION OF CHOLESTEROL BY PROBIOTICS

Although cholesterol plays a vital role in our body, elevated blood cholesterol has become a leading risk factor for the coronary heart diseases (CHDs) (Aloğlu and Öner, 2006). WHO has predicted that by 2030, cardiovascular diseases (CVD) will become the main cause of death. According to Yusuf et al. (2004), in Western Europe, 45% of heart attacks are due to hypercholesterolemia whereas in Central and Eastern Europe, it is contributed by 35%. Hypercholesterolemia has raised the risk of heart attack by three times as compared to those with normal blood lipid profiles. As per the statement by WHO, the increased risk of CVD is due to the intake of unhealthy diets that includes free sugar, the excess content of fat, salt, etc. According to Bliznakov (2002), cholesterol levels that are effectively reduced by pharmacological agents; however, they are expensive and have severe side effects. To overcome these major problems, scientists are showing interest towards probiotics. As per the paper published by Tsai et al. (2014), cholesterol levels can be lowered by LAB containing BSH by interacting with host bile salt metabolism. According to Begley et al. (2006), the survivability and colonization of the lower small intestine can be increased by the BSH activity of probiotics considering it as an important colonization factor. Since the pharmaceutical approach shows side effects, consumers are showing more interest towards non-pharmaceutical approach by regular intake of probiotic bacteria and yeast. Probiotics are becoming a source of treatment for CVD. A strong connection between low-density lipoprotein (LDL) cholesterol and high serum concentration of total cholesterol (TC) along with the increased risk of CVD have been observed through epidemiological and clinical studies (Guo et al., 2011).

10.3.4.1 BILE

Bile acids, cholesterol, biliverdin, and phospholipids are the major components of bile, which is a yellow-green aqueous solution. Bile is synthesized by the liver, and it is concentrated and stored in the gallbladder. After the intake of food, bile is finally released into the duodenum. Digesting fat is the most important function of bile, and it does that by solubilizing the lipids. Chenodeoxycholic acid and cholic acid are known to be the two primary bile acids that are synthesized in the human liver. Conjugation is the process by which liver metabolizes bile acids further to glycine or taurine. By the process of passive diffusion and active transport, both unconjugated and conjugated bile acids get absorbed along the gut and in the internal ileum, respectively. Deconjugation is one of the important transformation reactions that occur before any further possible modifications (Kumar et al., 2012). BSH enzymes catalyze deconjugation by hydrolyzing the amide bond resulting in the liberation of the taurine/glycine moiety from the steroid core.

10.3.4.2 REGULATING CHOLESTEROL SYNTHESIS

The amount of cholesterol synthesized per day in healthy adults is about 1 gram from which almost 0.3 gram is consumed. A consistent amount of cholesterol, i.e., 150–200 mg/dl is maintained by our body which is controlled by regulating the *de novo* synthesis. To some extent, the level of cholesterol synthesis is controlled by the intake of the dietary cholesterol. Most of this cholesterol is used for the bile acid synthesis.

The liver utilizes the cholesterol in two ways:

i. Part of it is utilized to produce bile salts. Ultimately the bile salt emulsifies the fat, and hence their absorption and ingestion; and
ii. Rest part of the cholesterol is utilized by the body for other requirements.

This is carried out by combining cholesterol with triglycerides covered with a particular protein to get dissolved in the blood. These large molecules are termed as VLDL (very LDLs) that is finally drained into the blood by the liver.

10.3.4.3 MECHANISMS OF CHOLESTEROL REDUCTION

The mechanisms that are involved in the cholesterol reduction are:

i. Bile deconjugation by the activity of BSH;
ii. The attachment or binding of cholesterol to the cellular surface of probiotics and incorporating into their cell membrane;
iii. Production of SCFA (short chain fatty acids) from oligosaccharides;
iv. Co-precipitation of cholesterol with deconjugated bile;
v. Cholesterol getting converted to coprostanol.

From the studies, it has been seen that most of the *Lactobacillus* and *Bifidobacterium* species consist of BSH genes that involve them in reducing cholesterol. To cleave the amide bond, probiotics produce BSH hydrolyzing conjugated bile acids resulting in the liberation of primary bile acids which are not reabsorbed efficiently and excreted in the faces (Lye et al., 2010a and Huang et al., 2013). In short, active BSH present in probiotics plays a vital role in reducing cholesterol by producing more amounts of bile salts from cholesterol-lowering the risk of CVD. According to Kumar et al. (2012), and Lye et al. (2010b), some *Lactobacillus* species like *Lactobacillus acidophilus* may use other mechanisms for lowering the cholesterol and they carry out this activity by binding tightly to phosphatidylcholine vesicles or exogenous cholesterol with the help of protease- sensitive receptors present on their cell surface that leads to the inclusion of cholesterol in their cell membranes.

As per the statement given by Guo et al. (2011), and Bordoni et al. (2013), probiotic cells reduce the cholesterol from culture media by secreting exopolysaccharides, peptidoglycan, and numerous amino acids that help them in maintaining strong interactive forces. These strong interactive forces help both dead and live probiotic cells in reducing cholesterol. In reducing cholesterol, SCFAs also plays a vital role. SCFAs are the result of probiotic metabolism on prebiotics. They carry out their activity by acting as ligands that bind to activate peroxisome proliferator-activated receptors (PPARs).

According to Guardamagna et al. (2014), probiotics have been successfully experimented on humans to control lipid levels resulting in the decrease of cholesterol. Still more high quality *in vivo* experiments is required to get more clear ideas about the cholesterol reduction mechanisms and get rid of the CVD. Also, the detailed mechanism of BSH should be studied. Although several investigations have been carried out about the cholesterol reduction mechanism of probiotics, the knowledge about the appropriate amount of dosage has yet to be studied. The researchers should also take care of the stability of the clinical trials.

10.3.5 ANTIOXIDATIVE ACTIVITY

Oxidative stress is a condition of increased levels of oxygen radicals resulting in the damage of lipids, proteins, and DNA (Schieber and Chandel, 2014) ultimately affecting the viability of the cells. Till now, the antimicrobial activity, cholesterol reduction, bile, and acid tolerance, and adhesion properties of probiotics are studied well. One more important property of probiotics which needs more investigation is antioxidant properties. Some of the highly active oxygen free radicals like hydroxyl radicals, superoxide anion radicals, and hydrogen peroxide are termed as reactive oxygen species (ROS). According to Mishra et al. (2015), most of the living organisms naturally possess enzymatic defense system such as glutathione peroxidase

(GPx), superoxide dismutase (SOD), glutathione reductase (GR) as well as non-enzymatic antioxidant defenses like vitamin C, vitamin E, thioredoxin, and glutathione that acts as a defense system against oxidative stress. Instead of the presence of the native antioxidant defenses, it has been found that these defense systems are not enough to fight against the oxidative stress. To protect the human body from oxidative stress, antioxidant additives are used, which can prevent or delay further oxidation of cellular substrates. The paper published by Luo and Fang (2008), states that butylated hydroxyanisole and butylated hydroxytoluene are some of the synthetic antioxidants whose safety have been questioned as they have been seen to promote carcinogenic effect and liver damage. To reduce these carcinogenic effects scientists are showing more interest in replacing the synthetic antioxidants with the natural antioxidants from bioresources. Here come the term probiotics that are gaining more attention due to their beneficial health effects. According to Lin and Yen (1999), Shen et al. (2011), and Persichetti et al. (2014), most of the probiotic bacteria have shown their antioxidant abilities both *in vitro* and *in vivo.*

10.3.5.1 REACTIVE OXYGEN SPECIES (ROS)

Oxygen is an essential element required by all aerobic organisms. The human body is affected by different types of stresses (exogenous and endogenous). When the concentration of oxygen goes below normal, the condition is referred to as hypoxia whereas when it is in high concentration, the condition is known to be the oxidative stress. ROS are produced by the normal oxygen-consuming metabolic processes (Wang et al., 2017). ROS being highly reactive can modify and damage other oxygen species, proteins, lipids, or DNA.

10.3.5.2 PROBIOTICS AND THEIR CONTRIBUTIONS AS ANTIOXIDANTS

As per the hypothesis, it is well understood about the activity of probiotics in antioxidation. Some of the contributions by various species and strains are discussed here. According to Stecchini et al. (2001), and Kullisaar et al. (2002), LAB have the skill to resist ROS including superoxide anions, hydroxyl, and peroxide radicals. Bao et al. (2012), hypothesized that *Lactobacillus plantarum* P-8 incorporated in high-fat diets of rats increased antioxidant activity. According to Martarelli et al. (2011), *Lactobacillus rhamnosus* shows its effect by increasing the level of antioxidants that neutralize the ROS in athletes with high oxygen concentration.

10.3.5.3 MODE OF ACTION OF PROBIOTICS IN REGULATING THE OXIDATIVE STRESS

During the past years, scientists have explored a lot about the probiotics. However, there are very few reviews depicting the antioxidant mechanisms of probiotics.

Wang et al. (2017), hypothesized that chelators like ethylene diamine tetraacetic acid (EDTA), penicillamine, bathophenanthroline disulfunic acid (BPS), and desferrioxamine plays a significant role in preventing the catalysis of the oxidation by the metal ions. Lin and Yen (1999), experimented 19 LAB strains for the chelating ability of their metal ions and found that *Streptococcus thermophilus* 821 showed the best result for chelating both Cu^{2+} and Fe^{2+}. Rest of the strains showed the chelating ability for either Cu^{2+} or Fe^{2+}. Similarly, according to Ahire et al. (2013), *Lactobacillus helveticus* CD6 showed higher Fe^{2+} ion chelation with its intracellular cell-free extract.

Probiotics also use a mechanism like antioxidant enzymatic systems in which enzyme SOD plays a vital role. Mitochondria producing superoxide is catalyzed by SOD into hydrogen peroxide and water and is therefore considered as the central regulator of ROS (Landis and Tower, 2005). Moreover, probiotics act as a support system for the host by stimulating their antioxidant system and increasing the antioxidase activity efficiently.

10.3.6 MODULATION OF IMMUNE HEALTH BY PROBIOTICS

Probiotics have contributed a lot towards the human health. Some of the functional properties have been discussed above. Another important functional property of the probiotics which is taken into consideration is the immunomodulation by probiotics.

By studying proteomic and genomic properties of probiotics, it has been found that there are several genes that are specifically derived from probiotic are responsible for immunomodulation. The immune response is of two types such as innate and adaptive immunity. The innate immunity comes into existence when the body is exposed to a foreign substance. However, this results in severe tissue damage and inflammation. Thus, disease-free state and good health can be maintained by manipulating the intestinal microbiota. According to Vanderpool et al. (2008), probiotics use Toll-like receptors that are responsible for the enchantment of the innate immunity along with modulation of the pathogen-induced inflammation.

10.3.6.1 PROBIOTIC GENES RESPONSIBLE FOR REGULATING HOST IMMUNE RESPONSES

With the help of metagenomic analysis, it has become possible to get deep knowledge about the probiotic genes that are responsible for modulating host immune responses. An experiment was conducted by taking forty-two *Lactobacillus plantarum* strains that were isolated and identified from different human and environmental sources where they experimented for their ability to stimulate interleukin (IL-10) and (IL-12). Six candidate genes were identified with their immunomodulatory abilities by analyzing strain specific cytokine responses and genome hybridization. Without these genes, *Lactobacillus plantarum* WCFS1 was unable to stimulate cytokine production (van Hemert et al., 2010).

10.3.6.2 REGULATION OF HOST IMMUNE SYSTEM BY PROBIOTICS

Probiotics play a crucial role in controlling the defense mechanisms along with the innate and adaptive immunity. They carry out this by interacting directly with the intestinal epithelial cells. The two major mechanisms of the host cells that occur due to the action of probiotics leading to immunomodulation are signaling pathways and regulation of gene expression. The intestinal epithelium plays a significant role by forming a barrier against pathogens and harmful substances that are present in the intestinal lumen (Yan and Polk, 2011). From the reviews, it was found that signaling pathways that were regulated by probiotics resulted in many responses such as the production of cytokine, production of cell protective proteins and antimicrobial substances including regulation of intestinal epithelial immune functions. *Lactobacillus johnsonii* N6.2 was found for regulating intestinal epithelial immunological functions using human Caco-2 cell monolayers. The most important function of probiotics is that they are responsible for preventing and treating immune diseases. They have the potential for preventing intestinal diseases, ulcerative colitis, diarrhea, irritable bowel syndrome, allergy, etc. However, more studies should be carried out to get further information.

10.3.7 PROBIOTICS PRODUCING ANTICARCINOGENIC AND ANTIMUTAGENIC PROPERTIES

Another functional property of bacteria that signifies them as probiotic is by the production of anticarcinogenic and antimutagenic substances. These mechanisms include procarcinogens binding, secretion of antimutagenic substances, procarcinogen enzymatic modulation carried out in the gut and tumor suppression. The LAB especially *Lactococcus*, *Lactobacillus*, *Enterococcus*, *Leuconostoc*, and *Pediococcus* species isolated from fermented products have been well studied *in vitro* for their antimutagenic and anticarcinogenic properties. As per the review published by Felgner et al. (2016), these properties are not just limited to probiotic bacteria but can be a property of both Gram-negative as well as Gram-positive bacterial species. According to Orrhage et al. (1994), *in vitro* mutagen binding is shown by some bacteria along with yeasts. Consumption of *Lactobacillus acidophilus* NCFB 1748 was studied using a human model which showed the lowering of urinary and fecal mutagenicity (Lidbeck et al., 1992), whereas consumption of *Lactobacillus* spp. (*Lactobacillus* GR-1) is well known for suppressing urinary mutagenicity (Brown and Valiere, 2004). Although many studies have been carried out *in vitro* as well as *in vivo* showing the antimutagenic and anticarcinogenic effects of probiotics, still it is a challenge for the scientists to understand the cause of cancer. So studies regarding these properties should be carried out more strongly to conclude.

10.4 TECHNOLOGICAL PROPERTIES OF PROBIOTICS

With the growing demand for the probiotics, it has become a great challenge for the researchers to come up with new ideas for the functional food which has been mentioned earlier from the beginning of this chapter. The food product along with the probiotic as well as prebiotic comprises the functional food. Today it has become a well-accepted product in the European market, and nowadays the interest in using the functional food in other countries is also increasing. To satisfy the need of the consumers, food industries need to produce functional food in large quantity. Several parameters are to be taken into consideration while production is carried out on an industrial level. Storage time is considered one of the most important parameters as it is really difficult to maintain the viability of the strain. It has been already mentioned above that to get benefited by probiotics it should be consumed at a concentration of 10^7 CFU per ml of live organisms per product. Since they lose their viability due to different stresses during storage, it is required to culture the organism at a very high concentration having a cell density of 10^{10} CFU per gram. From the literature, it has been found that probiotics especially *Bifidobacterium* are characterized by their very poor growth as they require very specific media along with anaerobic culture conditions. Due to this reason, it becomes difficult to incorporate *Bifidobacterium* and commercial strains of LAB together in functional food products (Sip and Grajek, 2010).

Before selecting probiotic strain some of the major criteria such as their ability to survive in the functional food, functionality of freeze-dried culture during storage, ease in manufacturing under industrial conditions, the conditions at which the functional food is to be stored as well as not producing any off flavors when incorporated into food are to be taken into consideration. The production of probiotic cultures at the industrial level is similar to that of the starter cultures consisting of certain stages:

i. Preparation of media followed by the inoculation;
ii. The growth of the culture in a bioreactor;
iii. Isolating the cells from the culture medium;
iv. Storing the cells in a freeze-dried condition followed by packaging in bulk.

To produce probiotics in large quantity, it is very much essential to obtain resistant strains that could provide the appropriate viability, genetic stability and proper metabolic activity (Girgis et al., 2003). Every single culture is produced with specific parameters. Before inoculation, the growth media is sterilized by using one of the two methods such as batch sterilization *in situ* which is mainly carried out in the fermentation tank or ultra-high temperature (UHT) treatment in order to get proper sterilization followed by cooling to the temperature at which inoculation can be carried out (approximately 30–40° C). Every single parameter like temperature, pH, concentration, and amount of inoculum, the rate of agitation, etc. are to be considered very seriously. For instance, LAB are known to produce acid during their fermentation period which may result in a reduction of the pH (Rhee et al., 2011). Excess production of acid is not acceptable as it affects the texture by producing off flavor in the food product (Barrett et al., 2010). To stabilize the pH condition, it is required to add a base (NaOH or NH_4OH). After a successful production of the cell biomass, it is collected and subsequently stored by freeze drying.

In some cases spray drying is also used but the main demerit of spray drying is that it promotes a very high temperature which becomes very hard for the probiotics to resist. As a result, it affects the survival rate of the cells. In that case, to protect the dried cell, some of the additives such as trehalose, skim milk, alginate solution, granular starch, may be applied. Another method used to protect the cell is through microencapsulation. Encapsulation increases the survival rate of probiotics by protecting against various stresses and mostly when it travels through the conditions of the GIT (Govender et al., 2014). The quality of the product, it specifically describes the performance of the strain. Quality of the product also depends on the absence of the contaminating organisms, physical appearance and water activity (a_w). This process is mainly considered after the process of production. It is a time-consuming process. Product quality is given special attention as it decides the performance and shelf life of the strain in the functional food product.

Some of the parameters that are to be taken into consideration while manufacturing the product in the industrial levels are discussed in the following subsections.

10.4.1 STORAGE STABILITY

Industrial processes are largely involved in manufacturing probiotic products in a huge quantity which results in reducing the viability of the strains. Manufacturers apply several strategies to maintain the stability and to improve the quality of the product during industrial processes. This is done by modifying the production parameters, as well as focusing on the strain of intrinsic resistance. Nowadays, the strains that can adapt stress and also the strains having the desired properties are gaining large attention. To manufacture probiotic products at the industrial level, it is necessary to understand the stress tolerance ability of probiotics. This knowledge may improve the stability of the final product. Stability not only includes the viability but also includes the functional and metabolic activity that is required to maintain the sensorial attributes.

Modification of the industrial processes may enhance the stability of the strains. This may also include suitable culture medium composition and the cell protectants that may influence the strain survival (Ross et al., 2005; Muller et al., 2011). Stability also depends on the chemical composition of the product. The presence of several antimicrobial compounds and certain food additives may change the composition of the product being detrimental to probiotics (Gueimonde and Sanchez, 2012).

10.4.2 ENHANCING STRESS TOLERANCE OF PROBIOTICS

Another important parameter to be taken into consideration when manufactured in industrial level is intrinsic stress tolerance ability of the strains. Hence, enhancing the stress tolerating ability of the probiotics is of great industrial demand which mainly includes the origin of the strain, their stress adaptation and most importantly the genetic manipulation resulting in a genetically modified organism (GMO) (Sánchez et al., 2012). The stress tolerance ability during their production and storage is different for each strain. Therefore, there is a need to screen the strains with better tolerance ability for increasing the viability and stability of the probiotics in the products. As per the hypothesis of Upadrasta et al. (2011), stress tolerance of the probiotics can be enhanced by pre-treating or pre-culturing the cells by exposing them to sublethal stress before exposed to more hard environmental conditions. Settachaimongkon et al. (2015), mentioned about two commercial strains, i.e., *Bifidobacterium animalis (*subsp. *Lactis*BB12) and *Lactobacillus rhamnosus* GG that developed adaptive responses by the pretreatment increasing their survivability. Stack et al. (2010), stated that the strains producing extra polysaccharide might show more tolerance towards stress. Makarova and Koonin (2007), clearly mentioned the reduction evolution. This phenomenon was used for LAB where they adapted to highly nutrient-rich environment leading to a loss of many genes, reducing the size of the genome. According to Saarela et al. (2011), *B. animals* subspecies *lactis* showed better stability in low pH products due to the mutagenesis induced by the use of different chemicals or UV light.

As mentioned earlier genetic engineering is another method for enhancing the probiotic stability during storage. This includes two ways, first by modifying the expression of homologous genes on the microorganisms, and second, the introduction of the heterologous genes.

The demand for growing probiotics in fermented food is increasing drastically. Since functional food incorporated with probiotics has been proved to be more healthy and nutritious than the normal food, food industries are concentrating on the production of functional food products in large number.

10.4.3 FUNCTIONAL FOOD

Today most of the food industries are predominantly producing functional food mainly the dairy products containing thc probiotics. Some of the examples of dairy products containing probiotics are cheese, yogurt, kefir, and ice cream. Nowadays non-dairy based functional foods are also available in the form of vegetable and fruit juices.

Probiotic products can be divided into two forms:

i. Products that are rich in probiotic bacteria on their own especially the traditional products such as cottage cheese, yogurt, kefir, etc.; and
ii. Foods incorporated with probiotic bacteria to get benefited by them.

Along with the nutritive value of the probiotic, sensory attributes are also given special attention. Addition of the probiotic bacteria to the fermented dairy milk and other vegetable and fruit juices modifies the taste of the product. Thus, to increase their sensory quality, aroma additives are frequently added. Most of the food industries producing functional foods launch their products by adding a prefix as "Bio" to promote their probiotic character or their origin. Nestle has produced probiotic yogurt that contains *Lactobacillus johnsonii* La1 (NCC533) which is one of the most known probiotic products in Europe. This product is known for stimulating the immune system. Along with this, probiotic bacteria are also being incorporated into confectionery like chocolates, and also into baby foods. Listed below are some of the probiotic foods:

10.4.3.1 YOGHURT

Many types of research have been done in past years to understand the ability of the bacterial strains to survive in yogurt. According to Gilliland et al. (2002), Donkor et al. (2006), and Martın-Diana et al. (2003), yogurt provides a very tough environmental condition for the probiotics to survive as compared to other dairy products. Still, many *Lactobacillus* species and strains have been successfully found to show good viability and storage stability in the yogurt conditions. As per the paper published by Vinderola et al. (2002a), the viability of the probiotic in yogurt is mainly affected by two factors:

i. Low pH; and
ii. Organic acids produced by the starter bacteria.

According to Kailasapathy et al. (2008), fruits are known for their good antimicrobial activity hence they may affect the viability when added to yogurt. To maintain the probiotic viability encapsulation in plain alginate beads can also be a good approach. Chitosan (CS)-coated alginate, alginate- prebiotic or alginate- pectin can also be used. The detailed techniques for encapsulation of probiotics for manufacturing yogurt are discussed by Krasaekoopt et al. (2003). After the successful manufacturing of the product, proper packaging is required for maintaining the shelf life of the product. Talwalkar et al. (2004), studied about the survival of the probiotic strain in yogurt concerning the effect of packaging materials by using different oxygen barrier. According to their study, they could not find any major difference in the viability of *Bifidobacterium infantis* CSCC 1912 and *Lactobacillus acidophilus* CSCC 2409 in yogurt in different packaging conditions and stated that oxygen affects the probiotic viability in some conditions.

10.4.3.2 CHEESE

Like yogurt, cheese has also been explored a lot as a potential carrier of probiotics. Mostly strains like *Bifidobacterium* and *Lactobacillus* are used as single cultures or in combination for producing cheese. The process of encapsulation has also experimented with the production of cheese. As compared to the fermented milk (yogurt), hard cheese has a higher pH value ensuring better survival conditions for probiotics. According to Gardiner et al. (1999), Boylston et al. (2004), da Cruz et al. (2009), and Roy (2005), hard cheese has a very high fat content which protects the probiotics from the stressful conditions of upper GI tract. Further, the probiotic viability of cheese is also affected by the biochemical and microbiological parameters that change due to the process of salting and ripening. According to da Cruz et al. (2009), as compared to the hard cheese, fresh cheese such as cottage cheese provides a better environment for probiotics as they are stored by refrigeration, and also they can have a very less salt content.

10.4.3.3 NON-DAIRY BASED PROBIOTIC PRODUCTS

When we talk about the non-dairy probiotic food, the products that are dominating the non-dairy based probiotic food market are mostly the vegetable and fruit juices. Vegetable juices can be produced in two ways both by fermentation and non-fermentation methods. Yoon along with his co-workers performed some experiments regarding the production of different vegetable juices by the help of the probiotic *Lactobacilli* and found that *Lactobacillus casei, Lactobacillus plantarum,* and *Lactobacillus acidophilus* could grow and produce acid in cabbage juice, tomato as well as beetroot and also confirmed their stability in different juices (Yoon et al., 2004–2006). In the experiment conducted by Savard et al. (2003), fermentation of mixed vegetable juice made of onions, carrots, and cabbage was carried out using nine different probiotic strains of *Lactobacillus* and *Bifidobacterium* that showed the growth of all the probiotic strains in the medium but showing a variable cell density. The initial pH of this fermented vegetable juice was 3.65 or 6.5. When this fermented vegetable juice was stored for 90 days at a temperature of 4°C, several probiotic strains were found to maintain their stability for about one month at pH 3.65 which declined rapidly after that period. Reid et al. (2007), hypothesized that *Lactobacillus rhamnosus* showed excellent stability in a mixed vegetable juice which was made of using several vegetables such as beets, carrots, tomatoes, celery, spinach, parsley, lettuce, and watercress at pH 4.35 for over two weeks. Several studies have also mentioned the growth of the probiotics in both non-supplemented and supplemented soy beverages. According to Vinderola et al. (2002b), probiotics do not find natural fruit juices to be a suitable environment to colonize due to their higher amount of organic acids and a very low pH. Champagne et al. (2005), hypothesized that *Lactobacilli* especially *Lactobacillus casei* and *Lactobacillus acidophilus* were found to show resistance to the acidic environment more as compared to *Bifidobacteria.* Since *Bifidobacteria* are sensitive to the pH of less than 4.6, they are unable to maintain their stability and viability in the fruit juices having pH values less between 3 and 4. Not only is the pH, but the viability of the probiotic strain is also affected by the presence of organic acids and the raw materials in the fruit juices. Therefore, to get probiotic products of good quality and high yield, it is very much essential to use a particular strain satisfying all the required parameters along with appropriate cell density.

10.4.4 PREBIOTICS

Till now it has been discussed about the stresses faced by the probiotics in several situations, may it be while passing through the GIT or when incorporated in different food products. These stresses affect the viability of the probiotics directly. To overcome this problem prebiotics were used as a supplementary. They maintain the balance of the intestinal microflora by selectively nourishing the growth of the probiotics. Prebiotics are not digestible and act as food for the probiotics. They show their beneficial effects by moving to the colon where they stimulate the growth of the probiotics. As per the hypothesis, Japan was the first to introduce prebiotics in the market. Some of the parameters that are needed to be fulfilled by the prebiotics are:

i. Should not be digested by our digestive system;
ii. Should be able to distinguish between the pathogenic and beneficial bacteria to selectively stimulate the beneficial bacteria;
iii. Should have the ability to produce SCFAs;
iv. Should be safe to be consumed by a human.

According to Rivero-Urgell and Santamaria-Orleans (2001), oligosaccharides were the first to be observed with the prebiotic properties. Along with oligosaccharides certain polysaccharides, peptides, fats, and proteins also exhibits prebiotic characteristics. Fructooligosaccharide, galactooligosaccharide, maltooligosaccharides, raffinose, and stachyose, are some of the well-known oligosaccharides found in the prebiotic properties. Finally, these saccharides move to the cecum where they interact with the probiotics and produce SCFAs.

Since the beneficial health effects along with their stability and viability of the probiotics are increasing by the addition of prebiotics, manufacturers are showing great interest in launching products consisting of both probiotics and prebiotics. These are known as the synbiotic products. Fermented dairy drinks are the best examples of the synbiotic products. It has been suggested by few scientists to produce synbiotics in the form of tablets, capsules containing one-third prebiotic and two third probiotics (Figure 10.1).

10.5 FUTURE TRENDS

Today European food markets are highly dominated by the functional foods. Maintaining a proper diet and a disease free health are the two targets that have been encouraging the scientists to carry out their research more on the probiotics since probiotics have been found to be one of the most important factors for preventing the rate of coronary diseases such as gastrointestinal disorders, cancer, and CVD. Considering consumer's health along with the growing interest for the healthy food, the food industries are producing functional food in large numbers.

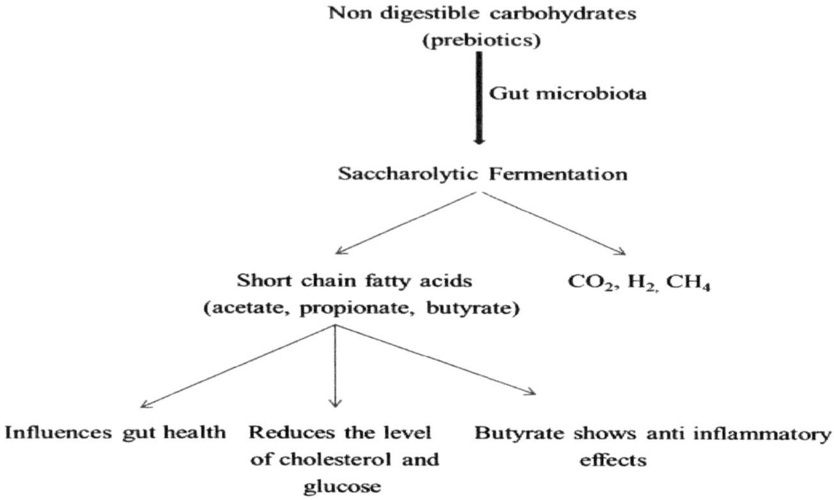

FIGURE 10.1 Synthesis and benefits of short-chain fatty acids (SCFAs) (Morrison and Preston, 2016; LeBlanc et al., 2017).

Looking towards their increasing beneficial health effects, Indian food industries are also improvising their skills for launching the functional food products in large numbers. The most crucial part in producing functional food is maintaining the stability and survivability of the probiotic bacteria (Heller, 2001). A lot of experiments have been carried out to maintain the viability of the strains in the food products. Encapsulation is one of the methods, which have been successfully implemented. Still, it requires a more advanced technological method that could enhance the shelf life of the product. From the literature search, it is known that some scientists prefer the selection of a suitable probiotic strain to be the best way to improve the shelf life of the functional food products. Despite several problems, our producers have successfully launched different types of functional food, tablets, capsules, health drinks, etc. Wagner et al. (2000), and Mottet and Michetti (2005) hypothesized that in future may be dead bacterial cells could be considered in the manufacture of functional food as scientists are showing some interest towards these dead cells due to their unseen health benefits. More experiments should be carried out regarding the activity of individual genes in the probiotic strains to modify the probiotic strains genetically.

Fermentation has always been the best method to store and improve the nutritional quality of the food since the ancient period. Day-by-day the consumption of homemade fermented food is getting replaced by the commercial food products. This non-commercial fermented food could also be a good carrier of a large number of unknown probiotic strains. Experiments should be carried out for isolating and identifying these strains before these products get replaced by the commercial fermented food products (Figure 10.2).

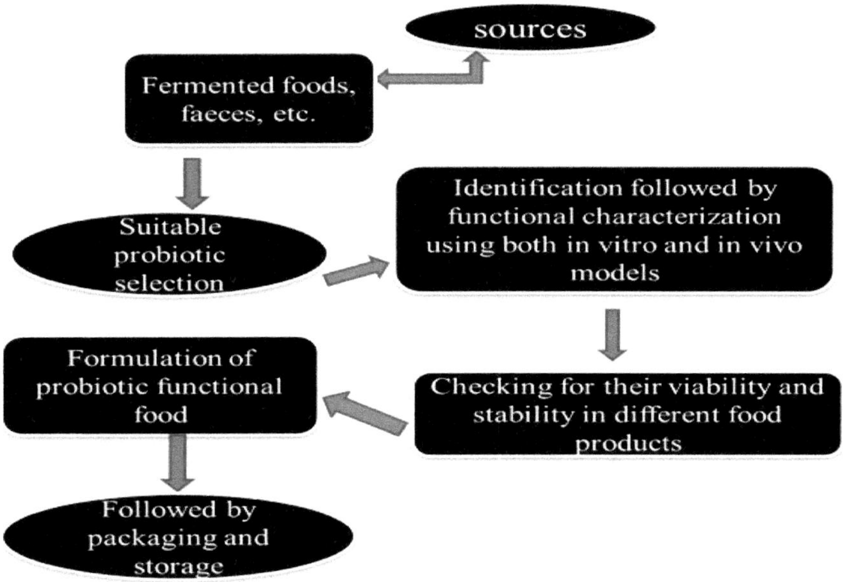

FIGURE 10.2 Schematic diagram representing the overall production of probiotic products (Sip and Grajek, 2010; Saarela et al., 2000).

KEYWORDS

- **anti-oxidative activity**
- **cholesterol**
- **functional foods**
- **prebiotics**
- **probiotics**
- **synbiotics**

REFERENCES

Adhikari, K., Mustapha, A., Grün, I. U., & Fernando, L., (2000). Viability of Microencapsulated Bifidobacteria in Set Yogurt During Refrigerated Storage1. *Journal of Dairy Science, 83*(9), 1946–1951.

AFRC, R. F., (1989). Probiotics in man and animals. *Journal of Applied Bacteriology, 66*(5), 365–378.

Ahire, J. J., Mokashe, N. U., Patil, H. J., & Chaudhari, B. L., (2013). The antioxidative potential of folate-producing probiotic Lactobacillus helveticus CD6. *Journal of Food Science and Technology, 50*(1), 26–34.

Aloğlu, H., & Öner, Z., (2006). Assimilation of cholesterol in broth, cream, and butter by probiotic bacteria. *European Journal of Lipid Science and Technology, 108*(9), 709–713.

And, H. C., & Hoover, D. G., (2003). Bacteriocins and their food applications. *Comprehensive Reviews in Food Science and Food Safety, 2*(3), 82–100.

Bao, Y., Wang, Z., Zhang, Y., Zhang, J., Wang, L., Dong, X., & Zhang, H., (2012). Effect of Lactobacillus plantarum P-8 on lipid metabolism in a hyperlipidemic rat model. *European Journal of Lipid Science and Technology, 114*(11), 1230–1236.

Barrett, D. M., Beaulieu, J. C., & Shewfelt, R., (2010). Color, flavor, texture, and nutritional quality of fresh-cut fruits and vegetables: Desirable levels, instrumental and sensory measurement, and the effects of processing. *Critical Reviews in Food Science and Nutrition, 50*(5), 369–389.

Begley, M., Hill, C., & Gahan, C. G., (2006). Bile salt hydrolase activity in probiotics. *Applied and Environmental Microbiology, 72*(3), 1729–1738.

Bliznakov, E. G., (2002). Lipid-lowering drugs (statins), cholesterol, and coenzyme Q10. The Baycol case–a modern Pandora's box. *Biomedicine & Pharmacotherapy, 56*(1), 56–59.

Bordoni, A., Amaretti, A., Leonardi, A., Boschetti, E., Danesi, F., Matteuzzi, D., & Rossi, M., (2013). Cholesterol-lowering probiotics: *In vitro* selection and *in vivo* testing of bifidobacteria. *Applied Microbiology and Biotechnology, 97*(18), 8273–8281.

Boylston, T. D., Vinderola, C. G., Ghoddusi, H. B., & Reinheimer, J. A., (2004). Incorporation of bifidobacteria into cheeses: Challenges and rewards. *International Dairy Journal, 14*(5), 375–387.

Brown, A. C., & Valiere, A., (2004). Probiotics and medical nutrition therapy. *Nutrition in Clinical Care: An Official Publication of Tufts University, 7*(2), 56.

Buck, B. L., Altermann, E., Svingerud, T., & Klaenhammer, T. R., (2005). Functional analysis of putative adhesion factors in Lactobacillus acidophilus NCFM. *Applied and Environmental Microbiology, 71*(12), 8344–8351.

Champagne, C. P., Gardner, N. J., & Roy, D., (2005). Challenges in the addition of probiotic cultures to foods. *Critical Reviews in Food Science and Nutrition, 45*(1), 61–84.

Da Cruz, A. G., Buriti, F. C. A., De Souza, C. H. B., Faria, J. A. F., & Saad, S. M. I., (2009). Probiotic cheese: Health benefits, technological and stability aspects. *Trends in Food Science & Technology, 20*(8), 344–354.

Donkor, O. N., Henriksson, A., Vasiljevic, T., & Shah, N. P., (2006). Effect of acidification on the activity of probiotics in yogurt during cold storage. *International Dairy Journal, 16*(10), 1181–1189.

FAO/WHO, (2006). Probiotics in Food. Health and nutritional properties and guidelines for evaluation. *FAO Food and Nutrition Paper no. 85, 25*(5), 700–704, ISBN 92-5-105513-0

FAO/WHO, (2006). *Probiotics in Food*. Health and nutritional properties and guidelines for evaluation. FAO food and nutrition paper no. 85, ISBN 92-5-105513-0.

Felgner, S., Kocijancic, D., Frahm, M., & Weiss, S., (2016). Bacteria in cancer therapy: Renaissance of an old concept. *International Journal of Microbiology, 2016*: 8451728, 14 pages.

Gardiner, G., Stanton, C., Lynch, P. B., Collins, J. K., Fitzgerald, G., & Ross, R. P., (1999). Evaluation of cheddar cheese as a food carrier for delivery of a probiotic strain to the gastrointestinal tract. *Journal of Dairy Science*, *82*(7), 1379–1387.

Gilliland, S. E., Reilly, S. S., Kim, G. B., & Kim, H. S., (2002). Viability during storage of selected probiotic lactobacilli and bifidobacteria in a yogurt-like product. *Journal of Food Science*, *67*(8), 3091–3095.

Girgis, H. S., Smith, J., Luchansky, J. B., & Klaenhammer, T. R., (2003). Stress adaptations of lactic acid bacteria. *Microbial Stress Adaptation and Food Safety*, 159–211.

Gomes, A. M., Pintado, M. M., Freitas, A. C., & Sousa e Silva, J. P., (2014). Gastrointestinal tract: microflora and transit aspects. In: Silva, J. P. S., & Freitas, A. C., (eds.), *Probiotic Bacteria: Fundamentals, Therapy and Technological Aspects*. Pan Stanford: Singapore, 18–56.

Govender, M., Choonara, Y. E., Kumar, P., Du Toit, L. C., Van Vuuren, S., & Pillay, V., (2014). A review of the advancements in probiotic delivery: Conventional vs. non-conventional formulations for intestinal flora supplementation. *Aaps. Pharm. Sci. Tech.*, *15*(1), 29–43.

Guardamagna, O., Amaretti, A., Puddu, P. E., Raimondi, S., Abello, F., Cagliero, P., & Rossi, M., (2014). Bifidobacteria supplementation: Effects on plasma lipid profiles in dyslipidemic children. *Nutrition*, *30*(7), 831–836.

Gueimonde, M., & Sanchez, B., (2012). Enhancing probiotic stability in industrial processes. *Microbial Ecology in Health and Disease*, *23*(1), 18562.

Guo, Z., Liu, X. M., Zhang, Q. X., Shen, Z., Tian, F. W., Zhang, H., & Chen, W., (2011). Influence of consumption of probiotics on the plasma lipid profile: A meta-analysis of randomized controlled trials. *Nutrition, Metabolism and Cardiovascular Diseases*, *21*(11), 844–850.

Hassanzadazar, H., Ehsani, A., Mardani, K., & Hesari, J., (2012). Investigation of antibacterial, acid and bile tolerance properties of lactobacilli isolated from Koozeh cheese. In: *Veterinary Research Forum* (Vol. 3, No. 3, p. 181). Faculty of Veterinary Medicine, Urmia University, Urmia, Iran.

Heller, K. J., (2001). Probiotic bacteria in fermented foods: Product characteristics and starter organisms. *The American Journal of Clinical Nutrition*, *73*(2), 374s–379s.

Holzapfel, W. H., & Schillinger, U., (2002). Introduction to pre-and probiotics. *Food Research International*, *35*(2/3), 109–116.

Huang, Y., Wang, X., Wang, J., Wu, F., Sui, Y., Yang, L., & Wang, Z., (2013). Lactobacillus plantarum strains as potential probiotic cultures with cholesterol-lowering activity. *Journal of Dairy Science*, *96*(5), 2746–2753.

Kailasapathy, K., Harmstorf, I., & Phillips, M., (2008). Survival of lactobacillus acidophilus and *Bifidobacterium animalis* ssp. lactis in stirred fruit yogurts. *LWT-Food Science and Technology*, *41*(7), 1317–1322.

Krasaekoopt, W., Bhandari, B., & Deeth, H., (2003). Evaluation of encapsulation techniques of probiotics for yogurt. *International Dairy Journal*, *13*(1), 3–13.

Kullisaar, T., Zilmer, M., Mikelsaar, M., Vihalemm, T., Annuk, H., Kairane, C., & Kilk, A., (2002). Two antioxidative lactobacilli strains as promising probiotics. *International Journal of Food Microbiology*, *72*(3), 215–224.

Kumar, M., Nagpal, R., Kumar, R., Hemalatha, R., Verma, V., Kumar, A., & Yadav, H., (2012). Cholesterol-lowering probiotics as potential biotherapeutics for metabolic diseases. *Experimental Diabetes Research, 2012*: 902917, 14 pages.

Landis, G. N., & Tower, J., (2005). Superoxide dismutase evolution and life span regulation. *Mechanisms of Ageing and Development, 126*(3), 365–379.

LeBlanc, J. G., Chain, F., Martín, R., Bermúdez-Humarán, L. G., Courau, S., & Langella, P., (2017). Beneficial effects on host energy metabolism of short-chain fatty acids and vitamins produced by commensal and probiotic bacteria. *Microbial Cell Factories, 16*(1), 79.

Lidbeck, A., Övervik, E., Rafter, J., Nord, C. E., & Gustafsson, J. Å., (1992). Effect of *Lactobacillus acidophilus* supplements on mutagen excretion in faces and urine in humans. *Microbial Ecology in Health and Disease, 5*(1), 59–67.

Lin, M. Y., & Yen, C. L., (1999). Antioxidative ability of lactic acid bacteria. *Journal of Agricultural and Food Chemistry, 47*(4), 1460–1466.

Luo, D., & Fang, B., (2008). Structural identification of ginseng polysaccharides and testing of their antioxidant activities. *Carbohydrate Polymers, 72*(3), 376–381.

Lye, H. S., Rahmat-Ali, G. R., & Liong, M. T., (2010a). Mechanisms of cholesterol removal by lactobacilli under conditions that mimic the human gastrointestinal tract. *International Dairy Journal, 20*(3), 169–175.

Lye, H. S., Rusul, G., & Liong, M. T., (2010b). Removal of cholesterol by lactobacilli via incorporation and conversion to coprostanol. *Journal of Dairy Science, 93*(4), 1383–1392.

Makarova, K. S., & Koonin, E. V., (2007). Evolutionary genomics of lactic acid bacteria. *Journal of Bacteriology, 189*(4), 1199–1208.

Martarelli, D., Verdenelli, M. C., Scuri, S., Cocchioni, M., Silvi, S., Cecchini, C., & Pompei, P., (2011). Effect of a probiotic intake on oxidant and antioxidant parameters in plasma of athletes during intense exercise training. *Current Microbiology, 62*(6), 1689–1696.

Martın-Diana, A. B., Janer, C., Peláez, C., & Requena, T., (2003). Development of a fermented goat's milk containing probiotic bacteria. *International Dairy Journal, 13*(10), 827–833.

Maurer, J. M., Schellekens, R. C., Van Rieke, H. M., Wanke, C., Iordanov, V., Stellaard, F., & Frijlink, H. W., (2015). Gastrointestinal pH and transit time profiling in healthy volunteers using the IntelliCap system confirms ileo-colonic release of ColoPulse tablets. *PloS ONE, 10*(7), e0129076.

Mishra, V., Shah, C., Mokashe, N., Chavan, R., Yadav, H., & Prajapati, J., (2015). Probiotics as potential antioxidants: A systematic review. *Journal of Agricultural and Food Chemistry, 63*(14), 3615–3626.

Morrison, D. J., & Preston, T., (2016). Formation of short chain fatty acids by the gut microbiota and their impact on human metabolism. *Gut Microbes, 7*(3), 189–200.

Mottet, C., & Michetti, P., (2005). Probiotics: Wanted dead or alive. *Digestive and Liver Disease, 37*(1), 3–6.

Muller, J. A., Ross, R. P., Sybesma, W. F. H., Fitzgerald, G. F., & Stanton, C., (2011). Modification of technical properties of *Lactobacillus johnsonii* NCC 533 by supplementing growth medium with unsaturated fatty acids. *Applied and Environmental Microbiology, 77*(19), 6889–6898, AEM-05213.

O'Hara, A. M., O'Regan, P., Fanning, A., O'Mahony, C., MacSharry, J., Lyons, A., & Shanahan, F., (2006). Functional modulation of human intestinal epithelial cell responses by *Bifidobacterium infantis* and *Lactobacillus salivarius*. *Immunology, 118*(2), 202–215.

Orrhage, K., Sillerström, E., Gustafsson, J. Å., Nord, C. E., & Rafter, J., (1994). Binding of mutagenic heterocyclic amines by intestinal and lactic acid bacteria. *Mutation Research/Fundamental and Molecular Mechanisms of Mutagenesis, 311*(2), 239–248.

Parada, J. L., Caron, C. R., Medeiros, A. B. P., & Soccol, C. R., (2007). Bacteriocins from lactic acid bacteria: Purification, properties and use as biopreservatives. *Brazilian Archives of Biology and Technology*, *50*(3), 512–542.

Pericone, C. D., Overweg, K., Hermans, P. W., & Weiser, J. N., (2000). Inhibitory and bactericidal effects of hydrogen peroxide production by *Streptococcus pneumoniae* on other inhabitants of the upper respiratory tract. *Infection and Immunity*, *68*(7), 3990–3997.

Persichetti, E., De Michele, A., Codini, M., & Traina, G., (2014). Antioxidative capacity of *Lactobacillus fermentum* LF31 evaluated in vitro by oxygen radical absorbance capacity assay. *Nutrition*, *30*(7), 936–938.

Picot, A., & Lacroix, C., (2004). Encapsulation of bifidobacteria in whey protein-based microcapsules and survival in simulated gastrointestinal conditions and in yogurt. *International Dairy Journal*, *14*(6), 505–515.

Prabhurajeshwar, C., & Chandrakanth, R. K., (2017). Probiotic potential of Lactobacilli with antagonistic activity against pathogenic strains: An *in vitro* validation for the production of inhibitory substances. *Biomedical Journal*, *40*(5), 270–283.

Reid, A. A., Champagne, C. P., Gardner, N., Fustier, P., & Vuillemard, J. C., (2007). Survival in food systems of *Lactobacillus rhamnosus* R011 microentrapped in whey protein gel particles. *Journal of Food Science*, *72*(1), M031–M037.

Rhee, S. J., Lee, J. E., & Lee, C. H., (2011). Importance of lactic acid bacteria in Asian fermented foods. In: *Microbial Cell Factories* (Vol. 10, No. 1, p. S5). *Bio. Med Central.*

Rivero-Urgell, M., & Santamaria-Orleans, A., (2001). Oligosaccharides: Application in infant food. *Early Human Development*, *65*, S43–S52.

Ross, R. P., Desmond, C., Fitzgerald, G. F., & Stanton, C., (2005). Overcoming the technological hurdles in the development of probiotic foods. *Journal of Applied Microbiology*, *98*(6), 1410–1417.

Round, J. L., & Mazmanian, S. K., (2009). The gut microbiota shapes intestinal immune responses during health and disease. *Nature Reviews Immunology*, *9*(5), 313.

Roy, D., (2005). Technological aspects related to the use of bifidobacteria in dairy products. *Le Lait*, *85*(1/2), 39–56.

Ruiz, L., Margolles, A., & Sánchez, B., (2013). Bile resistance mechanisms in Lactobacillus and Bifidobacterium. *Frontiers in Microbiology*, *4*, 396.

Saarela, M., Alakomi, H. L., Mättö, J., Ahonen, A. M., & Tynkkynen, S., (2011). Acid tolerant mutants of *Bifidobacterium animalis* subsp. lactis with improved stability in fruit juice. *LWT-Food Science and Technology*, *44*(4), 1012–1018.

Saarela, M., Mogensen, G., Fonden, R., Mättö, J., & Mattila-Sandholm, T., (2000). Probiotic bacteria: Safety, functional and technological properties. *Journal of Biotechnology*, *84*(3), 197–215.

Salminen, S., & Gueimonde, M., (2004). Human studies on probiotics: What is scientifically proven. *Journal of Food Science*, *69*(5), M137–M140.

Sánchez, B., Ruiz, L., Gueimonde, M., Ruas-Madiedo, P., & Margolles, A., (2012). Toward improving technological and functional properties of probiotics in foods. *Trends in Food Science & Technology*, *26*(1), 56–63.

Saranraj, P., (2013). Lactic acid bacteria and its antimicrobial properties a review. *International Journal of Pharmaceutical & Biological Archive*, *4*(6), 1124–1133.

Savard, T., Gardner, N., & Champagne, C. P., (2003). Growth of Lactobacillus and Bifidobacterium cultures in a vegetable juice medium, and their stability during storage in a fermented vegetable juice. *Sciences des Aliments (France)*, *23*(2), 273–283.

Schieber, M., & Chandel, N. S., (2014). ROS function in redox signaling and oxidative stress. *Current Biology*, *24*(10), R453–R462.

Settachaimongkon, S., Van Valenberg, H. J., Winata, V., Wang, X., Nout, M. R., Van Hooijdonk, T. C., & Smid, E. J., (2015). Effect of sublethal preculturing on the survival of probiotics and metabolite formation in set-yogurt. *Food Microbiology*, *49*, 104–115.

Shah, N. P., (2000). Probiotic bacteria: Selective enumeration and survival in dairy foods. *Journal of Dairy Science*, *83*(4), 894–907.

Shah, N. P., (2007). Functional cultures and health benefits. *International Dairy Journal*, *17*(11), 1262–1277.

Shehata, M. G., El Sohaimy, S. A., El-Sahn, M. A., & Youssef, M. M., (2016). Screening of isolated potential probiotic lactic acid bacteria for cholesterol lowering property and bile salt hydrolase activity. *Annals of Agricultural Sciences*, *61*(1), 65–75.

Shen, Q., Shang, N., & Li, P., (2011). *In vitro* and *in vivo* antioxidant activity of Bifidobacterium animalis 1 isolated from centenarians. *Current Microbiology*, *62*(4), 1097–1103.

Sip, A., & Grajek, W., (2010). Probiotics and prebiotics. *Functional Food Product Development* (pp. 146–177). Blackwell Publishing Ltd.

Stack, H. M., Kearney, N., Stanton, C., Fitzgerald, G. F., & Ross, R. P., (2010). Association of beta-glucan endogenous production with increased stress tolerance of intestinal lactobacilli. *Applied and Environmental Microbiology*, *76*(2), 500–507.

Stecchini, M. L., Del Torre, M., & Munari, M., (2001). Determination of peroxy radical-scavenging of lactic acid bacteria. *International Journal of Food Microbiology*, *64*(1/2), 183–188.

Talwalkar, A., Miller, C. W., Kailasapathy, K., & Nguyen, M. H., (2004). Effect of packaging materials and dissolved oxygen on the survival of probiotic bacteria in yogurt. *International Journal of Food Science & Technology*, *39*(6), 605–611.

Tennant, S. M., Hartland, E. L., Phumoonna, T., Lyras, D., Rood, J. I., Robins-Browne, R. M., & van Driel, I. R., (2008). Influence of gastric acid on susceptibility to infection with ingested bacterial pathogens. *Infection and Immunity*, *76*(2), 639–645.

Tsai, C. C., Lin, P. P., Hsieh, Y. M., Zhang, Z. Y., Wu, H. C., & Huang, C. C., (2014). Cholesterol-lowering potentials of lactic acid bacteria based on bile-salt hydrolase activity and effect of potent strains on cholesterol metabolism *in vitro* and *in vivo*. *The Scientific World Journal*, *2014*: 690752, 10 pages.

Upadrasta, A., Stanton, C., Hill, C., Fitzgerald, G. F., & Ross, R. P., (2011). Improving the stress tolerance of probiotic cultures: Recent trends and future directions. In: *Stress Responses of Lactic Acid Bacteria* (pp. 395–438). Springer, Boston, MA.

Van Hemert, S., Meijerink, M., Molenaar, D., Bron, P. A., De Vos, P., Kleerebezem, M., & Marco, M. L., (2010). Identification of *Lactobacillus plantarum* genes modulating the cytokine response of human peripheral blood mononuclear cells. *BMC Microbiology*, *10*(1), 293.

Van Tassell, M. L., & Miller, M. J., (2011). Lactobacillus adhesion to mucus. *Nutrients*, *3*(5), 613–636.

Vanderpool, C., Yan, F., & Polk, D. B., (2008). Mechanisms of probiotic action: Implications for therapeutic applications in inflammatory bowel diseases. *Inflammatory Bowel Diseases*, *14*(11), 1585–1596.

Vinderola, C. G., Costa, G. A., Regenhardt, S., & Reinheimer, J. A., (2002b). Influence of compounds associated with fermented dairy products on the growth of lactic acid starter and probiotic bacteria. *International Dairy Journal*, *12*(7), 579–589.

Vinderola, C. G., Mocchiutti, P., & Reinheimer, J. A., (2002a). Interactions among lactic acid starter and probiotic bacteria used for fermented dairy products. *Journal of Dairy Science, 85*(4), 721–729.

Von Ossowski, I., Reunanen, J., Satokari, R., Vesterlund, S., Kankainen, M., Huhtinen, H., & Palva, A., (2010). Mucosal adhesion properties of the probiotic *Lactobacillus rhamnosus* GG SpaCBA and SpaFED pilin subunits. *Applied and Environmental Microbiology, 76*(7), 2049–2057.

Wagner, R. D., Pierson, C., Warner, T., Dohnalek, M., Hilty, M., & Balish, E., (2000). Probiotic effects of feeding heat-killed *Lactobacillus acidophilus* and *Lactobacillus casei* to Candida albicans-colonized immunodeficient mice. *Journal of Food Protection, 63*(5), 638–644.

Wang, Y., Wu, Y., Wang, Y., Xu, H., Mei, X., Yu, D., & Li, W., (2017). Antioxidant properties of probiotic bacteria. *Nutrients, 9*(5), 521.

Yan, F., & Polk, D. B., (2011). Probiotics and immune health. *Current Opinion in Gastroenterology, 27*(6), 496.

Yoon, K. Y., Woodams, E. E., & Hang, Y. D., (2004). Probiotication of tomato juice by lactic acid bacteria. *The Journal of Microbiology, 42*(4), 315–318.

Yoon, K. Y., Woodams, E. E., & Hang, Y. D., (2005). Fermentation of beet juice by beneficial lactic acid bacteria. *LWT-Food Science and Technology, 38*(1), 73–75.

Yoon, K. Y., Woodams, E. E., & Hang, Y. D., (2006). Production of probiotic cabbage juice by lactic acid bacteria. *Bioresource Technology, 97*(12), 1427–1430.

Yusuf, S., Hawken, S., Ôunpuu, S., Dans, T., Avezum, A., Lanas, F., & Lisheng, L., (2004). Effect of potentially modifiable risk factors associated with myocardial infarction in 52 countries (the inter heart study): Case-control study. *The Lancet, 364*(9438), 937–952.

CHAPTER 11

Food Allergens: Detection and Management

JON JYOTI KALITA,[1] PAPORI BURAGOHAIN,[1]
PONNALA VIMAL MOSAHARI,[2] and UTPAL BORA[1,2]

[1]Department of Biosciences and Bioengineering, Indian Institute of Technology Guwahati–781039, Assam, India

[2]Center for the Environment, Indian Institute of Technology Guwahati–781039, Assam, India, E-mail: ubora@iitg.ernet.in

ABSTRACT

Food allergy is one of the public health concerns whose prevalence is now increasing globally. Food allergy was initially identified to be prevalent mostly in European countries, however, introduction of newly explored ingredients in food preparation and increased use of processed foods have given rise to it's prevalence in other parts of the world including India. New medical studies in a country like India has revealed the prevalence of food allergies in the Indian population, which can be referred to as just the tip of the iceberg. Food allergies can cause a mild immune response to severe cases of anaphylaxis and death. The presence of allergens in trace amount in food items by cross contamination or raw material adulteration possesses a risk for sensitized individuals. Several analytical methods based on immunological assays, mass-spectroscopy, DNA amplification and hybridization have been developed to detect allergenic peptides and proteins, but each method has its own advantages and disadvantages. Challenges lie in the detection of allergens hidden in the complex food matrix of raw materials and especially in processed foods. New generation analytical method like biosensors based on different transducers like optical, electrochemical, and piezoelectric have been developed to address the need for food process industries. Application of nanomaterial in biosensor enables the development of analytical methods with more sensitivity, selectivity, portability, and time and cost-effectiveness. Despite the regulation like food allergens labeling, undeclared allergens in food items cause trouble. Food allergen management can be done in food industries by applying standard guidelines based on GMP and HACCP, using risk assessment tactics and communicating the risk with consumers. In this book chapter, we discuss the advancements in various food allergens detection analytical methods and management of it in food manufacturing industries.

11.1 INTRODUCTION

Although we frequently consider allergy as a Western disease, in realism it is recklessly becoming a global ailment with increased prevalence in developed nations. In the last two to three decades food allergies have reportedly affected 5% of adult and 8% of children population of western countries (Sathe et al., 2016; Sicherer and Sampson, 2018). Food allergies are also not uncommon in Asia and Africa, with societies increasingly adopting a Western lifestyle, are noticing expanding rates of allergic syndrome across age groups especially in children. Even in India 1–2% of the total population is affected by food allergies (Gangal and Malik, 2003).

The growing use of processed and readymade food items, food safety issue has become a matter of concern. At the present situation, a cure for food allergy is not available, because the overall understanding of the underlying molecular mechanisms involving the immune reactions is lacking (Sathe et al., 2016b). The only treatment available currently is to avoid the foods causing allergy. In western nations food allergies are a matter of great concern with government agencies creating regulations on allergens and offending foods. Proper food labeling has become very important to maintain consumer safety on a worldwide basis (Prescott et al., 2013).

Allergies are hypersensitivity of the immune system characterized by the production of IgE antibodies against the allergens (antigen that cause allergic reactions) that may enter the human body through different exposure routes: oral, respiratory, skin, gastrointestinal, etc., (Sampson and Burks, 1996). The term "Food allergy," a type of hypersensitivity (type I hypersensitivity/immediate hypersensitivity) is the manifestation occurring due to an immune response towards some food proteins (potential allergens) causing vigorous immunologic reactions involving IgE antibodies in susceptible hosts. Although food allergies mostly involve IgE mediated immune reactions, but other immune mechanisms resulting from T-cell mediated inflammation or eosinophil-mediated reactions may also play some role. On the other hand, "Food Intolerance" doesn't involve immunologic reactions. Food allergy and Food intolerance can be grouped under "Adverse Food reactions," which is a broad term used both by a layman and a clinician and doesn't require the knowledge of the underlying pathophysiology (Johansson et al., 2001; Lammie and Hughes, 2016; S. L. Taylor and Hefle, 2001). Understanding these differences in immune and non-immune mediated allergic reactions is necessary for proper management and awareness of this illness. The following flow chart represents a classification scheme of Food allergies and Food Intolerances (Figure 11.1).

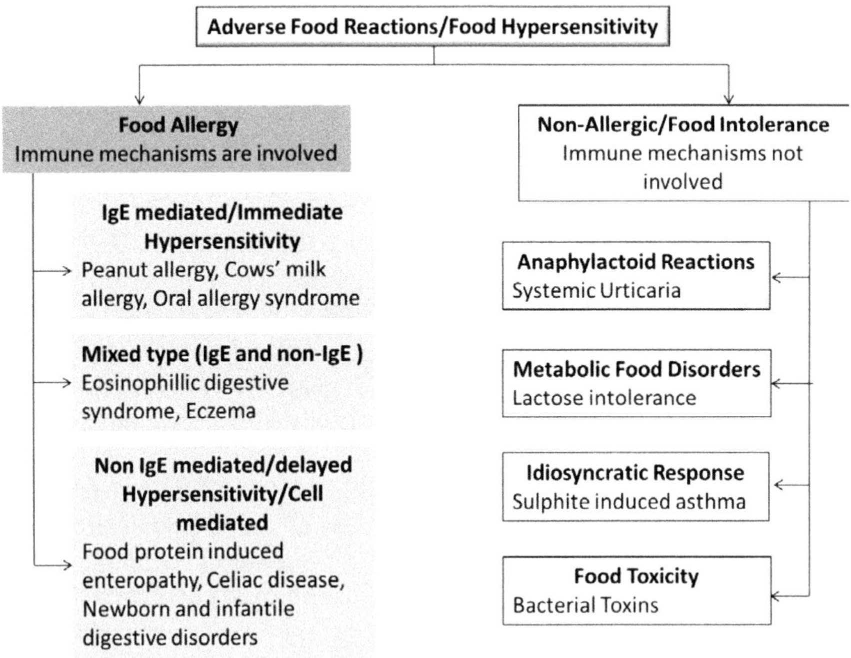

FIGURE 11.1 A schematic representation of food allergies and food intolerances. (*Source:* Burks et al. (2012); Flanagan (2014); Lee and Burks, (2006); Sathe, Liu and Zaffran, (2016); Sicherer and Sampson, (2018); Taylor and Lehrer, (1996); Valenta et al. (2010).

11.1.1 CHARACTERISTICS AND TYPES OF FOOD ALLERGENS

As per the definition, any edible or potable substance usually of animal or plant origin with nutritional benefits that people or animals consume in order to maintain life and growth is called food. It is consumed to provide nutritional support to an organism and at the same time most of us also find eating a pleasing experience. However, few individuals who are hypersensitive and susceptible to adverse food reactions fear the debilitating responses that may occur due to consumption of the offending food or food ingredients present therein. Therefore, food selection becomes tedious because very tiny amount of the offending substance (allergen) may cause allergic reactions that may cause a mild immune responses to life-threatening anaphylactic shock and eventual death of an individual in rare cases (Sampson and Burks, 1996; Steve Taylor and Baumert, 2010).

Allergens are mostly proteins and peptides. Similarly, Food allergens are also allergenic proteins with significant variability in their nature and structure. They may have multiple structural active sites or conformational epitopes to interact with the immune system (Flanagan, 2014). In the past decades, the structures of the common allergenic protein or peptides have been elucidated by molecular cloning (Valenta et al., 2010), which have helped in better understanding of their mode of action having clinical relevance in diagnosis, therapy, and prevention.

Mostly food allergens vary in their molecular weight between 10,000–70,000 Daltons where the size of 10 kDa possibly represents the lower limit of an allergen to cause allergic reactions and the upper limit may be because of the restricted mucosal absorption for bigger molecules (Sathe et al., 2016). Food allergens are commonly glycoproteins and their specific size range, water-solubility, and high stability required for their survival and transport in the host, which is sufficient to sensitize the susceptible organism (Lee and Burks, 2006; Pekar et al., 2018; Sathe et al., 2016). Modifications in their native form will change their antibody binding and crosslinking affinity and therefore, preservation of their unique three-dimensional conformations are critical in immune reactions. Furthermore, an improved understanding of the allergen proteins and their structural epitopes is necessary to understand this multifaceted issue of food allergy and provide supportive risk management guidance for the affected population (Sathe et al., 2016; Taylor and Lehrer, 1996; Steve Taylor and Baumert, 2010).

Several foods are now known to induce allergenic reactions, most of which are attributed to few major allergenic foods viz. milk, egg, peanut, soy, tree nuts (almonds, walnuts, hazelnuts, pistachios, pine nuts, etc.), wheat, fish, sesame, and mustard. Other more localized allergenic foods like allergic rice varieties and buckwheat also exist but yet to be characterized fully (Tables 11.1 and 11.2).

TABLE 11.1 List of Common Food Allergens and Their Sources*

Food	Allergen	IUIS name	Molecular weight
Bovine Milk	α-Lactalbumin	Bos d 4	14.2 kDa
	Immunoglobulin G	Bos d 7	160 kDa
	β-Lactoglobulin	Bos d 5	18.3 kDa
	Bovine serum albumin	Bos d 6	67 kDa
	αS1-Casein	Bos d 9	25 kDa
	αS2-Casein	Bos d 10	26 kDa
	β-Casein	Bos d 11	25 kDa
	κ-Casein	Bos d 12	21.2 kDa
Chicken Egg	Ovomucoid	Gal d 1	28 kDa
	Ovalbumin	Gal d 2	44 kDa
	Ovotransferrin	Gal d 3	78 kDa
	Lysozyme C	Gal d 4	14 kDa
	Serum albumin	Gal d 5	69 kDa
	YGP42	Gal d 6	35 kDa
Crustaceans	Tropomyosin	Cha f 1, Cra c 1, Hom a 1, Lit v 1, Mac r 1, Mel l 1, Met e 1, Pan b 1, Pan s 1, Pen a 1, Pen i 1, Pen m 1, Por p 1	34–39 kDa
	Myosin light chain 1	Art fr 5, Cra c 5	17.5 kDa
	Myosin light chain 2	Hom a 3, Lit v 3, Pen m 3	20–23 kDa
	Troponin C	Cra c 6, Hom a 6, Pen m 6	20–21 kDa
	Sarcoplasmic calcium-binding protein	Cra c 4, Lit v 4, Pen m 4, Pon l 4	20–25 kDa

TABLE 11.1 *(Continued)*

Food	Allergen	IUIS name	Molecular weight
	Arginine kinase	Cra c 2, Lit v 2, Pen m 2	40–45 kDa
	Triosephosphate isomerase	Arc s 8, Crac 8	28 kDa
	Troponin I	Pon l 7	30 kDa
Mollusks	Tropomyosin	Hel as 1, Tod p 1	36–38 kDa
Finfishes	β-Parvalbumin	Clu h 1, Cyp c 1, Gad c 1, Gad m 1, Lat c 1, Lep w 1, Onc m 1, Sal s 1, Sar sa 1, Seb m 1, Thu a 1, Xip g 1	11–12 kDa
	β-Enolase	Gad m 2, Sal s 2, Thu a 2	47.3–50 kDa
	Aldolase A	Gad m 3, Sal s 3, Thu a 3	40 kDa
	Vitellogenin	Onc k 5	18 kDa
	Tropomyosin	Ore m 4	33 kDa
Tree Nuts	2S albumin	Ana o 3, Ber e 1, Car i 1, Cor a 14, Jug n 1, Jug r 1, Pis v 1	7–16 kDa
	7S vicilin	Ana o 1, Cor a 11, Jug n 2, Jug r 2, Pis v 3	44–55 kDa
	11S legumin	Ana o 2, Ber e 2, Car i 4, Cor a 9, Jug r 4, Pis v 2, 5, Pru du 6	40–60 kDa
	Chitinase	Cas s 5	32 kDa
	nsLTP	Cas s 8, Cor a 8, Jug r 3, Pru du 3	9–13 kDa
	Bet v 1–related protein	Cor a 1	17 kDa
	Profilin	Cor a 2, Pru du 4	14 kDa
	Heat shock protein	Cas s 9	17 kDa
	Oleosin	Cor a 12, 13	14–17 kDa
	Manganese superoxide dismutase	Pis v 4	25.7 kDa

TABLE 11.1 *(Continued)*

Food	Allergen	IUIS name	Molecular weight
	60S acidic ribosomal protein P2	Pru du 5	10 kDa
Legumes	2S albumin	Ara h 2, 6, 7, Gly m 8	15–17 kDa
	7S/8S vicilin	Ara h 1, Gly m 5, Len c 1, Lup an 1, Pis s 1, Vig r 2	44–64 kDa
	11S legumin	Ara h 3, Gly m 6	60 kDa
	Profilin	Ara h 5, Gly m 3	14–15 kDa
	Bet v 1–related protein	Ara h 8, Gly m 4, Vig r 1, 6	16–18 kDa
	nsLTP	Ara h 9, 16, 17, Len c 3, Pha v 3	8.5–11 kDa
	Oleosin	Ara h 10, 11, 14, 15	14–17.5 kDa
	Defensin	Ara h 12, Ara h 13	5–9kDa
	TLP	Act d 2, Cap a 1, Mal d 2, Mus a 4, Pru av 2, Pru p 2	20–30 kDa
Lentil	Vicilin	Len c 1	48 kDa
	Seed biotinylated protein	Len c 2	66kDa
	Non-specific lipid transfer protein 1	Len c 3	9kDa
Cereals	Profilin	Hor v 12, Ory s 12, Tri a 12	14 kDa
	nsLTP	Tri a 14, Zea m 14	9 kDa
	2S albumin	Fag e 2, Fag t 2	16 kDa
	7S vicilin	Fag e 3	19 kDa fragment
	α-Amylase inhibitor	Hor v 15	14.5 kDa
	α-Amylase	Hor v 16	47.8 kDa
	β-Amylase	Hor v 17	57.3 kDa

TABLE 11.1 *(Continued)*

Food	Allergen	IUIS name	Molecular weight
	Agglutinin	Tri a 18	21.2 kDa
	Thioredoxin	Tri a 25, Zea m 25	13–14 kDa
	ω-5 gliadin	Tri a 19	65 kDa
	γ-Gliadin	Tri a 20	35–38 kDa
	High-molecular-weight glutenin	Tri a 26	88 kDa
	Low-molecular-weight glutenin	Tri a 36	40 kDa
	α-Purothionin	Tri a 37	12 kDa
	γ-Hordein	Hor v 20	34 kDa
	γ-Secalin	Sec c 20	70 kDa
Vegetables and Fruits	Bet v 1–related protein	Act c 8, Act d 8, Api g 1, Dau c 1, Fra a 1, Mal d 1, Pruar 1, Pru av 1, Pru p 1, Pyr c 1, Rub i 1, Sola l 4	9–18 kDa
	Profilin	Act d 9, Ana c 1, Api g 4, Cap a 2, Cit s 2, Cuc m 2, Dau c 4, Fra a 4, Lit c 1, Mal d 4, Mus a 1, Pru av 4, Pru p 4, Pyr c 4, Sola l 1;	13–15 kDa
	nsLTP	Act c 10, Act d 10, 11, Api g 2, 6, Aspa o 1, Bra o 3, Cit 3, Cit r 3, Cit s 3, Fra a 3, Lac s 1, Mal d 3, Mor n 3, Mus a 3, Pruar 3, Pru av 3, Pru d 3, Pru p 3, Pun g 1, Pyr c 3, Rub i 3, Sola l 3, 6, Vit v 1	7–11 kDa
	TLP	Act d 2, Cap a 1, Mal d 2, Mus a 4, Pru av 2, Pru p 2	20–30 kDa
Seeds	2S albumin	Bra j 1, Bran n 1, Ses i 1, 2, Sin a 1;	7–15 kDa
	7S vicilin	Sesi 3	45 kDa
	11S globulin	Sesi 6, 7, Sin a 2	51–57 kDa
	nsLTP	Hel a 3, Sin a 3	9–12.3 kDa

TABLE 11.1 *(Continued)*

Food	Allergen	IUIS name	Molecular weight
	Profilin	Sin a 4	13–14 kDa
	Oleosin	Sesi 4, 5	15–17 kDa
Soybean	Non-specific lipid transfer protein	Gly m1	18–71 kDa
	Defensin	Gly m2	8 kDa
	Pathogen-related protein (PR-10)	Gly m3	17 kDa
	Profilin	Gly m3	14 kDa
	Cupin (7S vicilin like globulin)	Gly m Bd28K;	28 kDa
	Thiol protease of the papain superfamily	Gly m Bd30K	34 kDa
	Lectin, SBA (agglutinin)	Gly m Lectin	14.5 kDa
	Cupin (7S vicilin like globulin)	Gly m Bd 60 K	63–67 kDa
	11S Glycinin	Gly m glycinin G1, Gly m glycinin G2, Gly m glycinin G4	22–62 kDa
	Kunitz trypsin inhibitor	Gly m TI	20 kDa
Mushroom	Manganese superoxide dismutase	NA	24 kDa
	Mannitol dehydrogenase	NA	27 kDa

*(Adapted from Sathe et al. (2016) and Ekezie et al. (2018), http://www.allergome.com, http://www.allergen.org).

Abbreviations: IUIS (International Union of Immunological Societies).

TABLE 11.2 List of Few Allergen Database

Database	URL	Features
Allergen Online	http://www.allergenonline.org/	Identification of allergenic cross-reactive proteins developed by genetic engineering and food processing
Informal	https://farrp.unl.edu/resources/gi-fas/informall	Information on allergenic foods
IUIS/WHO allergen Nomenclature	http://www.allergen.org/	Approved and officially recognized allergens database by WHO or the IUIS
Allermatch databases	http://www.allermatch.org/database.html	Sequences of known allergenic proteins
The Compare Allergen database	http://comparedatabase.org/	Clinically relevant protein allergens with description and amino acid sequence
Allergen database for food safety	http://allergen.nihs.go.jp/ADS/	Allergenic proteins and low-molecular-weight (LowMolWt) allergenic compounds
Structural database of allergen proteins (SDAP)	http://fermi.utmb.edu/	Sequence and structure of allergen epitopes
Allergome	http://www.allergome.org/	Sequence, structural, and immunological information of allergens

Abbreviations: IUIS (International Union of Immunological Societies), WHO (World Health Organization)

11.1.2 PATHOGENESIS

Immune responses towards allergenic foods occur due to the specific interaction of allergenic moiety with immune cells and triggering of cascading events involving multiple immune effector cells and their chemical mediators. The common mechanisms evoking food allergies occur mostly due to allergen-specific IgE-mediated immunogenic reactions with blood basophils and tissue mast cells that express high-affinity receptor (FcεRI) on the surface. When food allergen crosslinked IgE binds to these effector cells, as a result of consequent events the various food allergenic symptoms in individuals start appearing (Lee and Burks, 2006). Adverse reactions to food allergens can occur through the common routes like oral cavity (oral allergy syndrome), respiratory tract (inhalation airborne food allergens suspending in vapor and with powder, etc. in food processing area), gastrointestinal, and cutaneous routes (contact urticaria) and causes sensitization. Prevalence of food allergies is higher in children due to the immature immune system and the mucosal barriers especially gut. The prevalent food allergies occurring in young children are called Class I food allergy occurring from ingestion of food like cow milk, chicken egg, fish, and legumes etc. The Class 2 food allergy is caused by allergens entering through nasal tract. Sensitization causes generation of IgE antibodies specific to the allergen in mast cells and basophils. Later exposure to the allergens antibodies binding to antigens and immune system activated as immune mediators releases (Sampson et al., 2003; Sicherer and Sampson, 2018) (Figure 11.2).

Allergic reactions do not only depends upon the responses by antibodies, cellular mechanism also play a role. Certain proteins present in food items evade gut mucosal barrier are engulfed up by tissue dendritic cells, broken down into peptides. MHC (major histocompatibility complex) class II molecules present on the surfaces of dendritic cells present the peptides to native CD4+ T which triggers the native T cells to proliferate into TH2 cells, resulting in the release of cytokines (IL-4, IL-5, IL-9, and IL-13). These stimulatedTH2 cells further activate naive B cells and start class-switching to IgEbyIL-4 and IL-13. TH2 stimulates further proliferation of the antigen-specific activated B cell clones. Likewise, activated antigen-presenting B cells further activate new TH2 cells. IgE produced due to stimulated B cell binds to the IgE receptors (FcεRI) presented on the surface of the mast cells and basophils. Granulocytes presenting immunoglobin IgE recognizes the antigens upon any later exposure that results in degranulation releasing chemical mediators like histamines, lipids, chemokines, and cytokines which is the acute phase reaction. The chemical mediator will signal the eosinophils and TH2 cell to mobilize at the site of reaction which will release IL-5 and IL-9. These Interleukins will further activate mast cell and eosinophils for production of other mediators for the late phase reaction. (Burks et al., 2012; Lee and Burks, 2006; Sampson and Burks, 1996; Sathe et al., 2016; S L Taylor and Lehrer, 1996) (Figure 11.3).

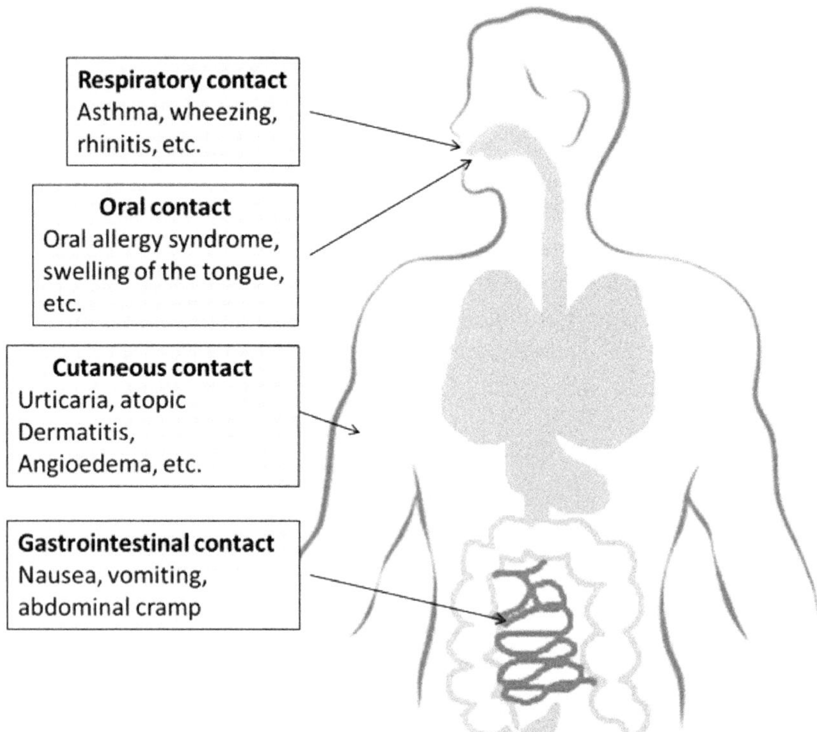

FIGURE 11.2 Diagrammatic representation of the exposure routes of common allergens.

11.2 ANALYTICAL METHODS FOR FOOD ALLERGEN DETECTION

Analytical methods are required to detect trace amount of allergen protein in food products to ensure the safety of the consumers and to comply with the food labeling regulation. There is a need for ideal analytical method which meets demands of optimal analytical performance parameters like sensitivity, specificity, reproducibility, precision, and accuracy. Also cost-effectiveness and ability performed in highly complex food matrix is also important criteria of analytical methods (Prado et al., 2015).

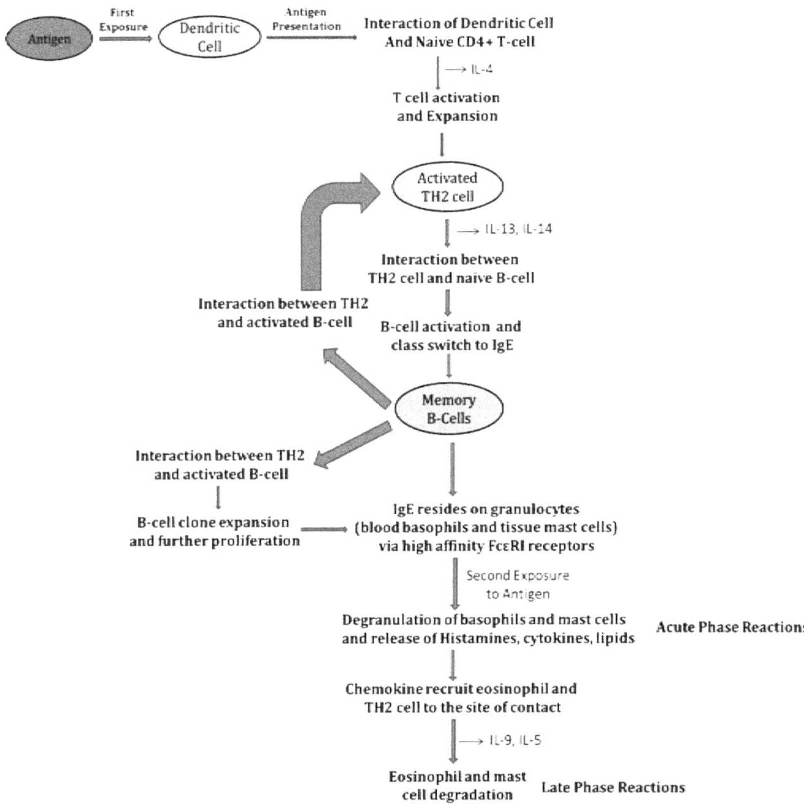

FIGURE 11.3 Mechanism of pathogenesis of IgE mediated food allergy.

11.2.1 CONVENTIONAL METHODS OF DETECTION

Allergens can be detected by different analytical methods which broadly falls under the followings categories: DNA-based methods, Immunoassay, Separation technique, Mass spectroscopic methods. These methods are currently used extensively in various laboratories and food industries. The methods adopted in organization are based on the types of food allergens present, complexity of food matrix, limit of detection required and economic viability.

11.2.1.1 IMMUNOASSAY

Immunoassay works on the principle of immune reaction of antibody against antigens which is the part of the natural defense mechanism of animals.

Immunoassays can detect the target allergens quantitatively using antibodies. It is a protein-based method that directly detect the allergen proteins or peptides by immune reactions. There are many variations of immunoassay developed at different point of time. Immunoassays include methods like radioallergosorbent test (RAST), enzyme allergosorbent test (EAST), rocket immuno-electrophoresis (RIE), immunoblotting, and enzyme-linked immunosorbent assay (ELISA). RIE and immunoblotting can detect food allergens semi-quantitatively while RAST, EAST, and ELISA detect quantitatively. Antibodies are central to immunoassays. Monoclonal antibodies recognize specifically one epitope of antigens, whereas polyclonal antibodies can recognize more than one epitopes in antigens. Polyclonal antibodies are cheaper than monoclonal antibodies, therefore it is widely used in food industries (Linda Monaci and Visconti, 2010). ELISA is most advanced and popularly used assay currently for food allergens detection (Poms, Klein, and Anklam, 2004).

RAST, EAST, and RIE are relatively old methods of allergen detection, usually use in clinical laboratories than the food industries. However in initial times, these methods were also used in allergens detection in food items (Poms et al., 2004). For allergens detection in food items, RAST, and EAST inhibition assay are performed where human IgE antibodies containing sera incubated with food extract sample may contain antigens (allergens). The incubated antibodies when added to solid phase immobilized antigens; presence of allergens in food sample will inhibit the binding of antibodies to immobilized antigens. Anti-IgE antibodies conjugated with a radioisotope (in RAST) will emit light indicating the binding of IgE to the immobilized antigens and this in turn implies the quantity of allergens (inversely related to intensity of emitted light) in food sample. In EAST, an enzyme like horseradish peroxidase is used instead of the radioisotope which forms Color on addition of a suitable substrate (Besler, 2001). RIE uses electrophoresis gel containing the antibody and antigens are allowed migrate through it; upon interaction, the antibody-antigen complex forms and precipitates in gel. Immunoblotting techniques involve the separation on proteins containing antigen in a SDS-PAGE gel. The protein bands resolved are transferred to a membrane where specific antibodies conjugated with enzyme or radioisotope detects the allergens (Besler, 2001; Poms et al., 2004).

1. **Enzyme-Linked Immunosorbent Assay (ELISA):** ELISA is commonly used old laboratory analytical technique. The reliability of ELISA has establish itself as the gold standard of allergen detection in clinical and food safety laboratories. This method is well known and

well-practiced among the food quality control technicians in industries and scientists in government regulatory laboratories around the world (Gasilova and Girault, 2015; Poms et al., 2004). Many ultrasensitive version of ELISA and advanced ELISA kits have been developed for ease of use by researchers and food safety technicians and their limit as low as in the scale of picograms. ELISA is very specific, highly sensitive and versatile technique. ELISA uses primary antibody raised against the allergen extract of food (Gasilova and Girault, 2015). ELISA is one of the most effective methods in detection of hidden allergens in food matrix. The antibodies are very specific in binding with antigens and this property enables the discriminating detection of allergens in complex food matrix (Poms et al., 2004). Quantification of the antigens can be done effectively as number of antigens binding to antibodies are directly related (Linda Monaci, Tregoat, Van Hengel, and Anklam, 2006). Gomaa and Ribereau, 2016 while evaluating commercial ELISA kits reported the limit of detection as low as 0.1 and 0.3 ppm for detecting egg and soy protein allergens respectively (Gomaa and Ribereau, 2016). ELISA has also been used in detection of food allergens in aerosol form that emitted in food processing setup (Kamath et al., 2014).

ELISA involves two types of strategies for food allergens detection, one is Sandwich ELISA and the other one is Competitive ELISA. In most of the allergen protein detection cases, sandwich ELISA has been used (Gasilova and Girault, 2015; Poms et al., 2004). In Sandwich ELISA, an antibody specific to the allergen, called capture antibody is coated in the inner surface of a microtiter plate. The sample solution may containing target allergen protein is added to the capture antibody. If the allergen is present in the analyzed sample, capture antibody will bind to it. Now another antibody conjugated with enzyme specific to the allergen, called detector antibody added to bind with the allergen bound to capture antibody. This will form a sandwich of antigenic allergen bound in between the capture and detector antibodies. The conjugated enzyme will catalyze a colorless substrate (usually tetramethylbenzidine) when added to it into a colored substrate, the measured absorbance of the color is proportional to the concentration of analyte. Limitation of sandwich ELISA is that the antigen should have more than one binding site to react with both the antigens (Poms et al., 2004; Immer and Lacorn, 2015).

Competitive ELISA is used when the analyte is relatively small size protein or a peptide or a hydrolyzed protein product. Antigenic allergen molecule is immobilized on the inner surface of a microtiter plate. The sample solution (food extract) may contain the target antigenic allergen is incubated with the primary antibody (Gasilova and Girault, 2015). After incubation, the primary antibody added to the antigen-coated microtiter plate. When the sample solution contains the food allergen, it will inhibit the binding of the primary antibody to the immobilized antibody. A secondary antibody conjugated with an enzyme specific to primary antibody added. The addition of substrate forms Color and the intensity of the produced Color is inversely proportional to amount of analyte present (Immer and Lacorn, 2015).

Although the ELISA is one of most used method for allergen detection, it also suffers from few limitations. ELISA in many cases was found to be not effective enough in allergen detection in thermally processed foods. Processing of food makes the detection of allergen protein complicated as the antigen binding site gets damaged by conformational change or hydrolysis, complex food matrix masks the binding site, alteration of the protein solubility, etc. (Koppelman et al., 2015; Linda Monaci and Visconti, 2009). ELISA also shows difference in result while using different manufacturer's kit or antibody. Using of ELISA in multiplex detection of large number of allergens are not yet very fruitful (Croote et al., 2017). ELISA method is very time consuming and labor intensive which is not well sought in industrial production setup (Zheng, Wang, Lu and Liu, 2012). ELISA may produce biased results as similar epitopic region present in other foods can cause cross-reactivity (García et al., 2017).

2. **Lateral Flow Devices (LFD):** These are portable, cost-effective version of immunoassay based on the principle of ELISA. LFD is semi-quantitative and mostly qualitative assay. LFD is membrane-based strip where analyte is added in sample zone, analyte is detected by antibodies in test zone after the movement of sample along the strip and a control zone indicating positive test. LFD assay can also be sandwich and competitive similar to ELISA. Many LFDs are available commercially for food allergen detection in food items like milk, soybean, hazelnut, almond, peanut, shellfish, gluten, and egg with reported LODs of 1 to 25 ppm (Prado et al., 2015).

11.2.1.2 SEPARATION TECHNIQUES AND MASS SPECTROSCOPIC BASED METHODS

Protein-based immunological methods are associated with several disadvantages like cross-reactivity leading to false positive results, interferences of some food matrix component with antibodies, the inability of antibody to recognize allergens due to structural alteration by various food processing operation leading to false negative results. Various separation techniques and mass spectroscopic methods have been traditionally utilized in proteomics offer advantages over immunological methods in food allergen detection (Koeberl, Clarke and Lopata, 2014). Separation techniques and mass spectroscopy based methods have molecular level sensitivity, unambiguous identification of allergens, reproducibility, and multiplex allergen detection in a single run (Pilolli et al., 2017). Two-dimensional polyacrylamide gel electrophoresis (2D-PAGE) and liquid chromatography (LC) are the two most popular separation techniques for allergens detection. 2D-PAGE separates the complex mixture of proteins into constituents proteins or peptides by independent orthogonal separations, first depending on the isoelectric points (pI) of protein and then by molecular weight. 2D-PAGE separates with high resolution, but direct detection of allergens suffers from low dynamic range of detection and prone to batch variations (Linda and Visconti, 2009). Liquid-phase separation methods, primarily liquid chromatography (LC) and others like capillary electrophoresis (CE) and field-flow fractionation (FFF) coupled with UV or fluorescence detectors can be utilized in direct detection and quantification of allergens. Although these methods offer higher dynamic range, resolution is not as good as 2D-PAGE as co-elution of other component in food matrix also possible (Gasilova and Girault, 2015). 2D-PAGE utilized as a prefractionilization methods before LC enables advantages of both the methods. Separation techniques coupled with mass spectrometer have been successful to a large extent in detection and quantification of allergenic proteins in various food items (Johnson et al., 2011). An ion source, a mass analyzer and a detector fundamentally construct the mass spectrometer. The most common ionization techniques used for biomolecules analysis are matrix-assisted laser desorption ionization (MALDI) and electrospray ionization (ESI). Four different types of mass analyzer are extensively used in MS are quadrupole (Q), ion-trap (IT; Q and Linear), time-of-flight (TOF) and Fourier-transform ion cyclotron resonance (FTICR) mass analyzers (Picariello et al., 2011). For application suitability, different combination of ionization technique and mass analyzer are available in MS systems such as

MALDI-TOF, ESI-IT, etc., (Koeberl et al., 2014). MS-based methods can be applied in food allergen identification, characterization, and quantification (Picariello et al., 2011). MS can identify the food allergen protein itself or the marker peptide of the allergen protein which may not be immunogenic. Selection of the marker peptide should be such that it is representative of the allergen, enough in quantity when allergen protein is hydrolyzed and should not be degraded in harsh food processing conditions (Gasilova and Girault, 2015). MS uses two strategies for allergen analysis, known as bottom up and top down. In bottom-up strategy, the extracted protein from food sample is digested with proteolytic enzyme like trypsin and obtained peptide fragments are analyzed in MS after a high throughput separation with method like LC. In top-down strategy, extracted protein in intact form is directly fragmented and analyzed inside MS. Identification and characterization of allergens in both the methods are done by using experimental data with referring specific protein databases like SEQUEST, MS-tag, and Mascot (Linda and Visconti, 2009; Prado et al., 2015). Quantification of the allergens can be done by intensities of the MS spectrum peak and also standard peptide can be added to protein extract as reference before running LC-MS for relative quantification (Gasilova and Girault, 2015). In bottom-up strategy, allergen protein can be identified based on the mass measured by MS technique like MALDI-TOF of the resultant enzyme hydrolyzed products which are unique to the allergen, known as peptide mass fingerprinting (PMF). Further fragmentation of digested peptide by MS techniques like MS^2 (e.g., LC-ESI-Q-TOF) provides the structural features related to amino acid sequence of the peptide fragment enabling characterization of the allergens, known as peptide-fragment fingerprint (PFF) (Linda and Visconti, 2009). New food allergens discovery has been made possible by the MS based proteomic studies of food proteins and this exercise is now called as 'Allergenomics.' By this strategy, new food allergens have been reported to be discovered in rice, clementine, kiwi, Nile perch and codfish (Gasilova and Girault, 2015). Various new methods of MS have been applied for allergen detection in a large number of food items. For instance, a MS/MS ion monitoring mass spectrometry has been standardized and used for rapid detection of fish allergen parvalbumin (Carrera et al., 2012). A micro-HPLC–ESI-MS/MS has been assessed for multi-allergen detection in a cookie food matrix with LODs of 0.1 μg/g for milk, 0.3 μg/g for egg and 2 μg/g for soy allergens (Monaci et al., 2014). A UHPLC–MS/MS method was reported for detection of milk (casein, whey protein), egg (yolk, white), soybean, and peanut allergens traces simultaneously in various complex and heat-processed food items like ice cream, chocolate, sauce, and processed cookies (Planque et al., 2016). MS produces very reliable results, but the equipment cost is high and requires specialized personnel to operate which makes it less suitable for routine analysis in small-scale food industries (Fernandes et al., 2015).

11.2.1.3 DNA BASED METHODS

DNA based method predominantly like PCR offers several advantages over the protein-based immunoassay. DNA are more stable and resistant to different harsh condition during food processing like thermal treatment, pH alteration, protease hydrolysis etc. than the protein molecules. Therefore, DNA based methods are used as alternative to immunoassays, where limitation in protein detection present. Although a very well established method in clinical and food safety research laboratory, DNA based methods are now gaining popularity in food processing industries. PCR methods are highly specific and sensitive which usually detects the amplified DNA sequence (gene) that codes the allergenic proteins or a marker DNA sequence of the gene (Prado et al., 2015; Słowianek and Majak, 2011). Use of designed Primers which are short complementary DNA sequences that binds to the start and end flanking regions of the target DNA sequence responsible for high specificity of PCR. But the PCR method does not detect the allergenic protein itself in the sample (Parker and Pereira, 2017).

The PCR method involves a thermostable DNA polymerase enzyme, the target DNA sequence called template, two primers specific to the target DNA, four deoxyribonucleotide triphosphates (dATP, dGTP, dCTP, and dTTP) and a buffer. The DNA sample from allergenic food sample is extracted and purified by suitable method based on food material whether it is plant or animal origin and this act as template DNA. The PCR follows three steps that are DNA melting (usually at 95°C), primer annealing (usually between 55 and 65°C) and DNA polymerization (usually between 60 and 72°C) in repeated cycles. The reactions are carried in a program automatically run device called thermocycler. The PCR is run for around 25–40 cycle to get sustainable amount of amplified DNA to be detected by various detection methods (Holzhauser and Röder, 2015). In simple PCR (End Point PCR), the amplified DNA can be separated using gel electrophoresis and visualized using UV light or fluorogenic DNA-intercalating dyes. This method of PCR is not quantitative, rather a qualitative technique. However, other advanced PCR methods like quantitative PCR and real-time PCR are generally used for quantification of food allergens (Prado et al., 2015; Fernandes et al., 2015; Holzhauser and Röder, 2015).

1. **PCR-ELISA or Enzyme-Linked Oligosorbent Assay (ELOSA):** PCR-ELISA method was developed in order to quantify the PCR products after simple PCR amplification. However this method is a semi-quantitative method. The amplified target DNA is labeled with biotin or digoxygenin. An immobilized DNA probe specific to the inner sequence of the PCR allowed to hybridize with the PCR products. A protein like avidin or streptavidin conjugated with enzyme binds to the biotin of the captured PCR products. The conjugated enzyme catalyzes a substrate to Color compound whose intensity is measured for the quantity of the DNA concentration. Advantage of this method is that the DNA probe allows the multiple detections of different allergens DNA when amplified together. (Poms et al., 2004; Prado et al., 2015).
2. **Quantitative PCR (qPCR):** It can quantify the amount of DNA amplified in real time. The qPCR method requires a sophisticated state of art real-time PCR setup for quantification of allergens. qPCR has several advantages as it allows to detect very small size product as electrophoresis gel visualization is not required here. It can be used for detecting highly fragmented DNA which is generally produced in highly processed foods. The chances of DNA contamination is less as it is confined in the reaction environment during amplification and analysis. Recent advancements in PCR technologies has made qPCR method to be highly precise, ultrasensitive, improved in dynamic range and resolution (García et al., 2017). qPCR in initial times, used a DNA intercalating dye like SYBR-Green which is cheaper for real-time quantification. But it has drawback of binding to both specific and non-specific PCR products. Probe-based quantification overcomes this problem. TaqMan probe used in qPCR is single-stranded nucleotide complementary to primers whose 5′ end contains a fluorescent reporter (donor) dye and 3′ end contains a quencher (acceptor) dye that works on the principle of FRET (Fluorescence resonance energy transfer). During amplification step, the Taq polymerase hydrolyzes the probe by 5′-3′ double-strand-specific exonuclease activity and this results in increase of fluorescence intensity (García et al., 2017; Poms et al., 2004; Prado et al., 2015). For quantification, qPCR express the amplification results as values in the plot of fluorescence intensity changes versus cycle number. The values are expressed as Cp (Crossing point), Ct (threshold cycle) or Cq (Quantification cycle), which are the PCR cycle when the amplified PCR products cross a threshold limit

beyond which fluorescence signal is detectable. Cq values have an inverse relationship with the target DNA starting copy number (Salihah et al., 2016). Quantification of allergens concentration from the DNA copy needs other strategies like the use of standard curve or use of internal reference DNA. In internal reference strategy, an internal standard material whose concentration known is added during DNA extraction or amplification process and concentration of the target allergens is calculated from the correlation of known internal standard material. However in most of the studies, authors used standard curve where known quantity of DNA is amplified in same rate parallel to the target DNA from allergenic food. Based on the known DNA to known allergen standard curve, the target allergens are quantified. However these methods do not certainly relate the quantity of target DNA to allergens in food matrix due to difference in reference material DNA size and food matrix composition (Linacero et al., 2016; Resea and Resea, 2011; Prado et al., 2015). qPCR also faces some problem with accurate quantification of allergens as in many processing operations of food like roasting, baking, pressure cooking etc., the detectable copy number of target DNA is lost. Also, in natural presence of some PCR inhibitors of some allergen food in the food matrix makes the inaccurate allergen quantification (Linacero et al., 2016). An advanced version of qPCR which uses ligation of bipartite hybridization probes can simultaneously detect and quantify multiple allergens DNA in single PCR assay which was demonstrated for peanut, cashew nut, pecan nut, pistachio nut, almond, hazelnut, walnut, sesame seeds, brazil nut and macadamia nut allergens (Linda and Visconti, 2010).

11.2.2 NEW METHODS OF DETECTION

New advanced techniques have been applied for development of detection devices that holds promises to overcome the existing drawbacks of the conventional methods of allergen detection. Methods like ELISA and PCR have been well established and reliable in current practice. But these methods are laborious, time-consuming, costly, prone to human error, requires state-of-art laboratory and expert technicians. In industrial scenario, quality control of food manufacturing needs online monitoring and real-time detection of allergens, less human error, reproducibility, portability of detection devices. For small and medium scale industries (SME) to maintain food safety, cost-effectiveness of the assays is crucial. Biosensors for food allergen detection have been developed in order to meet these demands from food processing industries. Advancements in nanotechnology have led to use of different nanomaterials in biosensing application for improved detection and biological signal amplification.

1. **Biosensors:** These are analytical devices which contain three basic parts, a biorecognition element such as antibody, aptamer, etc., a transducer which converts the biochemical activities to a measurable electrical signal and a display device along with signal processor that shows the end results of the assay. For detection of food allergens mainly antibody-based and aptamer-based biosensors have been developed and these are found effective against various food allergens. Researchers in recent years have devoted tremendous effort for bringing up biosensors for allergen detection which would be cost-effective, easy to perform, reproducible, real-time and fast, multiplex detection. Although challenges lie in achieving all the goals for a perfect biosensor, however success has been met in detecting various allergens to a great extent (Pilolli et al., 2013).

   Biorecognition elements in food allergen biosensors used mainly are antibodies, aptamers, and gene probes. It is called immunosensor when uses antibodies which are the same molecules that used in the immunoassays like ELISA and LFD. Based on the type of allergen molecule and sensitivity required, either monoclonal or polyclonal antibodies can be used as biosensors. Monoclonal antibodies interact with a single allergenic epitope, which increase their specificity and sensitivity to detect even the trace amount of the single allergen and reduced cross-reactivity with other food proteins. Monoclonal antibodies are produced by hybridoma technology which is costly. Polyclonal antibodies are raised in small animals against the protein extract of allergen food which recognizes more than one antigenic epitopes and their production cost is cheaper compared to monoclonal antibodies. The antibodies are generally immobilized on the sensor surface and antigen-antibody interaction is sensed by the transducer of the biosensor (Alves et al., 2016).

   Aptamers which offers some advantages over antibodies are now increasingly finding place in food allergen biosensors. Aptamers are small sequences of oligonucleotides (DNA and RNA), peptides or peptide nucleic acids (PNA) that can specifically binds to a diverse targets like small to large organic and inorganic molecules with high affinity (Sett et al., 2012). The aptamers form distinct secondary and tertiary structures by folding which enable them to interact with high specificity and affinity with the target molecule like antigen-antibody interaction. The aptamers can be selected against any target molecule by an *in-vitro* process called SELEX (Systematic Evolution of Ligands by EXponential enrichment). In SELEX process, the target molecule is allowed to interact with a library of oligonucleotide containing random sequences of order up to 10^{15}. The sequences having affinity for the target molecule are separated and amplified. The interaction, separation, and amplification process is then repeated in an iterative manner with increasing stringency during interaction. After several rounds, aptamer candidate with high affinity towards the target molecule is selected (Alves et al., 2016). Several aptamers have been selected against food allergens like egg lysozyme (Huang et al., 2009), Lup an 1(Nadal et al., 2012), Ara h1 (Tran et al., 2013), β-lactoglobulin (Eissa and Zourob, 2017). Aptamer are more stable, no batch to batch production variation, production process is simpler and can be modified with different other molecules unlike antibodies (Neethirajan et al., 2018).

Single-stranded oligonucleotide probes can also be used as biorecognition element which detects the complementary sequences of allergen DNA by hybridization. When used in biosensor, this type of sensor called genosensor. The advantages of using oligonucleotide probe that it is highly sensitive, more stable than antibodies, different types of modification can be done by labeling molecules and it allows multi allergens detection (Alves et al., 2016).

Biosensor developed for allergens in recent times mainly used optical, electrochemical, and piezoelectric transducers.

2. **Optical Biosensors:** These exploit the changes in optical properties to detect analyte. When an allergen binds to antibody or aptamer or any other biorecognition molecule immobilized on the sensor chip surface, then changes in the surface optical property like refractive index, reflection, light absorption of the incident light, or light scattering allows the transducer to produce respective electric signal indicating the quantity of allergen bound (Alves et al., 2016; Neethirajan et al., 2018). Several optical biosensors have been fabricated for food allergens detection using optical transducers on the principle of surface Plasmon resonance (SPR), localized surface plasmon resonance (LSPR), imaging SPR (iSPR), resonance enhanced absorption (REA), colorimetry, and fluorescent (Alves et

al., 2016). Detection of allergens by SPR has been demonstrated by many researchers. In SPR, biorecognition element (antibody or aptamer) are immobilized on a thin metal (mostly gold or silver) surface. A plane polarized light is incident through a prism on the metal surface and the light is reflected at certain angle. At certain resonance wavelength and angle (SPR angle), the incident light interacts with metal-free electrons (surface plasmon) and intensity of the reflected light decreases. The SPR angle is dependent on refractive index of the metal surface. When allergens bind to the biorecognition element on the metal surface and the refractive index changes which in turns changes the SPR angle. The change in angle is can be related amount of mass bound to the surface (Jonsson et Al., 2006). SPR is fully automated, highly sensitive, label-free, and produce results in real time, but SPR setup is not very cost-effective (Weng et al., 2016).

3. **Electrochemical Biosensors:** These uses transducers that senses the changes in the electrochemical property on an electrode surface and converts it to detectable electrical signal. The sensor system works as an electrochemical cell, which contains working and reference electrode. Oxidation and reduction due to presence of electroactive analyte can generate either a measurable current (amperometric) or a measurable potential or charge accumulation (potentiometric) based on the type of transducer used. Other electrochemical transducers also include conductometric, impedimetric, and field effect etc., (Grieshaber et al., 2008). The transducer mainly is working electrode upon which the biorecognition elements are attached by different surface modification and binding chemistry. The commonly used material of construction of working electrodes are platinum, gold, carbon, and silicon compounds (Alves et al., 2016). Electrochemical sensor offers the advantages of ease of use, low cost and ease of miniaturization (Pilolli et al., 2013). Amperometric biosensor monitors the current flow as a function of time due to oxidation and reduction of electroactive species associate with analyte of interest while maintaining a constant potential at working electrode. Current flow is the measure of analyte concentration (Grieshaber et al., 2008). Amperometric sensors when detecting allergens proteins uses redox mediators for electron transport. Advantage of amperometric biosensor is its high sensitivity and wide linear range (Wang, 2006). The current measurement of the working electrode can also be done while varying the potential and this type of measurement

is called voltammetry. The current peak value over linear potential proportional to the concentration of the analyte (Grieshaber et al., 2008). Several other strategies used in voltammetry are cyclic, linear sweep, square wave, stripping, and pulse voltammetry (Pilolli et al., 2013). Potentiometric biosensors work by measuring potential difference between working and reference electrodes when no current flows through them. The concentration of the analyte is found out by Nernst equation. Impedimetric biosensors have also been used in food allergen detection which measures the interfacial electron-transfer resistance at a functionalized electrode by applying a small sinusoidal voltage at a particular frequency and recording the respective current response (Alves et al., 2016).

4. **Piezoelectric Biosensor:** Piezoelectric or electromechanical biosensors use a transducer based on quartz resonator which generates electric signal due change in mass or thickness (Andreas et al., 2000). A piezoelectric transducer is known as quartz crystal microbalance (QCM) which generally resonates at a fundamental frequency. A biorecognition element is immobilized on the transducer surface and mass in the surface changes as it detects analyte which affects the resonant frequency of the microbalance. The change of resonant frequency can quantify the analyte mass. It has the advantage of high sensitivity and label-free detection. QCM based biosensors have been demonstrated in detection of shrimp allergen Pen a 1 (Xiulan et al., 2010).
5. **Application of Nanomaterial in Allergens Biosensing:** Nanotechnology has tremendously contributed in generation of nano-sized material with unique and improved physiochemical properties. These materials has been widely used in biosensor and other bioassay which immensely increases their specificity, sensitivity, rapidity, signal amplification, cost-effectiveness, and in situ detectability (Neethirajan et al., 2018). Nanomaterial are easy to synthesize, have high surface to volume ration and can be functionalized with different biomolecules like proteins, peptides, oligonucleotides etc. (Prado et al., 2015). Exploiting the advantages of unique optical, electrical, and electrochemical properties of Nanomaterial like metallic nanoparticles (NPs) of silver, gold, titanium etc., Iron oxide NPs, carbon, and graphene nanotubes and quantum dots (QD), food allergen biosensors with higher sensitivity than conventional biosensors have been developed (Neethirajan et al., 2018; Pilolli et al., 2013). Gold nanoparticles have been widely used in allergen

detection due advanced physical properties and high compatibility with biomolecules. In electrochemical biosensors, gold nanoparticles (AuNP) are used in transducers to facilitate effective electron transfer between the immobilized protein and the working electrode surface. Such strategy has be used in development of electrochemical biosensor for detection of peanut allergen Arh 2 and the sensitivity has been increased by 100 times than conventional methods (Liu et al., 2010). In many food allergen biosensor, graphene, and carbon tubes nanomaterial or their nanocomposites have been used in modification of electrode surface for better electrical and thermal conductivity and high mechanical strength (Eissa et al., 2017). A multilayer graphene-gold nanoparticle nanocomposite was deposited over a glassy carbon electrode (GCE) for development of an electrochemical stem-loop DNA biosensor which detects Ara h 1 gene with ultrasensitive detection limit of 0.041 fM in peanut milk beverages (Gómez-Arribas et al., 2018). In optical biosensors, use of NPs has greatly enhanced optical signal amplification. Several enhanced SPR based biosensors for allergen detection have been developed using different NPs. For instance, in an optical fiber SPR biosensor, magnetic nanobeads were used as label for secondary antibody for detecting the peanut allergen Ara h1 which enhanced the LOD of SPR biosensor by two orders of magnitude (from 9 to 0.09 μg/mL) (Pollet et al., 2011). Some metallic nanoparticle shows special phenomenon called LSPR, which has also been used in biosensors. LSPR can sensitively detect minute changes in dielectric environment around the NPs due to presence of the analyte (Sepúlveda et al., 2009). Based on LSPR induced optical process, several spectroscopic methods like REA, surface-enhanced fluorescence (SEF) and surface-enhanced Raman scattering (SERS) was developed which have been used in biosensors for food allergens detection (Gómez-Arribas et al., 2018). Due to high photoluminescence quantum yields, QD finds wide applications in biosensors. A fluorescent-linked immunosorbent assay (FLISA) was developed for detection of milk allergen α-lactoglobulin (α -La) by using CdSe/ZnS QDs conjugated monoclonal antibodies (Gómez-Arribas et al., 2018). Nanotechnology offers a lot of scope for innovation in developing nanobiosensor with superior detection capabilities, research carried out in this area certainly holds promises for a development of an ideal food allergen biosensor (Table 11.3).

TABLE 11.3 Example of Few Analytical Methods Applied for Detection of Few Food Allergens

Source Food	Allergenic Proteins/Peptides	Method of Detection	Sensitivity (Limit of Detection)	References
Cow's Milk	b-lactoglobulin (Bos d 5)	Peptide-based LC-MRM/MS	0.20 µg/ml	Ji et al., 2017
	Casein (Bos d 8)	SPR sensor	57.8 ng/mL	Ashley et al., 2017
	a-lactalbumin (Bos d 4)	Electrochemical magnetic beads-based immunosensor	11.0 pg/mL	Montiel et al., 2016
Egg	Ovomucoid (Gal d 1)	Electrochemical immunosensor	0.1 ng/ml	Benedé et al., 2018
	Ovalbumin (Gal d 2)	Graphene-based label-free voltammetric immunosensor	0.83 pg/mL	Eissa et al., 2013
Peanut	7S seed storage globulin (Ara h 1)	Electrochemical immunosensor	3.8 ng/ml	Alves et al., 2015
	2S albumin conglutin (Ara h6)	Anodic voltammetry gold immunosensor	0.27 ng/ml	Alves et al., 2017
Seafood	Tropomyosin from shrimp (Pen a 1)	Photoelectrochemical aptasensor	0.23 ng/mL	Tabrizi et al., 2017
Fish	Parvalbumin	Lateral flow immunoassay (LFIA)	5 µg/ml (Qualitative) 0.046 µg/ml (Quantitative)	Zheng et al., 2012
Tree nut	2S albumin from Brazil nut (Ber e 1)	TaqMan real-time PCR	2.5 mg/kg	de la Cruz et al., 2013
	Cor a 1 (PR-10/ Bet v 1 homolog) from hazelnut	ELISA	4 ng/ml	Trashin et al., 2011
Wheat	Gliadin	Graphite voltammetry immunosensor	7.11 µg/ml	Eksin et al., 2015
White mustard	2S albumin (Sin a 1)	LC–MS	5 ng/ml	Posada-Ayala et al., 2015

11.3 MANAGEMENT OF FOOD ALLERGENS

Current clinical practices recommend the total consumption restriction of the allergen food as treatment. For patients complete avoidance of the food sometimes is not achievable as processed food contains many ingredients and potential allergen contamination occurs during manufacturing, storage, and shipping. The food allergen management and control in industrial set up are therefore very necessary to mitigate the problem before reaching to the consumers. The current practice of food allergen management does not have an internationally agreed framework, but industries across countries applied their own standards. However standards based on universal common principles are desirable, few agencies in different countries have adopted such initiatives (Ward et al., 2010). Guidelines set by agencies and individual industries follow food allergens risk management based on GMP (Good Manufacturing Practice) and HACCP (Hazard Analysis and Critical Control Point) approach (Codex Alimentarius Commission, 1997). Food allergen risk management needs to be incorporated under standard food safety management system established for chemical and microbiological contaminants management (Crochane and Skrypec, 2014). One of the first requisite to apply HACCP in food allergen management is to form an allergen management team containing members from all the departments like quality control/assurance, manufacturing, engineering, warehousing, and regulatory affairs (Stone and Yeung, 2010). Principles of HACCP is applied to identify the food allergen hazards arise due to cross-contact in manufacturing operation and thus provide an opportunity to reduce the hazard to an acceptable level. Food allergen control should be applied to the whole product life cycles starting from providing training and awareness to employees, management at ingredients or raw materials supply and storage and handling level, at production area including raw material flow line to final product packaging line (Crochane and Skrypec, 2014). An allergen control plan needs to be set up in industries for effective management of food allergens. The allergen control plan ensures that no allergens contamination happens at different stages of production and it generally includes:

1. Formation of an allergen control team having members from various concern departments in the organization which would conduct risk assessment and designs an allergen management procedure to best suit the industry and safety of the customer.
2. Training and awareness program on food allergens for the staff in the industries.

3. Effective hygiene design of equipment and processing plant so that it allows proper monitoring and ease in elimination.
4. Design of product which avoids the uses of allergenic ingredients or raw material containing traces of allergens.
5. Zoning and segregation of allergenic foods during storage, handling, and processing so that chances of cross-contamination reduce.
6. Selection of supplier which has own allergen control plan that guarantees allergen free raw material.
7. Measures to be taken for prevention of allergen cross-contamination during the process by effective scheduling for processing of different food items at different time, use of dedicated equipment for specific item production, minimizing the reuse of processing water or oil etc.
8. Use of validated cleaning procedure to remove allergens from processing equipment.
9. Use of proper labeling to communicate information on allergens to customer. (Crevel, 2015; Jackson et al., 2008; Adams, 2018; Crochane and Skrypec, 2014).

One of the most critical points in food allergen management is the cleaning of the common equipment or processing lines used for manufacturing two or more different food products. Cleaning method includes dry cleaning and wet cleaning of the equipment to remove the residual allergens from the surface. The wet cleaning is found to be more effective as most allergenic proteins solubilize in hot water (Jackson et al., 2008).

Food allergens are relatively recent among the many issues of food safety, recognized only around thirty years back. With the global increase in the consumption of propriety, packaged, and processed food, it has become an unavoidable risk, which needs to be assessed and managed. There is still a lack of knowledge base on the risk factors to be considered and the susceptibility of the population across the various group, ages, and geographical distribution. Work spanning decades, have contradicted the notion that uniqueness in response of each allergens to each individual cannot be assessed with established risk assessment techniques (Crevel et al., 2014). However, due to its complexities in response mechanism, a simple safety assessment based on absolute population thresholds is inadequate. Risk assessment of the impact of hazard, forms the basis of risk management and decision-making. Risk assessment if not followed by management and communication, leads to ineffective decision making.

The basic elements in assessing the risk exposure include hazard identification, dose-response effect, exposure assessment and risk characterization.

Hazard Identification describes the toxic effects of the allergens, in dose-response effect, the dose that results in toxic effect is considered; in the exposure assessment, the intensity and duration of both direct and indirect exposure is considered. Risk assessment sums up the above three elements, and assess whether the exposure will cause any adverse effect to the exposed subject or population. Risk assessment in food allergens is more complex as not every individual is allergic towards one particular allergen and the reaction may vary among individuals. The susceptibility of individuals towards any food allergens vary across population and majority may not be affected even at high dosage. This makes it difficult to perform an extensive risk assessment. The dose exposure-response, exposure time, age, and frequency of exposure is difficult to document. The protein causing the risk of eliciting a response may be different than sensitizing protein (Madsen, 2001). For new proteins originating from genetically modified food and processed food, certain novel techniques are required. Hazard identification identifies and links the symptoms to the allergens. Unlike in chemical risk assessment, where it is easy to identify the exact cause of response, most of the times the whole particular food is considered to cause the allergy. It becomes important to identify the particular protein that causes allergy to the particular individual for proper hazard identification. Dose-response for individuals may vary over time, and hence more emphasis is given on establishing the cause of allergy. Hazard identification can be easily done, but dose-response will be unique for particular group or individuals. Identifying the proteins responsible is difficult. To carry out risk characterization for risk analysis and management, threshold for reaction needs to be identified, although not enough for detailed characterization. More the knowledge of threshold level across population, more accurately we can determine the safety level of any ingredient. In New Zealand and Australia, Voluntary Incidental Trace Allergen Labeling (VITAL) was constituted to help the food manufactures in managing cross-contamination of allergens in industries. Reference Doses ED01 (milk, egg, and peanuts) and ED05 (soybean) are the lower confidence interval for the safety of 99% and 95% of allergenic people, respectively. VITAL specifies, based on these references doses, food containing not more than 2.5 mg milk proteins, 5 mg peanut proteins, 0.75 mg egg proteins, or 25 mg soybean proteins per kilogram (portion size: 40 g) can be labeled as "allergen-free." VITAL advisory are not regulatory. VITAL risk assessment tool utilizes no observed adverse effect levels (NOAELs) and lowest observed adverse effect levels (LOAELs) data from patients' low dose oral clinical challenge studies. The VITAL risk assessment tool has attracted international interest and certain manufacturers were encouraged to use it (Zurzolo et al., 2012).

The effective communication to customer is crucial in food allergen management. Different regulations have been made in this regard. Although 160 different food materials are reported as allergenic, only few food items are responsible for 90% of all allergens. Milk, egg, fish, crustacean shellfish, tree nuts, wheat, peanut, and soybeans are identify as 'big eight' which causes most severe and frequent allergenic reactions. FDA's Food Allergy Labeling and Consumer Protection Act (FALCPA), 2004 in the USA requires labeling of the food product if it contains any of the eight food items. Labeling Directive (European Commission, 2000) and its amendments (European Commission, 2003, 2007) in European also issued directive for labeling the eight food items along with other six which are celery, mustard, sesame seeds, sulfur dioxide or sulfites, lupin, and mollusks. This food items may be intentionally added by food manufacturer and declared in labeling so that patients can avoid these foods. However, problem arises due to presence of undeclared allergens in trace amount in food items. Undeclared allergens or hidden allergens may be present in food due to cross-contaminating in handling of food, shared food processing equipment with other products in same factory and in ingredient switching or cross-reactivity due to similar protein from other ingredients (Prado et al., 2015; Puglisi and Frieri, 2007). FALCPA, 2004 regulations was not extended to the accidental cross-contamination of allergens from one food to another. In such cases precautionary advisory labeling (PAL) labeling ('may contain') has been applied by manufacturer to manage and communicate the consumers when manufacturer shares the equipment to produce more than on food. However, PAL was not declared mandatory by FALCPA stating that use of PAL indiscriminately should not be misleading while compromising with GMP (FDA, 2006). In 2011, FDA's Food Safety Modernization Act (FSMA) included provisions to control accidental food allergen cross-contact in manufactured foods (FSMA, 2011) (Adams, 2018).

11.4 CONCLUSION

Control of food allergens is very crucial as it is hazardous for consumer health and the producers face huge economic loss due to food recall. As demand for processed foods is massively increasing not only in developed countries, but in developed countries too, the challenges of food allergy increases. Detection and quantification of the food allergen is the core to total food allergens management. Different analytical methods developed have both advantages and disadvantages and one method may have advantages for particular allergens while others may be totally ineffective. Food industries uses ELISA to large extent, but other methods are also gaining popularity. SME require analytical method development which would be cost-effective and ensures reliable food allergens detection. Development of biosensors with applications from nanotechnology certainly holds promises for affordable, reliable, and portable food allergen detection devices. Such devices developed as a point-of-care device will also enable to personal allergen management devices in patients. Food allergen management requires attention of all the stakeholders from manufacturers, researchers, clinicians to policymakers. Use of processing methods to reduce allergenicity, exploration of substitute of allergens ingredients, development of clinical therapy of food allergy cure (like immunotherapy) and allergen management guidelines and regulatory practices together in a holistic approach food allergen management can be effectively achieved.

KEYWORDS

- **analytical methods**
- **food allergens detection**
- **food allergens management**
- **food allergies**

REFERENCES

Adams, T., (2018). *Allergen Management in Food Processing Operations: Keeping What is Not on the Package Out of the Product* (pp. 117–130). https://doi.org/10.1007/978-3-319-66586-3 (Accessed on 14 June 2019).

Alves, R. C., Barroso, M. F., González-García, M. B., Oliveira, M. B. P. P., & Delerue-Matos, C., (2016). New trends in food allergens detection: Toward biosensing strategies. *Critical Reviews in Food Science and Nutrition, 56*(14), 2304–2319. https://doi.org/10.1080/10408398.2013.831026 (Accessed on 14 June 2019).

Alves, R. C., Pimentel, F. B., Nouws, H. P. A., Marques, R. C. B., González-García, M. B., Oliveira, M. B. P. P., & Delerue-Matos, C., (2015). Detection of Ara h 1 (a major peanut allergen) in food using an electrochemical gold nanoparticle-coated screen-printed immunosensor. *Biosensors and Bioelectronics, 64*, 19–24. https://doi.org/10.1016/j.bios.2014.08.026 (Accessed on 14 June 2019).

Alves, R. C., Pimentel, F. B., Nouws, H. P., Silva, T. H., Oliveira, M. B. P., & Delerue-Matos, C., (2017). Improving the extraction of Ara h 6 (a peanut allergen) from a chocolate-based

matrix for immunosensing detection: Influence of time, temperature and additives, *Food Chem., 218,* 242–248.
Andreas, J., Hans-Joachim, G., & Claudia, S., (2000). Piezoelectric mass-sensing devices as biosensors—an alternative to optical biosensors? *Angewandte Chemie International Edition, 39*(22), 4004–4032. https://doi.org/doi:10.1002/1521–3773(20001117)39:22<4004:AID-ANIE4004>3.0.CO,2–2 (Accessed on 14 June 2019).
Ashley, J., Piekarska, M., Segers, C., Trinh, L., Rodgers, T., Willey, R., & Tothill, I. E., (2017). An SPR based sensor for allergens detection. *Biosensors and Bioelectronics, 88,* 109–113.
Benedé, S., Montiel, V. R. V., Povedano, E., Villalba, M., Mata, L., Galán-Malo, P., & Pingarrón, J. M., (2018). Fast amperometric immune platform for ovomucoid traces determination in fresh and baked foods. *Sensors and Actuators B: Chemical, 265,* 421–428.
Besler, M., (2001). *Determination of Allergens in Foods, 20*(11), 662–672.
Burks, A. W., Tang, M., Sicherer, S., Muraro, A., Eigenmann, P. A., Ebisawa, M., & Sampson, H. A., (2012). ICON: Food allergy. *Journal of Allergy and Clinical Immunology, 129*(4), 906–920. https://doi.org/10.1016/j.jaci.2012.02.001 (Accessed on 14 June 2019).
Cochrane, S., & Skrypec, D., (2014). Food allergen risk management in the factory–from ingredients to products. *In Risk Management for Food Allergy* (pp. 155–166)
Crevel, R. W. R., (2015). Food allergen risk assessment and management. In: *Handbook of Food Allergen Detection and Control* (pp. 41–66).
Crevel, R. W., Baumert, J. L., Baka, A., Houben, G. F., Knulst, A. C., Kruizinga, A. G., & Madsen, C. B., (2014). Development and evolution of risk assessment for food allergens. *Food and Chemical Toxicology, 67,* 262–276.
De La Cruz, S., López-Calleja, I. M., Alcocer, M., González, I., Martín, R., & García, T., (2013). TaqMan real-time PCR assay for detection of traces of Brazil nut (*Bertholletia excelsa*) in food products. *Food Control, 33*(1), 105–113.
Eissa, S., & Zourob, M., (2017). *In vitro* selection of DNA aptamers targeting beta-lactoglobulin and their integration in graphene-based biosensor for the detection of milk allergen. *Biosens. Bioelectron., 91,* 169–174. https://doi.org/10.1016/j.bios.2016.12.020 (Accessed on 14 June 2019).
Eissa, S., L'Hocine, L., Siaj, M., & Zourob, M., (2013). A graphene-based label-free voltammetric immunosensor for sensitive detection of the egg allergen ovalbumin. *Analyst, 138*(15), 4378–4384.
Ekezie, F. G. C., Cheng, J. H., & Sun, D. W., (2018). Effects of nonthermal food processing technologies on food allergens: A review of recent research advances. *Trends in Food Science & Technology, 74,* 12–25.
Eksin, E., Congur, G., & Erdem, A., (2015). Electrochemical assay for determination of gluten in flour samples. *Food Chemistry, 184,* 183–187.
Fernandes, T. J. R., Costa, J., Oliveira, M. B. P. P., & Mafra, I., (2015). An overview on fish and shellfish allergens and current methods of detection. *Food and Agricultural Immunology, 26*(6), 848–869. https://doi.org/10.1080/09540105.2015.1039497 (Accessed on 14 June 2019).
Gangal, S. V., & Malik, B. K., (2003). Food allergy–how much of a problem really is this in India? *Journal of Scientific and Industrial Research, 62*(8), 755–765.
García, A., Madrid, R., García, T., Martín, R., & González, I., (2017). *Food Allergens, 1592,* 95–108. https://doi.org/10.1007/978-1-4939-6925-8 (Accessed on 14 June 2019).
Gasilova, N., & Girault, H. H., (2015). Bioanalytical methods for food allergy diagnosis, allergen detection and new allergen discovery. *Bioanalysis, 7*(9), 1175–1190. https://doi.org/10.4155/bio.15.49 (Accessed on 14 June 2019).

Gomaa, A., & Ribereau, S., (2016). Detection of allergens in a multiple allergen matrix and study of the impact of thermal processing. *Journal of Nutrition & Food Sciences*, *9*, 1–6. https://doi.org/10.4172/2155–9600.S9–001 (Accessed on 14 June 2019).

Gómez-Arribas, L., Benito-Peña, E., Hurtado-Sánchez, M., & Moreno-Bondi, M., (2018). Biosensing based on nanoparticles for food allergens detection. *Sensors*, *18*(4), 1087. https://doi.org/10.3390/s18041087 (Accessed on 14 June 2019).

Grieshaber, D., MacKenzie, R., Vörös, J., & Reimhult, E., (2008). Electrochemical biosensors–sensor principles and architectures. *Sensors*, *8*(12), 1400–1458. https://doi.org/10.3390/s80314000 (Accessed on 14 June 2019).

Flanagan, S. (Ed.). (2014). In *Handbook of Food Allergen Detection and Control.* Elsevier.

Jackson, L. S., Al-Taher, F. M., Moorman, M., DeVries, J. W., Tippett, R., Swanson, K. M. J., & Gendel, S. M., (2008). Cleaning and other control and validation strategies to prevent allergen cross-contact in food-processing operations. *Journal of Food Protection*, *71*(41), 445–458. https://doi.org/10.4315/0362–028X-71.2.445 (Accessed on 14 June 2019).

Ji, J., Zhu, P., Pi, F., Sun, C., Sun, J., Jia, M., & Sun, X., (2017). Development of a liquid chromatography-tandem mass spectrometry method for simultaneous detection of the main milk allergens. *Food Control, 74*, 79–88.

Johansson, S. G., Hourihane, J. O., Bousquet, J., Bruinzeel-Koomen, C., Dreborg, S., Haahtela, T., & Van Cauwenberge, P., (2001). A revised nomenclature for allergy. An EEACI position statement from the EEACI nomenclature task force. *Allergy*, *56*, 813–814.

Kamath, S. D., Thomassen, M. R., Saptarshi, S. R., Nguyen, H. M. X., Aasmoe, L., Bang, B. E., & Lopata, A. L., (2014). Molecular and immunological approaches in quantifying the air-borne food allergen tropomyosin in crab processing facilities. *International Journal of Hygiene and Environmental Health*, *217*(7), 740–750. https://doi.org/10.1016/j.ijheh.2014.03.006 (Accessed on 14 June 2019).

Koeberl, M., Clarke, D., & Lopata, A. L., (2014). Next generation of food allergen quantification using mass spectrometric systems. *J. Proteome Res.*, *13*(8), 3499–3509. https://doi.org/10.1021/pr500247r (Accessed on 14 June 2019).

Koppelman, S. J., Söylemez, G., Niemann, L., Gaskin, F. E., Baumert, J. L., & Taylor, S. L., (2015). Sandwich enzyme-linked immunosorbent assay for detecting sesame seed in foods. *Bio. Med. Research International*, https://doi.org/10.1155/2015/853836 (Accessed on 14 June 2019).

Lammie, S. L., & Hughes, J. M., (2016). Antimicrobial resistance, food safety, and one health: The need for convergence. *Annual Review of Food Science and Technology*. https://doi.org/10.1146/annurev-food-041715–033251 (Accessed on 14 June 2019).

Lee, L. A., & Burks, A. W., (2006). Food allergies: Prevalence, molecular characterization, and treatment/prevention strategies. *Annual Review of Nutrition*. https://doi.org/10.1146/annurev.nutr.26.061505.111211 (Accessed on 14 June 2019).

Linacero, R., Ballesteros, I., Sanchiz, A., Prieto, N., Iniesto, E., Martinez, Y., & Cuadrado, C., (2016). Detection by real-time PCR of walnut allergen coding sequences in processed foods. *Food Chemistry*, *202*, 334–340. https://doi.org/10.1016/j.foodchem.2016.01.132 (Accessed on 14 June 2019).

Madsen, C., (2001). Where are we in risk assessment of food allergens? The regulatory view. *Allergy*, *56*(67), 91–93.

Monaci, L., & Visconti, A., (2009). Mass spectrometry-based proteomics methods for analysis of food allergens. *TrAC–Trends in Analytical Chemistry*, *28*(5), 581–591. https://doi.org/10.1016/j.trac.2009.02.013 (Accessed on 14 June 2019).

Monaci, L., & Visconti, A., (2010). Immunochemical and DNA-based methods in food allergen analysis and quality assurance perspectives. *Trends in Food Science and Technology, 21*(6), 272–283. https://doi.org/10.1016/j.tifs.2010.02.003 (Accessed on 14 June 2019).

Monaci, L., Pilolli, R., De Angelis, E., Godula, M., & Visconti, A., (2014). Multi-allergen detection in food by micro-high-performance liquid chromatography coupled to a dual cell linear ion trap mass spectrometry. *Journal of Chromatography A, 1358*, 136–144. https://doi.org/10.1016/j.chroma.2014.06.092 (Accessed on 14 June 2019).

Monaci, L., Tregoat, V., Van Hengel, A. J., & Anklam, E., (2006). Milk allergens, their characteristics and their detection in food: A review. *European Food Research and Technology* (Vol. 223). https://doi.org/10.1007/s00217-005-0178-8 (Accessed on 14 June 2019).

Montiel, V. R. V., Campuzano, S., Torrente-Rodríguez, R. M., Reviejo, A. J., & Pingarrón, J. M., (2016). Electrochemical magnetic beads-based immunosensing platform for the determination of α-lactalbumin in milk. *Food Chemistry, 213*, 595–601.

Neethirajan, S., Weng, X., Tah, A., Cordero, J. O., & Ragavan, K. V., (2018). Nano-biosensor platforms for detecting food allergens–New trends. *Sensing and Bio-Sensing Research, 18*, 13–30. https://doi.org/10.1016/j.sbsr.2018.02.005 (Accessed on 14 June 2019).

Picariello, G., Mamone, G., Addeo, F., & Ferranti, P., (2011). The frontiers of mass spectrometry-based techniques in food all ergonomics. Jou*rnal of Chromatography A, 1218(42), 7386–7398. https://doi.org/1*0.1016/j.chroma.2011.06.033 (Accessed on 14 June 2019).

Pilolli, R., De Angelis, E., & Monaci, L., (2017). Streamlining the analytical workflow for *multiplex MS/MS allergen d*etect*i*on in processed foods. Food Chemistry, 221, 1747–1753. https://doi.org/10.1016/j.foodchem.2016.10.110 (Accessed on 14 June 2019).

Pilolli, R., Monaci, L., & Visconti, A., (2013). Advances in biosensor development based on integrating nan*otechnology an*d *app*lied to food-allergen management. TrAC–Trends in Analytical Chemistry, 47, 12–26. https://doi.org/10.1016/j.trac.2013.02.005 (Accessed on 14 June 2019).

Planque, M., Arnould, T., Dieu, M., Delahaut, P., Renard, P., & Gillard, N., (2016). Advances i*n ultra-high performance liquid chr*om*at*ography coupled to tandem mass spectrometry for sensitive detection of several food allergens in complex and processed foodstuffs. Journal of Chromatography A, 1464, 115–123. https://doi.org/10.1016/J.CHROMA.2016.08.033 (Accessed on 14 June 2019).

Platteau, C., De Loose, M., De Meulenaer, B., & Taverniers, I., (2011). Detection of allergenic i*ngredients using real-time* PC*R: A* case study on hazelnut (Corylus avellena) and soy (Glycine max). Journal of Agricultural and Food Chemistry, 59(20), 10803–10814.

Pollet, J., Delport, F., Janssen, K. P. F., Tran, D. T., Wouters, J., Verbiest, T., & Lammertyn, J., (2011). Fast and accurate peanut allergen detect*ion with nanobead enhanced optical fiber SPR b*iosensor. Talanta, 83(5), 1436–1441. https://doi.org/10.1016/j.talanta.2010.11.032 (Accessed on 14 June 2019).

Poms, R. E., Klein, C. L., & Anklam, E., (2004). Methods for Allergen Analysis in Food: A Review: Food Additives *and Con*tam*i*nants (Vol. 21). https://doi.org/10.1080/02652030310001620423 (Accessed on 14 June 2019).

Posada-Ayala, M., Alvarez-Llamas, G., Maroto, A. *S., Maes, X., Muñoz-Garcia, E., Villalba, M., & C*u*esta-Herranz, J., (2015). Nove*l liquid chromatography-mass spectrometry method for sensitive determination of the mustard allergen Sin a 1 in food. Food Chemistry, 183, 58–63.

Prado, M., Ortea, I., Vial, S., Rivas, J., Calo-Mata, P., & Barros-Velázquez, J., (2015). Advanced DNA- and protein-based methods for the detection and investigation of food allergens. *Critical Reviews in* Food Science and Nutrition, 56(15), 2511–2542. https://doi.org/10.1080/10408398.2013.873767 (Accessed on 14 June 2019).

Prescott, S. L., Pawankar, R., Allen, K. J., Campbell, D. E., Sinn, J. K., *Fiocchi, A., & Lee, B.-W., (2013). A global s*urv*e*y of changing patterns of food allergy burden in children. The World Allergy Organization Journal, 6(1), 21. https://doi.org/10.1186/1939–4551–6–21 (Accessed on 14 June 2019).

Puglisi, G., & Frieri, M., (2007). Update on hidden food allergens and food labeling. Allergy and *Asthma Proceedings, 28(6), 634–639. h*ttps://doi.org/10.2500/aap.2007.6.3066 (Accessed on 14 June 2019).

Salihah, N. T., Hossain, M. M., Lubis, H., & Ahmed, M. U., (2016). Trends and advances in food ana*lysis by real-time polymerase* cha*i*n reaction. Journal of Food Science and Technology, 53(5), 2196–2209. https://doi.org/10.1007/s13197-016-2205-0 (Accessed on 14 June 2019).

Sampson, H. A., & Burks, A. W., (1996). Mechanisms of Food Allergy. *Annual Review of Nutrition, 16*(1), 161–177.

Sathe, S. K., Liu, *C., & Zaffran, V. D., (2016). Food al*lergy. Annual Review of Food Science and Technology. https://doi.org/10.1146/annurev-food-041715–033308 (Accessed on 14 June 2019).

Sepúlveda, B., Angelomé, P. C., Lechuga, L. M., & Liz-Marzán, L. M., (2009). LSPR-based n*anobiosensors. Nano Today, 4(3), 244–251. h*ttps://doi.org/10.1016/j.nantod.2009.04.001 (Accessed on 14 June 2019).

Sett, A., Das, S., Sharma, P., & Bora, U., (2012). Aptasensors in health, environment and food safety monitoring. *Open Journ*al of Applied Biosensor, 01(02), 9–19. https://doi.org/10.4236/ojab.2012.12002 (Accessed on 14 June 2019).

Sharma, G. M., Khuda, S. E., Parker, C. H., Eischeid, A. C., & Pereira, M. (2016). Detection of A*llergen Markers in Food: Analytic*al *M*ethods. Food Safety: Innovative Analytical Tools for Safety Assessment, 65–121.

Sicherer, S. H., & Sampson, H. A., (2018). Food allergy: A review and update on epidemiology, pathogenesis, diagnosis, prevention, and management. *Journal of Allergy and Clinical Immunology*, *141*(1), 41–58. https://doi.org/10.1016/j.jaci.2017.11.003 (Accessed on 14 June 2019).

Stone, W. E., & Yeung, J. M., (2010). *Principles and practices for allergen management* an*d* control in processing. Allergen Management in the Food Industry, 145–165.

Słowianek, M., & Majak, I., (2011). Methods of Aller*gen Detection Based on DNA Analysis, 75(*2), 39–44.

Tabrizi, M. A., Shamsipur, M., Saber, R., Sarkar, S., & Ebrahimi, V., (2017). A high sensitive visible light-driven photoelectrochemical aptasensor for shrimp allergen tropomyosin detection using graphitic carbon nitride-TiO2 nanocomposite. *Biosensors and Bioelectronics, 98*, 113–118.

Taylor, S. L., & Baumert, J. L., (2010). Cross-contamination of foods and implications for food allergic patients. *Current Allergy and Asthma Reports*, *10*(4), 265–270. https://doi.org/10.1007/s11882-010-0112-4 (Accessed on 14 June 2019).

Taylor, S. L., & Hefle, S. L., (2001). Food allergies and other food sensitivities. *Food Technology*, *55*(9), 68.

Taylor, S. L., & Lehrer, S. B., (1996). Principles and characteristics of food allergens. *Critical Reviews in Food Science and Nutrition*, *36*(2013), S91–S118. https://doi.org/10.1080/10408399609527761 (Accessed on 14 June 2019).

Tran, D. T., Knez, K., Janssen, K. P., Pollet, J., Spasic, D., & Lammertyn, J., (2013). Selection of aptamers against Ara h 1 protein for FO-SPR biosensing of peanut allergens in food matrices. *Biosensors and Bioelectronics*, *43*(1), 245–251. https://doi.org/10.1016/j.bios.2012.12.022 (Accessed on 14 June 2019).

Trashin, S. A., Cucu, T., Devreese, B., Adriaens, A., & De Meulenaer, B., (2011). Development of a highly sensitive and robust Cor a nine specific enzyme-linked immunosorbent assay for the detection of hazelnut traces. *Analytica Chimica Acta, 708*(1/2), 116–122.

Valenta, R., Ferreira, F., Focke-Tejkl, M., Linhart, B., Niederberger, V., Swoboda, I., & Vrtala, S., (2010). From allergen genes to allergy vaccines. *Annual Review of Immunology*, *28*(1), 211–241. https://doi.org/10.1146/annurev-immunol-030409–101218 (Accessed on 14 June 2019).

Wang, J., (2006). Electrochemical biosensors: Towards point-of-care cancer diagnostics. *Biosensors and Bioelectronics, 21*(10), 1887–1892.

Ward, R., Crevel, R., Bell, I., Khandke, N., Ramsay, C., & Paine, S., (2010). A vision for allergen management best practice in the food industry. *Trends in Food Science and Technology*, *21*(12), 619–625. https://doi.org/10.1016/j.tifs.2010.09.004 (Accessed on 14 June 2019).

Weng, X., Gaur, G., & Neethirajan, S., (2016). Rapid detection of food allergens by microfluidics ELISA-based optical sensor. *Biosensors*, *6*(2), 1–10. https://doi.org/10.3390/bios6020024 (Accessed on 14 June 2019).

Zheng, C., Wang, X., Lu, Y., & Liu, Y., (2012). Rapid detection of fish major allergen parvalbumin using superparamagnetic nanoparticle-based lateral flow immunoassay. *Food Control*, *26*(2), 446–452. https://doi.org/10.1016/j.foodcont.2012.01.040 (Accessed on 14 June 2019).

Zurzolo, G. A., Mathai, M. L., Koplin, J. J., & Allen, K. J., (2012). Hidden allergens in foods and implications for labeling and clinical care of food allergic patients. *Current Allergy and Asthma Reports*, *12*(4), 292–296. https://doi.org/10.1007/s11882-012-0263-6 (Accessed on 14 June 2019).

CHAPTER 12

Cereals and Pseudocereals: General Introduction, Classification, and Nutritional Properties

NISAR AHMAD MIR,[1] BASHARAT YOUSUF,[2] KHALID GUL,[3] CHARANJIT SINGH RIAR,[1] and SUKHCHARN SINGH[1]

[1]*Department of Food Engineering & Technology, Sant Longowal Institute of Engineering and Technology, Longowal, Punjab–148106, India*

[2]*Department of Post Harvest Engineering and Technology, Faculty of Agricultural Sciences, Aligarh Muslim University, Aligarh–202002, India*

[3]*Food Process Engineering Laboratory, School of Applied Life Sciences, Gyeongsang National University, 900 Gajwa-Dong, Jinju, 660701, Republic of Korea*

ABSTRACT

Cereals are the crops, which belong to the Gramineae family and are composed of endosperm, germ, and bran. According to FAO cereals crops refer to those crops, which are harvested for dry use and can be utilized in different food formulations. They are generally classified according to their genus but when two or more genera are grown and harvested together they are usually referred to as mixed grains. However, some plant species like buckwheat, quinoa, and chia belong to a dicotyledonous family and they are referred to as pseudocereals. Cereals usually contain 12–14% moisture on dry weight basis and apart from moisture content cereals contain inedible substances such as cellulose along with traces of minerals, vitamins, carbohydrates mainly starches (comprising 65–75% of their total weight), as well as proteins (6–12%) and fat (1–5%). They are considered as the staple foods for mankind worldwide and represent the main constituent of animal feed. Most recently, cereals have been additionally used for energy production, for example by fermentation yielding biogas or bioethanol. Among the cereal crops, the major ones are wheat, corn, rice, barley, sorghum, millet, oats, and rye. These major cereal crops occupy nearly 60% of the cultivated land in the world. Among these major cereal crops wheat, corn, and rice share the major portion of land for cultivation and hence produce the largest quantities of cereal grains.

12.1 INTRODUCTION

Cereals occupy an important place in the world due to their nutritional properties and are widely cultivated in order to obtain the edible portion of their fruit seeds. In botanical terms, the fruits of cereal crops are termed as 'caryopsis' and are composed of endosperm, germ, and bran. As they contain high amounts essential nutrients and provide energy greater than other types of crops they are cultivated in huge quantities, therefore, they are known as staple crops. They are rich sources of complex carbohydrates that is the reserve source of starch and they also help to combat various diseases like cancer, constipation, colon disorders, and high blood sugar levels. They also enrich our overall health with abundant proteins, fats, lipids, minerals, vitamins, and enzymes. Cereals are rich sources of niacin, iron, riboflavin, and thiamine, and in addition to this most of the cereal grains contain abundant amounts of dietary fiber contents, especially barley, oat, and wheat. Moreover, cereal grains are also good sources of soluble bran that assists in lowering blood cholesterol levels thus prevents different cardiac ailments. The consumption of cereals is also associated with high intake of protein likewise breakfast cereals are often eaten with milk which makes for a protein-rich meal. Cereal grains have gained importance for the formulation of infant foods. For example, iron-fortified cereals developed for infants are termed as premium solid foods. Apart from their excellent nutritional profile, cereals grains have gained popularity all over the world due to their prolific growth and plentiful production in most countries. Their consumption also varies across the globe this is due to their diverse nature and response to different environmental conditions. For example, wheat is the most significant cereal in the diets of most European countries and India. Regarding China, Japan, South East Asia, Bangladesh, Pakistan, Brazil, Myanmar, Vietnam, and US rice is the primary grain used in these countries for edible purposes. On the other hand maize or corn is the staple food in northern, Central America and Africa, and millets and sorghum are widely consumed in India and Africa as well. A wide range of factors like climate, population, and production quality are responsible for these global differences of cereal consumption in tropical and subtropical regions. Some people know cereals by the name "staff of life" but this name is not totally justified in a true sense because we cannot live on cereals solely lifetime and retain optimum health. Whole cereal grains have an outer bran coat, a starchy endosperm, and a germ which are briefly discussed as follows:

1. **Bran:** The outer layers of the kernel are called bran, which is made of about 5% of the kernel. The kernel is rich in fiber and minerals while the bran is rich in thiamine and riboflavin.
2. **Aleuron:** While refining, the bran layer is removed and the aleurone layer is exposed, which lies just below the bran. This layer is also rich in phosphorous, proteins, fat, and thiamin.
3. **Endosperm:** This large central part of the kernel acts as a source of starch and protein and is low in vitamin or mineral content.
4. **Germ:** The small structure at the rear part of the kernel is known as the germ. Rich in protein, fat, minerals, and vitamins, this germ is a storehouse of nutrients for the seeds during germination.

The general parts of cereal grains are shown in Figure 12.1.

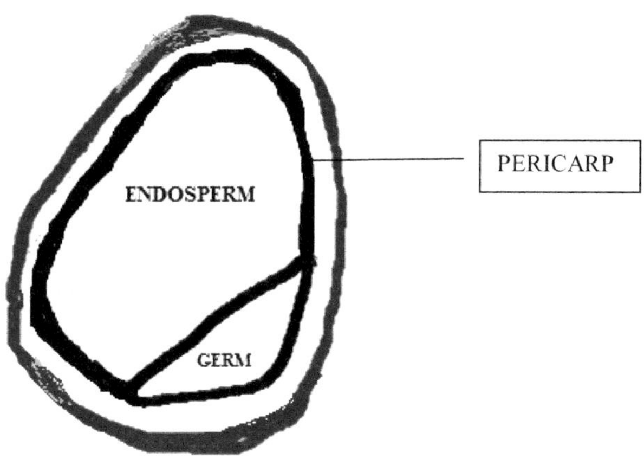

FIGURE 12.1 General parts of cereal grains.

12.1.1 CLASSIFICATION

Grains may be classified into two different categories which include cereal (true grasses) and pseudocereals (non-grasses). These are the primary sources of carbohydrates for the world's population. Nearly half of the annual cereal production is used for human food. Moreover, they also serve as a primary food for dairy and draft animals, poultry, and wild birds and also serve as source for the production of alcohol. The primary cereals include wheat, rice, corn, sorghum, millets, oats, barley, and triticale. Wheat and rice provide nearly 50% of the world's food energy. Millet is a term that refers to small-seeded grain and has been applied to many unrelated species. Pseudocereals are the second category of plants that produce seeds or fruits that are consumed as grains however botanically they are neither grasses nor true cereals and are therefore placed in dicotyledonous family. The main pseudocereals include quinoa, amaranth, buckwheat, and chia. Despite being abandoned by the traditional world crops the pseudocereals have regained their position due to their excellent nutritional and functional profile. They are well balanced with essential amino acids as compared to traditional cereal crops and contain a myriad of other essential nutrients like phenolic phytochemicals. In this chapter, we will be discussing about the major cereal crops, which include wheat, rice, maize, barley, and sorghum and some pseudocereals, which have gained popularity due to their excellent nutritional profile. The general classification of these grains is shown in Table 12.1.

TABLE 12.1 Classification of Cereal Crops

True Cereals	Pseudocereals
Wheat (*Triticum* spp.)	Album (*Chenopodium album*)
Maize, corn (*Zea mays* ssp. mays)	Quinoa (*Chenopodium quinoa*)
Rice (genus *Oryza*)	Kaniwa (*Chenopodium pallidicaule*)
Barley (*Hordeum vulgare*)	Amaranth (*Amaranthaceae*)
Sorghum (*Sorghum vulgare*)	Kiwichae (*Amaranthus caudatus*)
Millet (*Pennesetum glaucum*).	Chia (*Salvia hispanica* L)
Rye (*Secale cereale*)	Buckwheat (*Fagopyrum esculentum*).
Oats (genus *Avena*)	Cockscomb (genus *Celosia*)
Triticale (X *Triticosecale*).	
Teff (*Eragrostis* tef.)	
Wild rice (*Zizania*).	
Spelt (*Triticum spelta*).	

12.1.2 CEREALS

1. Rice: (*Oryza sativa*) is one of the important cereal crops and widely consumed foods in the world and provides basic nutrition to more than half of the population of the world. Rice not only feeds the largest number of people around the world, it forms the bulk of food of a large number of poor people. The importance of rice to the dietary requirements of world population is evident from the presence of rice in the diet of a quarter of world's population (Gul et al., 2016a; Gul et al., 2015).

Rice serves as a staple food for over 39 countries. However, the culinary traditions and preferences for rice vary from region to region, yet rice is particularly common in areas where population density is high and availability of arable land is very low. For this reason, rice is widespread in Asia and Africa. Low labor costs also contribute to rice's popularity throughout Asia and Africa, since cultivation of rice is labor-intensive. Dietary intake surveys from China and India show an average adult intake equivalent to about 300 grams of raw rice per day (Popkin et al., 1993).

Rice is a semi-aquatic annual grass plant and nearly 20 species of the genus *Oryza* have been recognized, but almost all cultivated rice is *Oryza sativa* L. The domestication of rice perhaps occurred independently countries such as China, India, and Indonesia, thereby giving rise to three different races of rice which include sinica (also called japonica), indica, and javanica (also called bulu). There are some evidences that rice was cultivated in India between 1500 and 2000 B.C. and in Indonesia around 1648 B.C.

Rice is basically a tropical crop but its production extends to subtropical and temperate regions also. More than 90% of the rice across world is grown in Asia alone (Figure 12.2) and yet it feeds the largest number of people in the world. Over 700 MT of rice is produced worldwide (Figure 12.3), with China, India, Indonesia, Bangladesh, Vietnam, Thailand, Myanmar, Philippines, Brazil, and Japan the top ten rice producing countries of the world (Figure 12.4).

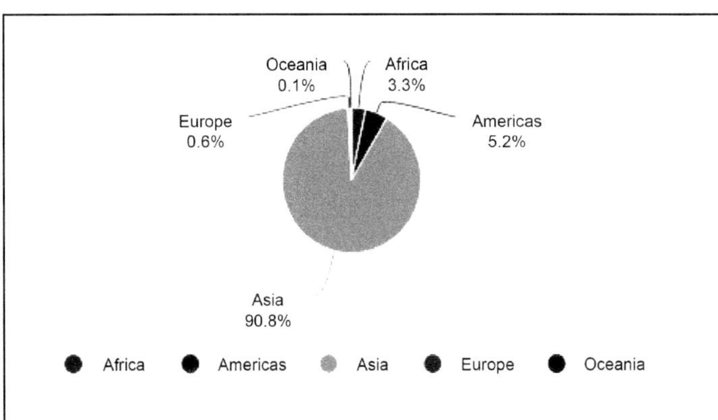

FIGURE 12.2 Production share of rice by regions (FAOSTAT, 2016).

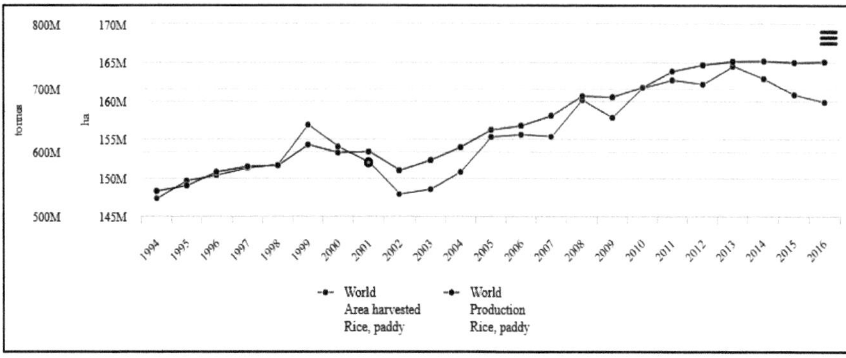

FIGURE 12.3 Production/yield quantities of rice in world (FAOSTAT, 2016).

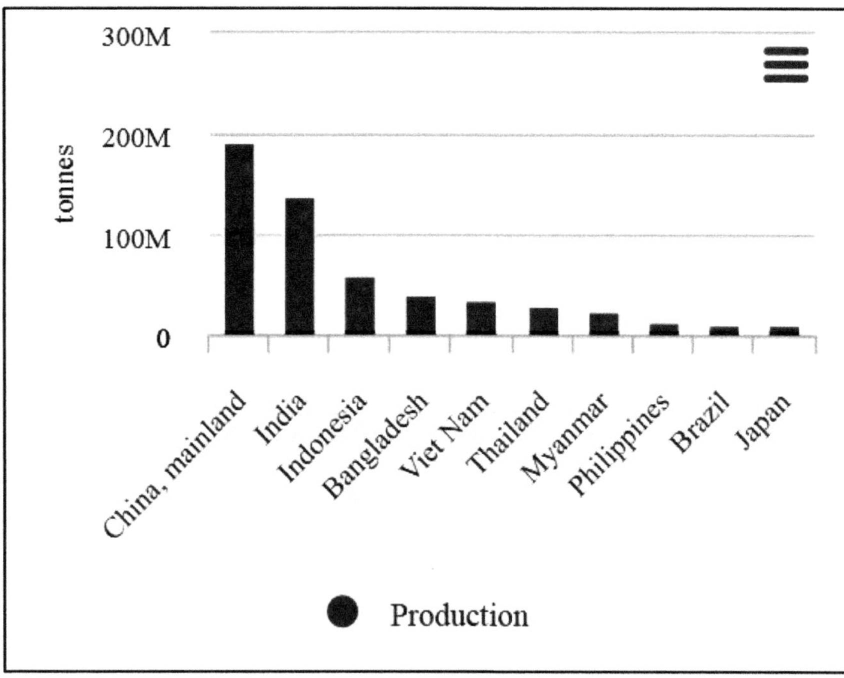

FIGURE 12.4 Top 10 rice producing countries and their production (FAOSTAT, 2016).

2. Rice Structure: The rice grain (paddy) is comprised of outer protective covering called hull (husk) and brown rice. Brown rice, in turn, contains bran which comprises the outer layer and the edible portion. Most rice varieties are composed of roughly 20% rice hull or husk, 11% bran layers, and 69% starchy endosperm, also referred to as the total milled rice. Brown rice, i.e., paddy without hull, comprise of outer layers of pericarp, seed-coat, and nucellus; the germ; and the endosperm. The endosperm is made up of the aleurone layer and the endosperm proper, consisting of the sub-aleurone layer and the starchy or inner endosperm (Figure 12.5). The aleurone layer encloses the embryo (Juliano and Bechtel, 1985).

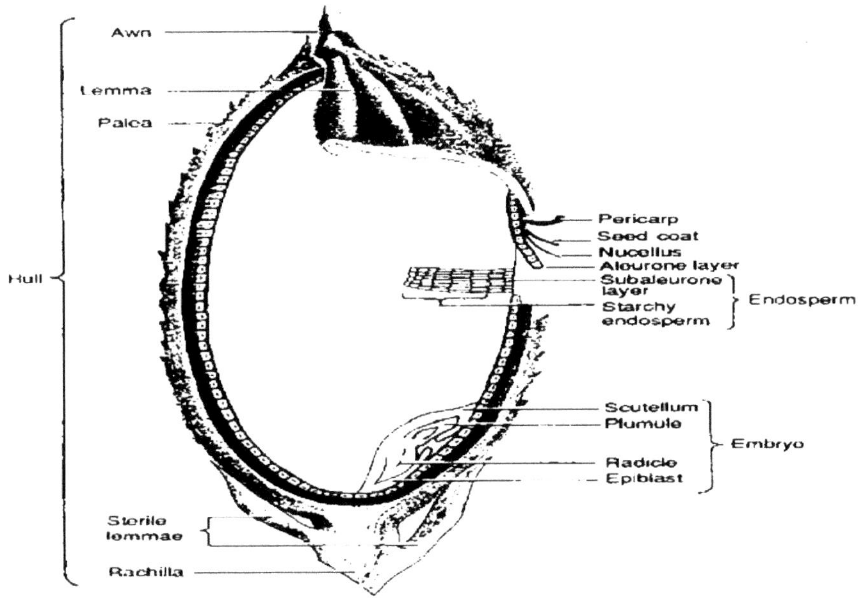

FIGURE 12.5 Longitudinal section of rice grain (FAO, 1993).

Rice is consumed after milling to obtain the edible portion for consumption as the entire paddy grain is not edible. One-fifth of its weight is roughly woody-siliceous covering called the husk or the hull which is inedible and must be removed and the process is called dehusking. The resulting grain after the removal of husk is called brown rice which has some fibrous and fatty covering left, preventing its easy cooking. This layer is called bran, therefore needs to be removed at least partially by a process of attrition or abrasion, this process being called whitening or pearling or simply milling or polishing. The entire process of producing milled rice or white rice from paddy is referred to as milling (Figure 12.6).

The process is carried out with care to avoid excessive breakage of the kernels and to improve recovery of rice. The extent of recovery during milling is influenced by many factors like variety of paddy, degree of milling required, the quality of equipments used, the operators, etc.

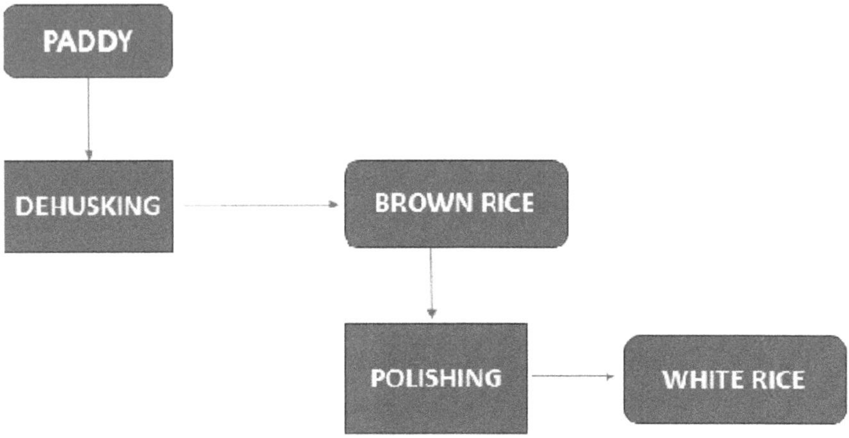

FIGURE 12.6 Schematic of the rice milling.

Rice is marketed according to three-grain size and shape classes (long, medium, and short). Kernel dimensions are primary quality factors in most phases of processing, drying, handling equipment, breeding, and grading. Rice is classified as extra-long, >7.50 mm; long, 6.61 to 7.50 mm; medium, 5.51 to 6.60 mm; and short, <5.50 mm. The grain shape is characterized on the basis of length-to-width ratio: slender, >3.0; medium, 2.1 to 3.0; bold 1.1 to 2.0; and round, < 1.0. Methods for measuring rice grains include use of photographic enlargers to magnify kernels or simply measuring with a ruler the length, width, and thickness of several grains placed in adjacent positions for particular measurements (USDA, 1989).

3. Composition of Rice: Human beings consume relatively large amount of carbohydrates, proteins, and lipids, small amounts of macro elements and trace amounts of some microelements (Fe, Zn, Cu, I, and Se) and vitamins. Rice grain contains most of these beneficial components some of which are present in large amounts, while the rest are only in small quantities. Rice provides up to 75% of the dietary energy and protein for 2.5 billion people in Asia. In South Asia, rice provides 68% of the total energy contribution, wheat 10% and maize 2.5%; rice also provides 69% of the total dietary protein (Juliano, 1985). The chemical composition of rice varies widely, depending on environment, soil, and variety. The nutrient composition of rice (per 100g) is presented in Table 12.2. Though rice contains essential minerals and vitamins in addition to carbohydrates, proteins, and lipids, only the major constituents have been discussed in this chapter.

TABLE 12.2 Nutritional Composition of Rough, Brown, and White Rice

Nutrient	Rough rice	Brown rice	White rice
Proximate composition (g)			
Protein	5.8–7.7	7.1–8.3	6.3–7.1
Total lipid (Fat)	1.5–2.3	1.6–2.8	0.3–0.5
Ash	2.9–5.2	1.0–1.5	0.3–0.8
Crude fiber	7.2–10.4	0.6–1.0	0.2–0.5
Carbohydrates	64–73	73–87	77–89
Vitamins and minerals			
Thiamine (mg)	0.26–0.33	0.29–0.61	0.02–0.11
Riboflavin (mg)	0.06–0.11	0.04–0.14	0.02–0.06
Niacin (mg)	2.9–5.6	3.5–5.3	1.3–2.4
Calcium (mg)	10–80	10–50	10–30
Phosphorous (g)	0.17–0.39	0.17–0.43	0.08–0.15
Iron (mg)	1.4–6.0	0.2–5.2	0.2–2.8
Amino acids			
Histidine	1.5–2.8	2.3–2.5	2.2–2.6
Isoleucine	3.0–4.8	3.4–4.4	3.5–4.6
Leucine	6.9–8.8	7.9–8.5	8.0–8.2
Phenylalanine	9.3–10.8	8.6–9.3	9.3–10.4
Valine	4.6–7.0	4.8–6.3	4.7–6.5
Tryptophan	1.2–2.0	1.2–1.4	1.2–1.7
Threonine	3.0–4.5	3.7–3.8	3.5–3.7

Adapted and modified from Pedersen and Eggum, (1983); Juliano (1985); USDA, (2009).

It shows that rice is rich in energy and is a good source of protein as well. Rice also contains a reasonable amount of thiamine, riboflavin, niacin, and other nutrients. Because of the quantity consumed, it is the principal source of energy in Asian diets. Carbohydrates are the major component of the rice grain. Starch is the major constituent of rice and constitutes over 90% of the dry matter (DM) in milled rice, and 72 to 82% of brown rice (Frei et al., 2003). Starch is present as granules which are partially crystalline particles composed mainly of two homopolymers of glucopyranose with different structures: amylose, which is composed of units of D-glucose linked through α–D–(1–4) linkages and amylopectin, the branching polymer of starch, composed of α–D–(1–4)-linked glucose segments containing glucose units in α–D (1–6) branches. The apparent amylose content, or amylose/amylopectin ratio, serves as one of the most important indices for evaluation of the quality of rice products. Based on amylose content, rice is classified as waxy (1 to 2 percent), very low amylose (2 to 12 percent), low amylose (12 to 20 percent), intermediate (20 to 25 percent) and high (25 to 33 percent) (Bao et al., 2006). During digestion, these links are broken and the resulting glucose is absorbed into the body. Amylopectin contains branches and is less resistant to digestion whereas amylose is a straight chain molecule and harder for the digestive system to break up. This means that rice varieties with a greater proportion of starch in the form of amylose tend to have a lower glycemic index (GI).

Cooking characteristics, texture, water absorption capacity, stickiness, expansion, hardness, and even the whiteness of the cooked rice are also affected by the amylose content (Juliano, 1985b). Waxy varieties, for example, produce only amylopectin and these starches are non-gelling because of the lack of amylose. Both the amylose contents and amylopectin branch chain-length distributions strongly affect the pasting properties of starch (Jane et al., 1999).

Cereal lipids are a chemically diverse group and separated into neutral lipids, glycolipids, and phospholipids (Mano et al., 1999). The ratio of these lipid classes do not differ in japonica and indicarices (Mano et al., 1999), but their distribution is not uniform within the grain. The endosperm lipids contain higher proportion of polar lipids (Fujino and Mano, 1972; Choudhury and Juliano, 1980). The lipid content of brown rice ranges from 2.76 to 3.84% on a dry weight basis and is strongly dependent on the varieties and the growing environments (Kitta et al., 2005). The lipids mainly are monoacyl lipids (fatty acids and phospholipids) complexed with amylose and the major fatty acids identified by Kitta et al. (2005) include oleic (18:1) and linoleic (18:2) acids, followed by palmitic acid (16:0) which together account for more than 90% of the total fatty acid content. The lipid or oil content of rice is concentrated in the bran fraction specifically as lipid bodies. Starch lipids (bound together with starch molecules) can be extracted by more polar alcoholic solvents at elevated temperatures. The selection of solvent is very important for the effective extraction of lipids; and both water and heat are necessary to swell the native starch granule sufficiently to permit the solvent to penetrate and extract the lipids (Morrison, 1988). Rice bran oil is commercially available and is unique among edible oils due to its rich source of nutritionally important phytochemicals such as oryzanol, lecithin, tocopherols, and tocotrienols. However, most of these phytochemicals are removed from the oil as waste byproducts during the refining process (Patel and Naik, 2004). Antioxidants from rice bran can potentially satisfy the demand of finding effective and economical natural antioxidants, and is one of the interesting areas of research (Gul et al., 2016b).

Proteins, after starch, are the major components of rice, with an approximate composition of 8% (Marshall and Wordsworth, 1994). Rice protein is valuable and unique because it has hypoallergenic properties and ranks high in nutritive quality (rich in the essential amino acid lysine) among the cereal proteins (Ju et al., 2001). Based on Osborne classification of proteins, rice grain has about 10% albumin, 5% globulin, 20% prolamin, and 65% glutelin. However, the protein contents decrease with an increase in the degree of polishing, i.e., the removal of bran, as these constituents are mainly concentrated in the peripheral layers of the kernel (Pal et al., 1999). Rice proteins are most abundant in the sub-aleurone layers but are also present in aleurone cells (Azhakanandam et al., 2000). The hypoallergenic property and the high nutritive value make rice an ideal choice for the development of protein concentrates or isolates. However, an understanding of the physicochemical and functional properties of rice proteins is very important for its application in food industries.

4. Importance of Rice Bran: Rice bran is a mixture of bran (brown layer) and germ which is produced as a by-product during the milling process in the production of white rice from brown rice (Gul et al., 2016a). It is a good source of proteins, minerals, fatty acids, and dietary fiber and its beneficial properties, which make it a health-promoting functional food product, have widely been recognized (Kahlon and Smith, 2004; Cicero and Derosa, 2005). The increasing demand for rice bran and rice bran oil has led to the emergence of defatted rice bran as a potential by-product obtained as rice bran meal after the extraction of oil from rice bran (Chan et al., 2013).

Rice bran is rich in dietary fiber and in view of its therapeutic potential, the addition of rice bran can contribute to the development of value-added foods or functional foods that are currently in high demand (Gul et al., 2016a). Since most of the phytochemicals are deposited in the bran layer, compared to brown rice, bran contains most of the antioxidants and has, therefore, higher values of antioxidant capacity. Rice bran proteins are complete proteins, hypoallergenic, and easily digestible with a high nutritional value. However, applications of rice bran in food industry are limited by its instability due to rancidity caused by exposure of oil to lipases during milling. Therefore, stabilization of bran is carried to prevent rancidity and protect its nutritional value.

12.2 FUTURE PROSPECTS

Rice grain contains different bioactive compounds and there is growing interest in the potential health benefits of these substances. In future, rice will remain an important and a staple food for billions of people around the world. To help ensure that rice can contribute to the healthy diets of rice consumers worldwide, studies on the GI of rice and development of healthier rice are promising areas of research.

12.3 WHEAT

Generally, it is believed that wheat was one of the first cereal grains to be cultivated. Considering the global scenario, wheat is one of the most important cereal crops accounting a major share to the world cereal production. There are many species of wheat which together make the genus *Triticum* but the most widely grown species *Triticum aestivum*. Wheat covers nearly 25% of the total global area devoted to cultivation of cereal crops and its production is being considered second to maize among the cereal crops. Wheat is one of the major grains in the diet of large number of the world's population. It is a staple source of food for billions of people worldwide. It gives a good nutritional quality of the diet and human health. Wheat and wheat-based food ingredients are ideal for development of functional foods. Wheat bran used for enhancing health & reduction of risk of the chronic diseases.

Wheat can be classified into different types on the basis of the (a) texture of endosperm and (b) the protein content. The texture of the grain is related to the way the grain breaks down during the milling process while as the protein content is related to the flour quality and its suitability for different purposes. A brief disruption of wheat types according to different classifications is given ahead.

There are two types of wheat based on the nature or type of endosperm-vitreous wheat and mealy wheat. This vitreous or mealy character is genetic however; it may be influenced by environment. The vitreous kernels are translucent and they appear bright against a strong light. In contrast, the mealy kernels are opaque, and they appear dark against a strong light. On the basis of milling character or how the endosperm breaks down during milling, there are two types of wheat-hard wheat and soft wheat. In case of hard wheat, the fracture of endosperm during milling occurs along the lines of cell boundaries whereas in case of soft wheat fractures occur in a random manner. This theory suggests that there are areas of mechanical strength and weakness in hard wheat while as uniform areas of mechanical weakness in soft wheat (Kent, 1994). Hardness is defined as a mechanical property of the individual wheat grain or resistance to deformation or crush. The importance of wheat grain hardness lies in the fact that it determines the end use of the wheat that is suitability of the flour milled from that wheat for various purposes. It affects the force required to fracture the grain, manner of fracture, resultant size of fragments, and sifting behavior of the flour. Wheat yielding flour, which has high protein content is known as "strong wheat." Such wheat flour has the ability to large loaf volume. Wheat yielding flour which has low protein content is "weak wheat," forming only a small loaf volume. Strong flour is suitable for bread making while as weak flour is suitable for biscuits and cakes. Protein content of wheat may vary over a wide range as it is affected by a number of factors including fertilizer treatment, soil, and climatic conditions during growth.

12.3.1 STRUCTURE AND CHEMICAL COMPOSITION OF WHEAT KERNEL

Wheat is a single-seeded fruit consisting of three principal parts–germ and endosperm and bran (Onipe et al., 2015). Wheat grain size varies enormously as in other cereals, and generally, a wheat kernel is about 5–8 mm in length and 2.5–4.5 mm in width. Inherent characteristics are considered to be the major source for this wide variation. The kernels vary considerably in color, form, and size. The color of wheat may vary from red to white, which is dependent on various factors such as species or the genetic make-up. The size may also vary with these factors and also the soil, environmental, and nutritive conditions during its development. Germ is typically removed in milling process. Typical representational diagram of wheat kernel is shown in Figure 12.7.

The chemical composition of wheat grain is dependent on genetic and environmental factors. In addition, the physical and chemical effects acting on the grain during its storage and processing also influence the final chemical composition.

Chemical composition varies with species to species. Wheat grain contains a number of chemical constituents, the main ones being starch and proteins. The primary quantitative component of wheat is starch. Besides, starch the wheat grain and especially the grain coat, the aleurone layer and the embryo, contain other carbohydrates, such as cellulose, hemicelluloses, and sugars. Pentosans, although their content is low (2–3%) are important owing to their water absorbing capacity, which is ten times their mass. In combination with other hemicelluloses, they form the basic structure of endosperm cell walls. Percentage of different chemical components of whole wheat grain and its different fractions are presented in Tables 12.3.

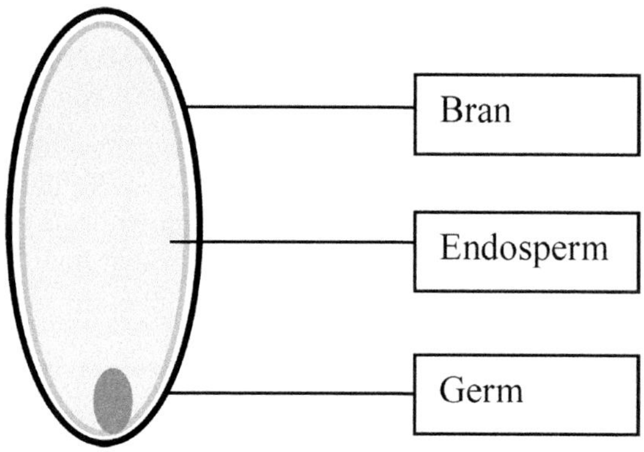

FIGURE 12.7 Typical representational diagram of wheat kernel showing different parts.

TABLE 12.3 Chemical Composition of Wheat Grain and its Different Parts (% Dry-Weight Basis)

Chemical components	Whole grain	Grain coat with endosperm	Embryo	Endosperm
Carbohydrates (%)				
Starch	60.0–70.0	0	0	78.0–84.0
Sugar	3.0–6.0	3.0–5.0	22.0–28.0	3.0–4.0
Pentosans	6.0–9.5	30.0–40.0	9.0–11.0	2.5–3.0
Cellulose	2.5–3.3	12.0–20.0	3.0–5.0	0.13–0.18
Protein (%)	10.0–12.0	23.0–33.0	36.0–42.0	9.0–14.0
Fat (%)	2.0–2.5	7.0–8.5	12.0–16.0	0.5–0.7
Minerals (%)	1.4–2.3	9.0–11.0	5.0–6.0	0.3–0.5

12.3.2 WHEAT BRAN: CHEMICAL COMPOSITION AND PHYSIOLOGICAL EFFECTS

Wheat Bran is the hard outside layer of the kernel, which is usually separated from the other parts of the kernel by milling. Wheat bran is a richest source of natural food fiber, recognized for its role to help maintain bowel regularity. It is an edible broken seed coat, or protective outer layer of wheat, which is separated from the kernel. Wheat bran consists of the outer coats of the wheat grain, which can be separated from the germ and various grades of flour during the milling process. Wheat bran, the outer layer of the grain, is rich in fiber and other nutrients. Wheat bran is produced globally in large quantities, as an important by-product of cereal industry.

Wheat bran consists of 16% protein, 11% natural fiber, and 50% carbohydrate. It is water-insoluble fiber consists mainly of cell wall components such as cellulose, lignin, and hemicelluloses. Insoluble fiber shortens bowel transit time, increases fecal bulk, and renders feces softer. It also contains antioxidants, starch, protein, vitamin E, vitamin B, and minerals. Wheat bran represents not only a good source of dietary fiber, but also of phenolic acids, known to contribute significantly to the total antioxidant activity of wheat (Yu et al., 2003).

Wheat Bran acts as a functional food due to its medicinal effects on various diseases and disorders. It provides many important physiological effects. There is scientific evidence demonstrating the positive impact on health due to dietary fiber present in wheat bran. Wheat bran helps to prevent:

- colorectal cancer;
- constipation;
- diverticulosis;
- irritable bowel syndrome;
- obesity;
- hypertension;
- cholelithiasis; and
- diabetes mellitus.

12.4 MAIZE/CORN

Corn, *Zea Mays*, is grown in most countries throughout the world. It requires, however, warmer climates than found in the temperate zones to grow to maturity. Corn is the major cereal crop of the United States and the second most important cereal crop worldwide. This cereal is a staple food for large groups of people in Latin America, Asia, and Africa (Boyer and Hannah, 2001). It has a wide adaptability under varied climatic conditions and is cultivated in more than 160 countries around the world. Corn is considered as one of the highest yielding among world's major crops. It gives higher average grain yield as compared to other major cereals such as wheat and rice. Due to its worldwide distribution and relatively lower price corn is believed to have many advantages over other cereals. Almost, each part of corn plant is being utilized for one or the other purpose and nothing goes waste, therefore it is known as "queen of cereals."

In India, corn occupies a prominent position and is the third most important crop after wheat and rice. The main driving forces for the emerging importance of maize crop in India is the increasing use of maize as feed, increasing interest of the consumers in nutritionally enriched products, and rising demand for maize seed. It serves as a primary ingredient in number of industrial products. Its importance lies in the fact that it is not only used for human food and animal feed but is also important corn starch industry, corn oil production, baby corns and so on. The higher demand of corn is tackled by technological interventions like adoption of high yielding single cross hybrids seeds, biotechnological intervention and improved package of practices. The improvement of crops with the use of genetics has been occurring for years. Such approaches will lead to the improvement in input use efficiency, productivity, and sustainability.

12.4.1 CHEMICAL COMPOSITION

Corn grain is composed of several chemicals constituents. The mature kernel is composed of 70% to 75% starch, 8% to 10% protein, and 4% to 5% oil. Starch is the predominant carbohydrate while other carbohydrates such as simple sugars like glucose, fructose, and sucrose are also present in small amounts. After starch, Protein is the next largest chemical component of the corn. Generally, it varies between 8%–11% in the case of different common varieties. The most abundant protein in corn is known as zein, which contributes approximately 44–79% of the total protein content (Table 12.4). Corn kernel is comprised of different parts (Figure 12.8). The two major parts of the kernel are endosperm and germ (or embryo) which constitute approximately 80% and 10% of the mature kernel dry weight, respectively (Boyer, and Hannah, 2001). Corn endosperm is largely composed of starch (approximately 90%) while as the germ has high levels of oil (30%) and protein (18%). Corn contains a high percentage of essential amino acids. Yellow corn is the richest source of Vitamin-A. It contains more riboflavin than wheat and rice, and is also rich in phosphorous (Sapna et al., 2013). The oil and protein are of commercial value and are often obtained as by-products from the production of starch from corn. Germ is the major contributor of the oil content of the corn kernel. The immature corn kernels have relatively high levels of sugars and lesser amounts of starch, protein, and oil, which accumulate during development phase. Corn also contains a fair amount of fiber, though the content may vary among the different varieties. Insoluble fibers such as hemicellulose, cellulose, and lignin are the predominant types of fiber present in corn. In addition, corn also contains various bioactive components. These may include ferulic acid, anthocyanins, zeaxanthin, lutein, and phytic acid (Figure 12.8).

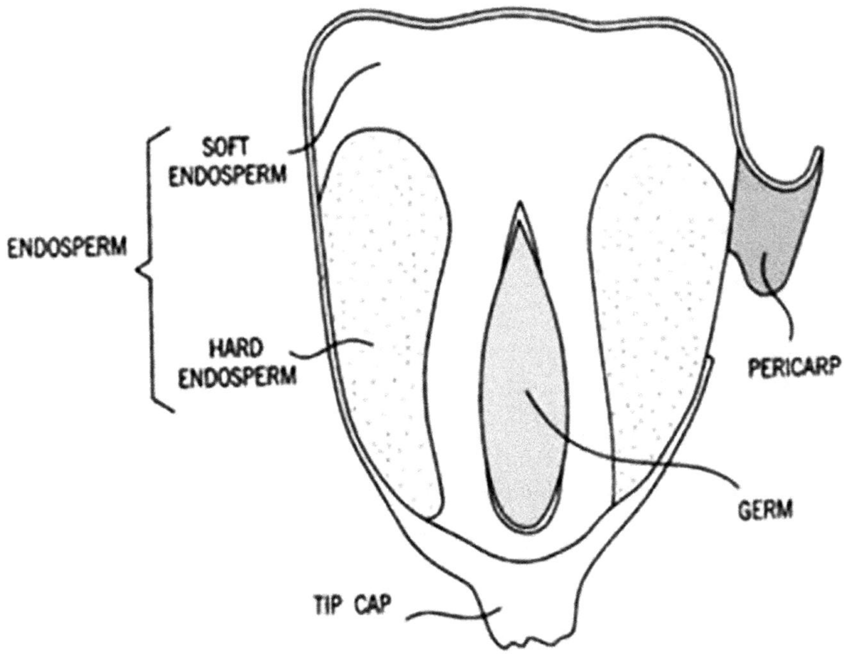

FIGURE 12.8 Structure of corn kernel (Source: Reprinted from Burger and Rheeder, 2017. Open access).

TABLE 12.4 Proximate Composition of Various Parts of Maize Kernel

Chemical constituent	Whole grain	Pericarp	Endosperm	Germ
Protein (%)	11.1	3.7	8.0	18.4
Crude fiber (%)	2.7	86.7	2.7	8.8
Ash	1.5	0.8	0.3	10.5
Starch	70–75	7.3	87.6	8.3
Sugar	-	0.34	0.62	10.8

Source: Watson, S. A. (1987); http://www.fao.org.

12.4.2 DIFFERENT VARIETIES OF CORN

There are different varieties of corn each having some special feature and subsequently have some specific uses. These are described briefly ahead:

Dent corn (*Zea Mays Indentata*), also called "field" corn is a variety with kernels that contain both hard and soft starch and become indented at maturity. It is popularly known as dent corn because of dent formation on the top of the kernel. It is largely used to make food, animal feed, and industrial products. This variety is also used for manufacturing of cornstarch.

Flint corn (*Zea Maysindurate*), is a variety of corn having hard, horny, round, or short and flat kernels. The soft and starchy endosperm is completely enclosed by a hard outer layer. It is mostly grown in South America. This type of corn has an early maturity and has kernels that are round on the top.

The grains of waxy corn have a waxy appearance when they are cut. It is grown to make special kind of starches for thickening of different foods items. Its origin is believed to be in China but many waxy hybrids developed in the United States of America are being grown commercially.

Sweet corn, also known as green corn is eaten fresh, canned or frozen. It is occasionally considered a distinct species, a subspecies, or a specific mutation of dent corn. The kernels of sweet corn contain a high percentage of sugar in the milk stage. The sugar content is the most apparently recognizable constituent of sweet corn quality due to which sweet corn is eaten at an immature stage of development. In United States and Canada, Sweet corn is one of the most popular vegetables. It is popular both as fresh and processed vegetable. Its consumption is also increasing in eastern Asia, and Europe.

Popcorn (*Zea Mayseverta*) is indigenous to the Americas. Popcorn has been a commercial crop in the U.S. for more than 100 years. It has round kernels with very hard corneous endosperm. The endosperm forms a white starchy mass many times the size of the original kernel on exposure to dry heat. The endosperm pops out by the expulsion upon exposure to heat. The formation of large flakes after the kernels explode in response to heating is the major character that distinguishes popcorn from other types of corn. This character/trait is called "popping expansion."

Flour corn, also called soft corn (*Zea Maysamylacea*), has flint corn shaped kernels which are almost entirely composed of soft starch. Generally, it has 8 to 12 rows of grain with a soft, floury endosperm without dents or wrinkles. It is crushed using stone mills and sifting to remove some of the very coarse pericarp to produce refined meal and flour. Flour corn is used to make tortillas, chips, and baked goods. The popularity of floury corns may be due to their relative easy reduction into flour or reduced cooking times.

Baby corn, is young and unfertilized ear of the corn plant which is harvested when the corn silks have either not emerged or just emerged. It is a highly nutritious vegetable. The requirements of baby corn for the fresh market or processing include (1) ear size of 4 to 9 cm length and 1.0 to 1.5 cm diameter and (2) good quality, i.e., yellow color, straight ovary row arrangement, and unfertilized and unbroken ear (Boyer and Hannah, 2001).

12.4.3 USES OF CORN

Corn as a crop has multiple uses. Three main uses of corn are as food, feed for livestock, and raw material for industry. Corn has a range of uses due to its worldwide distribution and comparatively lower price. It is used directly for human consumption, in industrially processing foods, as Live-stock feed and has a number of industrial and non-food applications such as starches, acids, and alcohols. Even corn has been employed for production of ethanol, which is used as a substitute for petroleum-based fuels. In India, over approximately 85% of the corn production is used as food. Some of the main and popular products from corn include, corn starch, corn flakes, dextrose powder, corn gluten, liquid glucose, edible corn oil, etc. Global utilization pattern and use of corn India is presented in Figure 12.9.

12.5 MILLETS

Millet is an important crop grown in the semiarid tropics of Asia and Africa, especially in India, Mali, Nigeria, and Niger. Millets are a group of highly variable small-seeded grasses which are widely grown around the world as a cereal crop. There are different varieties of millet grown around the world, but the most common and important cultivar is Pearl millet (*Pennisetum glaucum*) (Clotault et al., 2011). It is a major cereal crop in West Africa. Pearl millet is the most widely grown of all millets and has the property of being tolerant to drought and heat. It has the highest yield potential of all millets under drought and heat stress. Generally, millets can survive in areas with as little as 300 mm or less of seasonal rainfall. These advantages that have made it the traditional staple cereal crop in subsistence or low-resource agricultural regions, especially hot semiarid regions. In addition to Pearl millet, finger millet (*Eleusine coracana*), proso millet (*Panicum miliaceum*), and foxtail millet (*Setaria italica*) are also important millet types.

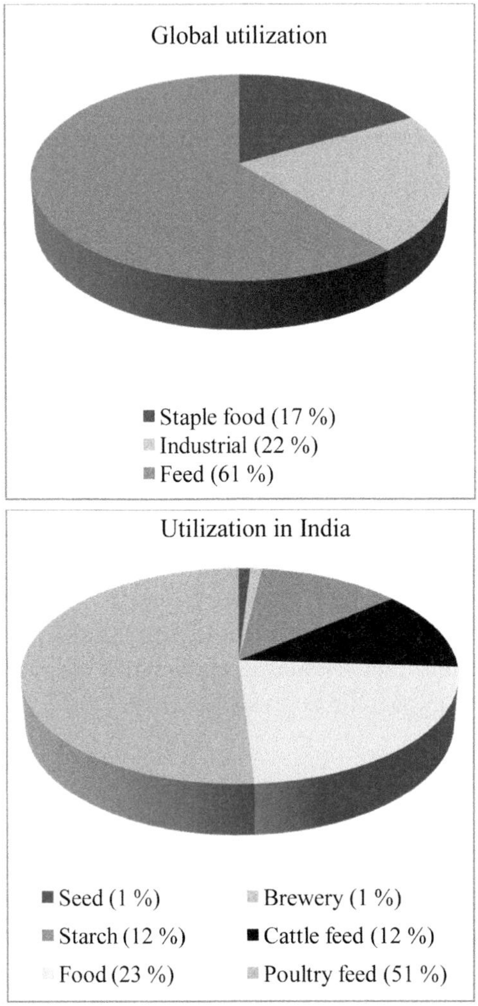

FIGURE 12.9 Global utilization and utilization pattern of corn India (Dass, 2013).

Millet is a good source of different nutrients, vitamins, minerals, and organic compounds that may be beneficial to human health. They are rich in calcium and polyphenols. In addition, millet contains significant levels of protein and dietary fiber. Furthermore, millet is gluten-free, thus celiac disease sufferers can use millet-based products instead of relying on wheat. Millet has a high nutrient content. However, presence antinutritional factors such as phytic acid, polyphenols, and tannins limits the bioavailability of nutrients present in millet (Elyas et al., 2002). The antinutrients are present in considerable amounts, limit protein, and starch digestibility, hinder mineral bioavailability and inhibit proteolytic and amylolytic enzymes. Fermentation is one of the processes known to reduce these antinutrients. Fermentation increases millet protein digestibility, decreases phytic acid and polyphenols and improves availability of minerals. Millet has a well-balanced protein, except for its lysine deficiency, with high concentration of threonine.

12.5.1 MILLET GRAIN STRUCTURE AND CHEMICAL COMPOSITION

Shape of millet grain can vary considerably from globular to lanceolate. The color of grain also varies widely as a result of various factors. Pearl millet grain comprises of approximately 75% endosperm, 17% germ, and 8% bran (Abdelrahman and Hoseney, 1984). The germ is firmly embedded in the endosperm and may not be completely removed by milling. Pericarp of the millet grain is comprised of three layers of different cell structures. The epicarp has one or two layers of thick cubic cells with a thin layer of cutin on the outer surface and is considered to play important role in resisting "weather" damage Figure 12.10.

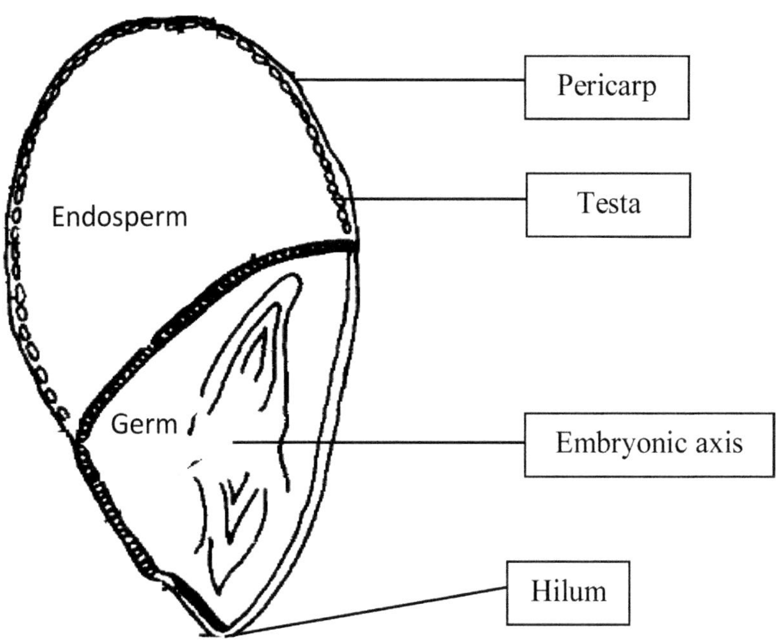

FIGURE 12.10 General diagram of pearl millet grain.

Pearl millet indicate approximately 12% protein content, 69%carbohydrates, 5% lipids, around 2.5% fiber and ash each, and the remainder being moisture. Among carbohydrates, starch represents 56% to 65% of the grain. Free sugars range from 2.6% to 2.8% of the grain. The main sugar present in different millet types is sucrose. Protein content of pearl millet may, however, range from 8 to 19% depending upon different factors. Millets generally contain significant amounts of essential amino acids especially the sulfur-containing amino acids (methionine and cysteine). Generally, it is considered that pearl millet has higher tryptophan levels than in other cereal (Chung and Pomeranz, 1985). Millets are also higher in fat content than maize, rice, and sorghum (Obilana and Manyasa, 2002). About 75% of the fatty acids in pearl millet are unsaturated, and linoleic acid is particularly high (46.3%).

Millet also contains dietary fiber, most of which is insoluble. Apart from important nutrients, millets also contain antinutritional components such as polyphenols and tannin, phytic acid and phytate, goitrogens, and oxalic acid (Léder, 2004). Polyphenols and tannin compounds are concentrated in the bran.

12.5.2 USES OF MILLET

Millets are small-seeded cereals used as food, feed, and forage. The principal use of millet grain is for food (85%), and about 9% is used for feed (Kent, 1978). They are widely cultivated in the tropics and consumed in several forms after cooking, particularly by people in developing countries. It is a staple food in Namibia and Uganda. Several food preparations are made from millet which varies from country to country. It has been reported to possess many nutritional and medical functions (Obilana and Manyasa, 2002). Millet can be used as a traditional cereal, and can also be used in porridge, snacks, and other types of bread. Millets are often ground into flour, rolled into large balls, parboiled, and then consumed as porridge with milk. Porridges and flat unleavened bread are two principal types of food that are traditionally made from pearl millet. Pearl millet is primarily grown for human food in the drier tropical regions of the world where agricultural production is at most risk from pest, disease, and highly variable weather conditions (Andrews and Kumar, 1992). The emerging uses of millets as an industrial raw material include production of biscuits and confectionery, beverages, weaning foods and beer (Anukam and Reid, 2009). Millet is also used as forage crop especially in United States, Australia, and southern

Africa. Pearl millet gives a productive pasture for grazing, especially with dwarf varieties, and silage is easily made.

In addition, millet is more than just an interesting alternative to the more common cereals. The grains are rich in various phytochemicals which are associated with reduced risk of diseases. Thus, besides nutritional value, millets promote different health benefits. For people suffering from celiac diseases, millet is the best alternative to wheat and other more common cereal grains which contain gluten. Millet is gluten-free, therefore provides nutrition and health benefits to celiac disease sufferers.

12.6 BARLEY

Barley *(Hordeum vulgare L.)* is one of the most important cereal crop and ranks 4th among cereal crops in production around the world behind maize, wheat, and rice. It belongs to Poaceae family and is usually grown in temperate climates around the globe. Archaeological evidences shows that it was being cultivated in Egypt along the River Nile around 17,000 years ago (Badr et al., 2000). It outperforms other cereal under various environmental stresses due to its winter-hardy, drought resistant and early maturing nature and is generally more economical to cultivate. About 90% of the cultivated barley is used for feeding stock and for alcoholic beverage production. Besides this, the demand for barley has increased considerably due to its other nutritional benefits like high portion of biologically active compound like β-glucan, tocols including tocopherols and tocotrienols and phenolic compounds which are having a multitude of health benefits in comparison with rice, wheat, and maize. The proximate composition of different varieties of barley is presented in Table 12.5.

TABLE 12.5 Proximate Composition of Different Barley Varieties

Variety	Crude protein (%)	Moisture content (%)	Ash (%)	Fiber (%)	Fat (%)	Carbohydrate (%)	Energy (kcal)
Shege	11.61	13.07	2.16	10.00	3.60	51.31	293.60
Dimtu	10.54	12.18	1.96	11.47	4.50	52.36	300.28
Harbu	11.88	12.98	1.43	14.42	5.01	48.72	296.00
Setegen	12.92	12.19	2.34	14.16	2.75	51.87	292.39
Belemi	12.61	12.39	2.21	10.13	4.90	55.14	324.30
Cross 41/98	14.10	12.42	2.22	14.30	3.10	48.04	285.74

Adapted and modified from: Abeshu and Abrha, (2017).

Regarding composition the major portion of barley includes starch which is around 62–77% of the total weight of the grain; protein content is 8–20% with the preferred brewing range being 10–12% Figure 12.11. This protein is desirable for brewers for using a liquid adjunct or for those using a low protein solid adjunct such as rice such in which the protein content is round about 12–13%. The starch digestibility is important in brewing, animal feed and human food with varied effects hence the degradation of barley starch is major determinant of functional properties of fermentable sugars during brewing, in human gut health and providing energy for livestock. Out of the total protein content, the major portion of the barley protein includes 50% of prolamin (hordein) with other three protein fraction comprising of albumins, globulins, and glutelins making up the remaining protein portion. The protein synthesis commences early after anthesis although there are different reports on the actual timing of these protein components. A number of enzymes are responsible for the synthesis of barley proteins have been identified with wide range of individual genes and allelic variations.

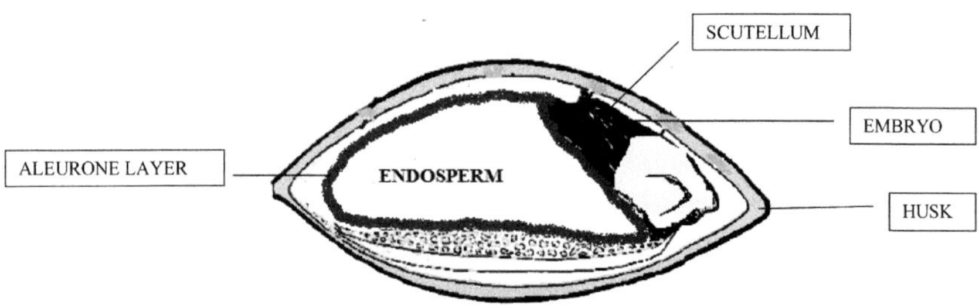

FIGURE 12.11 General structure of barley kernel.

In addition to this barley has gained its importance duo its high portion of β-glucan which is a polymer of β-D-glucose through glycosidic linkages (1,4) and (1,3), belonging to the hemicellulose plant polysaccharides (soluble fiber) (Baik and Ullrich, 2008). Meanwhile, different studies have proved the positive impact of β-glucan on several health-related claims like reduction of cholesterol levels and glycemic responses. Moreover, The United States Food and Drug Administration (USFDA) have also accepted these claims. Further, the European Food Safety Authority (EFSA) issued a note indicating that the consumption of β-glucan from barley as part of a meal contributes to the reduction of the rise in blood glucose after the meal. These effects are produced when the food contains at least 4 g of β-glucan from oats or barley for each 30 g of carbohydrates. Moreover, the EFSA assumes that β-glucan maintains normal blood LDL cholesterol concentrations (EFSA, 2011). However, the average content of β-glucan in barley is not well defined and mainly related to the cultivar. Data reported in the literature show contents of β-glucan ranging between 2 and 11% in dry grain (Izydorczyk and Dexter, 2008). Waxy barley varieties have been associated with higher β-glucan content, although results reported in literature are not clear-cut (Baik and Ullrich, 2008). Some researchers suggest that β-glucan when comes in contact with water results in an increase in viscosity due to which is associated with an increased binding of bile acids and their subsequent excretion. Bile acids are synthesized de novo by the cholesterol-7α-hydroxylase, which is the key enzyme of bile acid synthesis. Hereby, plasma cholesterol serves as substrate for newly synthesized bile acids, which leads to a reduction of blood cholesterol levels. Other mechanisms suggest that β-glucan is fermented by intestinal bacteria in the colon leading to the formation of short chain fatty acids (SCFA) such as acetate, propionate, and butyrate. These SCFA serve as signal molecules and modulate the glucose and cholesterol metabolism by distinct receptors (e.g., Ffr 2/3, free fatty acid (FFA) receptors). Via these receptors, SCFA can increase the concentration of gastrointestinal hormones such as GLP-1 (glucagon-like-peptide 1) and PYY (peptide YY). PYY induces glucose intake in muscle- and fat tissue and GLP-1 indirectly reduces blood glucose concentration by increasing the concentration of insulin and reducing the glucagon production in the pancreas.

The tocopherols and tocotrienols (Vitamin E) also known as tocols are having potential impacts on human health. They are associated with various chemical and physiological properties. Their biological activity is generally believed to be due to their antioxidant action where they inhibit lipid peroxidation in biological membranes. The vitamin E family includes eight compounds:α-, β-, γ- and δ-tocopherol (T) and α-, β-, γ- and δ-tocotrienol (T3). Different studies have shown that barley grain contains all four tocotrienol and the four tocopherol vitamers in different proportions (Ryynanen et al., 2004). Some studies also indicate that a high intake of α-tocopherols decreases lipid peroxidation, platelet aggregation, as well as functioning as a potent anti-inflammatory agent (Jialal and Devaraj, 2005; Tiwari and Cummins, 2009). Apart from their antioxidant properties, the tocol content of cereals such as barley can offer a multitude of health benefits including prevention and reduction of many degenerative diseases like cancer and cardiovascular diseases (CVD) (Tiwari and Cummins, 2009).

12.7 SORGHUM

Sorghum (*Sorghum bicolor L.* Moench) is one of the most important and world's fifth cereal crop, which is drought-tolerant, resistant to waterlogging and grows in various soil conditions cereal. It is a hardy cereal grain mostly found in semi-arid regions such as northern Africa and western Asia (Taylor and Shewry, 2006). The seeds are small in contrast with the kernels of corn and wheat. The pericarp consists of several tissue layers and it may be pigmented and its outer surface is covered with waxy cuticle. A central layer of large thin-walled cells composed of starch and constitutes the mesocarp. The testa is a tissue attached to the pericarp next to the endosperm and is absent in many sorghum varieties. The embryo near the base of the seed consists of a large scutellum, a terminal plumule, a short central axis, and a primary root. Endosperm of sorghum seed is composed of outer single-cell layer of aleurone tissue, a region of horny endosperm, and an inner floury endosperm (Figure 12.12). Aleurone cells are rich in oil and protein.

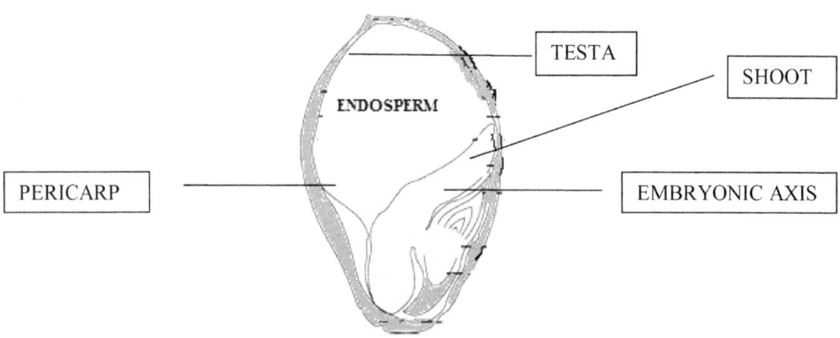

FIGURE 12.12 Structure of sorghum grain.

In Australia, the production of sorghum grain averaged 275000 t through the year 1960s, but increased to an average of 1.035×10^6 t in 1970s. This dramatic rise in production is due to its increased use in beef feedlot industry. Sorghum grain is the leading cereal grain on African continent and Nigeria is the world second largest producer of the grain (ICRISAT, 2002).

It has got excellent nutrient profile which is due to the high concentration of dietary fibers, carbohydrates, and proteins however its nutritional profile is hindered by some interfering molecules present in it (Awika and Rooney, 2004). Its flour is used in many food formulations like gluten-free breads, composite breads, and extruded snack like foods, breakfast cereals, and instant porridges. Absence of gluten makes it a good candidate for the formulation of gluten-free products; however, due to poor rheological properties, it can be used in combination with other flours to overcome this problem. Proximate composition of sorghum is shown in the Table 12.6.

TABLE 12.6 Proximate Composition of Sorghum

Parameter	% age
Protein	11.6
Starch	75.9
Fat	3.3
Fiber	1.9
Ash	1.3

Source: Adapted and modified from: Jones and Beckwith, (1970).

Sorghum proteins are classified into albumins, globulins, and glutelins accounting for 30%, while the remaining 70% are alcohol soluble prolamins called kafirins (Belton et al., 2006). Kafirins are classified as α-kafirin, β-kafirin and γ-kafirin with molecular weight of 25 and 23 kDa, 20, 18 and 16 kDa and 28 kDa and 50 kDa, respectively, and the less characterized δ-kafirin with molecular weight of 13 kDa (Mokrane et al., 2010). Sorghum also contains significant amounts of dietary fiber, minerals, and bioactive compounds such as condensed tannins which are present in varieties with pigmented tests (Dykes and Rooney, 2006; Paiva et al., 2017). The presence of tannins in sorghum has got number of benefits like reduction of bird damage and bird predation in pre and post-harvest. The tannins present in sorghum has antioxidant and radical scavenging functions like immunomodulatory, anticancer, anti-inflammatory, cardioprotective, vasodilating, and antithrombotic effects (Wu et al., 2012). On the other hand, tannins are known to bind to proteins and carbohydrates, reducing caloric availability and may bring health benefits for special human diets aimed at weight loss (Awika and Rooney, 2004) but this property may cause a decrease in the feed efficiency (Dykes and Rooney, 2006). The proteins and starch present in sorghum can be processed into grits, starch flour, and flakes, which can be further utilized in number of industrial applications. Moreover, sweet sorghum has the potential of producing bioethanol which has got tremendous applications (Dar et al., 2018)

12.8 PSEUDOCEREALS

The American Heritage Dictionary of English language defines pseudocereals as, "Plant species that does not belong to the grass family but produces fruits and seeds used as flour for bread and other staple foods." The most common pseudocereals known so far are grain amaranth (Amaranth caudatus, Amaranth cruentus, Amaranth hypochondriacs; family Amaranthaceae), quinoa (Chenopodium quinoa sub sp. quinoa; Chenopodiaceae), album (Chenopodium album), Chia (Salvia hispanica), and buckwheat (Fagopyrum esculentum; Polygonaceae) (Mir et al., 2018). Figure 12.2 shows the images and seeds of different pseudocereals. Pseudocereals are underutilized crops; they are gluten free and contain innumerable essential nutrients. The pseudocereals belong to the most important food for the majority of mankind consumed in the form of bread, breakfast cereals, or cereal bars in developed countries (Kockovaet al., 2013). They are a good source of saccharides (especially starch and fibers), proteins with good quality of amino acids, lipids (essential fatty acids), vitamins, and minerals (Fletcher, 2004). The values for pseudo-cereal proteins are definitively higher when compared to cereals and are close to those of casein (Gorinstein et al., 2002). The high amount of dietary fiber in pseudocereals helps in improving lipid metabolism and in the prevention of LDL-C oxidation. Pseudocereals have gained their importance due to their excellent nutritional profile, phytochemical content and their use in gluten-free products. These plant species has a more important role in the development and diversification of agricultural products and food. Due to the excellent nutritional profile of pseudocereals, their production has increased considerably. The production of *Chenopodium quinoa* is projected at 74, 382 MT in Bolivia, 3711 MT in Ecuador, and 114,725 MT in Peru in the year 2014 (FAOSTAT, 2014). However, until today cultivation of *Chenopodium* is still low and even production data on *Chenopodium album* is not listed in the FAO statistics yet. The yield of amaranth grain strongly depends on environment, weather conditions, species, genotype, and production techniques, and varies in a wide range from 500 to 2,000 kg grain per ha (Mlakar et al., 2009). With the advent of modern agricultural practices and appropriate varieties, it is expected that production yields of 1,500 to 3,000 kg grain per ha can be obtained (Williams and Brenner, 1995). According to Jamriska (1990) and Kaul et al. (1996), grain yields in Europe ranged between 2,000 in 3,800 kg per hectare.

Regarding buckwheat, the worldwide production has reached 2.5 Mha providing 2 MT. China, Russian Federation, Ukraine, and Kazakhstan are the largest producers of buckwheat (Li and Zhang, 2001; Bonafaccia et al., 2003).

In addition to this, buckwheat is also produced in Slovenia, Poland, Hungary, and Brazil (Kreft et al., 1999a). China produced 564,900 MT, Russia 661,764 MT and Ukraine 167,440 MT (FAOSTAT, 2014). In Europe, 83,499 MT were produced in Poland, 111,300 MT in France and little in Hungary, Slovenia, Latvia, and Lithuania. The production of buckwheat is projected at 83, 499 MT in Poland, 1934 MT in Republic, and 67 MT in Slovakia in the year 2014 (FAOSTAT, 2014). Additionally, due to their different morphology and functional properties, the commonly available processing methods are not applicable for their processing (Figure 12.13). The seeds of pseudocereals possess a new, unknown taste compared to conventional cereals (Berghofer and Schoenlechner, 2010).

12.9 AMARANTH

Amaranthus collectively known as amaranth belongs to Amaranthaceae family and cosmopolitan genus of annual or short-lived perennial plants. The scientific plant name amaranth signifies in Greek "immortal," "everlasting" or "non-wilting" (Mlakar et al., 2010). In botanical terms, amaranth is not true cereals; it is dicotyledonous plants as opposed to most cereals (e.g., wheat, rice, barley) and referred as pseudocereal, as their seeds resemble in function and composition those of the true cereals. The amaranth grains are very small in size and they occur in huge numbers, sometimes more than 50,000 to a plant (Berghofer and Schoenlechner, 2010). Amaranth seeds are small (1–1.5 mm diameter), they are lenticular in shape and weight per seed is 0.6–1.3 mg. In addition, the percentage of bran fraction (seed coat and embryo) in amaranth seeds are higher in comparison with common cereals, such as maize and wheat, which explains the higher levels of protein and fat present in these seeds (Bressani, 2003).

Amaranth was used as a staple food throughout history in Inca, Maya, and Aztec civilizations. At present amaranth is grown for commercial purposes in Mexico, South America, the United States, China, Poland, and Austria (Milán-Carrillo et al., 2012). In India amaranth is grown in Himalaya regions from Kashmir to Bhutan, and also in south Indian hills. There are approximately 60 species, which according to the uses for human consumption can be divided into grain and vegetable amaranths. The leaves of the young *Amaranthus blitus*, *Amaranthus tricolor*, *Amaranthus cruentus*, *Amaranthus dubius*, *Amaranthus edulis, and Amaranthus hypochondriacus* plants are used in salads and soups. The grains of *Amaranthus caudatus*, *Amaranthus hypochodriacus*, *Amaranthus cruentus*, *Amaranthus*

Pseudo cereal	Plant	Seed
Album		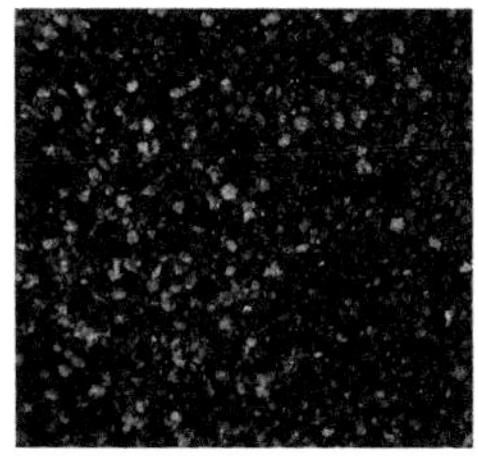
Quinoa		
Amaranth		
Chia		
Buckwheat		

FIGURE 12.13 Images of different pseudocereal plants and their seeds.

hybridus, and *Amaranthus mantegazzianus* are used into breads, cakes, cookies, confectionery, and soups, whereas some species are not safe for consumption by either humans or livestock such as *Amaranthus retroflexus, Amaranthus Viridis*, and *Amaranthus spinosus (*Caselato-Sousa and Amaya-Farfan, 2012). Nowadays, amaranth classified as a new, forgotten, neglected, and an alternative crop of great nutritional value. In India amaranth used in beverages, sauces, porridges, tortillas (also with maize flour), popped grains like maize, and for various medicinal uses. Besides in the diet, amaranth had an important position also in Indian religion (Mlakar et al., 2009).

12.10 BUCKWHEAT

Tartary buckwheat (*Fagopyrum tataricum* Gaertn.) is an important underexploited pseudocereal belonging to the Polygonaceae family. It is an annual plant, about 0.6–1.3 m tall, having reddish stems and white to pink flowers. Due to its fast growth rate, it is grown as a cover crop which helps in binding soils and subsequently checks soil erosion during rainy seasons. The plant is especially tolerant to poor quality, sandy or acidic soils. It is a crop with multiple uses such as the tender shoots are consumed as leafy vegetables, the green leaves have medicinal uses, e.g., to promote circulation, and the grains are processed into flour for human consumption as well as livestock feed. Common buckwheat is used as a cereal but does not belong to the family of the grasses (where the common cereals, such as rice and wheat, belong), thus it is known as a pseudocereal. It is a good source of protein and is processed and consumed in the form of many dishes depending upon local cultures. For instance, the flour is used in the preparation of noodles in China and Japan, pancakes, and biscuits in Europe and North America, and porridge and soup in Russia and Poland. While as, it is used to make unleavened bread chapattis in India.

There are some botanical and physiological similarities between buckwheat and weeds, one of them being the ability to correct growth without the use of artificial fertilizers or pesticides (Kreft et al., 1996). Moreover, buckwheat absorbs less water and lower amounts of nutrients from soil than other main crops (Li and Zhang, 2001). The amino acid composition of buckwheat proteins is well balanced and of a high biological value (Kato et al., 2001), although the protein digestibility is relatively low (Liu et al., 2001). Buckwheat grains are an important source of microelements, such as: Zn, Cu, Mn, Se (Stibilj et al., 2004), and macroelements: K, Na, Ca,

Mg (Wei et al., 2003). With 80% unsaturated fatty acids more than 40% are constituted by polyunsaturated fatty acid (PUFA) (Krkošková and Mrázová, 2005). The significant contents of rutin, catechins, and other polyphenols, as well as their potential antioxidant activity, are also of significance to the dietary value. Moreover, buckwheat grains are a rich source of TDF (total dietary fiber), soluble dietary fiber (SDF), and are applied in the prevention of obesity and diabetes (Brennan, 2005).

12.11 *CHENOPODIUM* SPECIES

Chenopod (*Chenopodium sp.*) is one of the largest genera of the family *Chenopodiaceae* and comprises over 250 species of which about 8 species are found in India. *Chenopodium* spp. has been cultivated for centuries as leafy vegetables and as a subsidiary grain crop in different parts of the world (Bhargava et al., 2003). The Chenopod is an important vegetable and grain crop for the hill region and is largely consumed mixed with other cereals. The Himalayan grain chenopod is comparable to Andean quinoa in nutrient composition and is much better than wheat, barley, maize, and rice (Pachauri et al., 2013). The chenopod grains are sometimes used as a staple food consumed in the form of porridge, pudding, and also cooked with rice.

12.12 QUINOA

Quinoa (Chenopodium quinoa) an excellent pseudocereal is a dicotyledonous seed crop native to the Andean region, where it has been used as a food staple for thousands of years (Abugoch James, 2009). Quinoa has received considerable attention due to its high protein content (14–16%) (Föste et al., 2015). Its protein consists of two major protein fractions: 11S (globulins) and 2S (albumins) types (Brinegar et al., 1996; Brinegar and Goundan, 1993). It has a balanced pattern of essential aminoacids and is rich in lysine (4.8 g/100 g protein), threonine (3.7 g/100 g protein), histidine, isoleucine, and methionine (Wright et al., 2002; Repo-Carrasco et al., 2003). Its high lysine content is attributable to albumins and globulins, which are 44–77% of the total protein. According to joint Food and Agriculture Organization/World Health Organization/United Nations University (FAO/WHO/UNU) expert consultation, the recommended protein profile for adults could potentially be provided by raw quinoa. In recent years, the chemical composition, nutritional, and functional applications and processing of quinoa have increased dramatically (Comai et al., 2007; Nongonierma et al., 2015; Valenzuela et al., 2013). This fact has been linked to its agricultural properties, gluten-free nature, and nutritional value. Quinoa seed has been recognized as a very nutritious grain because of the quantity and quality of its proteins (compared to traditional cereals), and the content of fatty acids, dietary fiber, vitamins, and minerals. Moreover, multiple phytochemicals present in quinoa provide it a remarkable advantage over other grains in terms of human health (Vilcacundo and Hernández- Ledesma, 2017). It is known that flavonoids, phenolic acids, and saponins in quinoa contribute to its biological functions. Although some studies have reported a good correlation between the total phenolic content and the antioxidant activity, other authors have suggested the role of non-phenolic compounds such as ascorbic acid, phytic acid, tocopherols, sterols, carotenoids, and ecdysteroids on the antioxidant activity of this seed (Nsimba, Kikuzaki, and Konishi, 2008). Recently, polysaccharides extracted from quinoa with water and alkali has also demonstrated to contribute on the antioxidant activity attributed to this plant (Hu et al., 2017). Quinoa proteins have been suggested to exert some beneficial effects by themselves and as a source of bioactive peptides. Quinoa proteins have previously shown 2,2-diphenyl-1-picrylhydrazyl (DPPH) radical scavenging capacity when hydrolyzed with Alcalase® (Aluko and Monu, 2003). Moreover, due to their excellent functional properties, quinoa seed proteins could also be used to produce novel healthy foods.

12.13 ALBUM

Chenopodium album botanically assigned to the *Dicotyledonae* is long forgotten under-explored plant which produces starch-rich seeds like cereals. The seeds are small with diameter ranging from 1 to 2.5 mm and it can grow in areas with low rainfall of 300 to 400 mm. *Chenopodium album* essentially on account of it's nutritional, functional, agricultural, and technological potential could prove the focus of attention in many scientific studies (Berghofer and Schoenlechner, 2002). The plant is characterized by great diversity of species, and green parts of some species are used as a vegetable. The grains contain no gluten and are safe to consume for individuals with celiac disease. The pseudo-cereal grown by simple methods may be potential and prospective nutritional source for the "starving world." This can be good complements to the nutritional values of traditional corn-based food products because of their high nutrient composition as well as some agricultural benefits such as relative high grain yield, resistance to drought and short production time.

12.14 CHIA

Chia (Salvia hispanica) is an annual plant of Lamiaceae family commonly known as Chia and is native to Mesoamerica. The plant it is approximately a meter tall, with opposite, petiolate, and serrated leaves that are 4 to 8 cm long and 3 to 5 cm wide. The flowers are hermaphrodite and grow in numerous clusters in a spike protected by small bracts with long pointed tips. The plant has quadrangular stems that are ribbed and hairy. It grows in light to medium, clay, and sandy soils, and even in arid soils that have good drainage but are not too wet. Chia is grown mainly in mountainous areas and has little tolerance to abiotic phenomena, such as freezing and sunless locations. Morphologically, wild, and domesticated plants differ very little, and today chia has been classified within the cultivated lands of Mesoamerica. Chia was the most important crop for the pre-Colombian people besides corn, bean, and amaranth (Ayerza and Coates, 2005). The diet of the Columbians was considered superior as to what we consume today because as chia was an important part of their diet and it consist of superior quality of proteins, high amount of omega-3-fatty acids, high extent of dietary fiber, vitamins, minerals, and antioxidants which helps in increasing the shelf life of seeds by protecting it from microbial and chemical breakdown. Chia holds future perspective for food, pharmaceutical, and cosmetic sector because it is a rich source of natural antioxidants, proteins, and dietary fiber, and also contains high amount of unsaturated fatty acids (da Silva Marineli et al., 2014). The seed protein is rich in essential amino acids which makes it a valuable candidate for supplements. It is an excellent and vegan source of omega three fatty acids and has benefits for prevention of coronary heart diseases (CHDs). It forms mucilaginous coats as soon as it comes in contact with water. The mucilage is mostly made up of insoluble and soluble fibers which can help in assisting the regulation of bowel movement. The mucilage can be extracted and used as an emulsifying agent in bakery products moreover the phytochemical composition is also appreciable due to this reason it can assist in prevention of rancidity in fats thus prolonging their shelf life. The leaves contain essential oils that act as insect repellents, thus the plant can be grown without pesticides or other chemical compounds. In Mexico it grows easily and mainly in juniper, oak, pine, and pine-oak forests, spreading by seed dispersal, with the wild type an average height of 1.9 meters. In the state of Jalisco, Mexico, chia is grown on farmland from late spring to early summer. The chia seed is a package of important nutrients in comparison with other seeds like pumpkin, quinoa, and basil. The dietary fiber is also in good amounts as compared to other seeds such as quinoa and basil. It also contains natural phenolic compounds as caffeic acid, chlorogenic acid, myricetin, and kaempferol. Chia is an important source of different vitamins and minerals also. The mucilage present in chia is made up of polysaccharides. The mucilage of chia seed comprises of 93.8% carbohydrate, mucilage can be used as a food stabilizer in food systems due to its weak gel characteristics and high pseudoplasticity. The mucilage contains 95% nonstarch polysaccharide, i.e., 35% soluble fraction and 65% insoluble fraction. When mucilage is freeze-dried it shows weak gel and strong shear depth properties at low concentration due to this property it can be used as an effective food stabilizer. This property is desirable in low-fat products where it can aid in stabilizing the large particles and also helps in improving the mouthfeel. Besides this several in vivo studies confirms the beneficial effects of chia seed for prevention and reduction of CVD. Therefore extensive research should be carried out to explore the other compounds present in chia seeds and their possible impacts on humans. Proximate composition and amino acid profile of different pseudocereals is presented in Tables 12.7 and 12.8.

TABLE 12.7 Proximate Composition of Different Pseudocereals

Parameter (%)	Pseudocereals				
	Amaranth	**Buckwheat**	**Quinoa**	**Album**	**Chia**
Moisture	12.31	13.86	12.05	9.43	12.45
Fat	5.08	2.52	3.44	6.50	30.41
Ash	1.59	1.67	2.10	3.25	3.02
Protein	11.00	13.07	11.32	13.12	20.04
Fiber	4.80	17.44	10.55	13.09	37.65
Carbohydrates	56.12	56.32	57.11	54.61.23	37.11

(Adapted and Modified from: Collar and Angioloni, (2014), Chauhan et al. (2015), Jan et al. (2016), Xiao et al. (2017))

TABLE 12.8 Amino Acid Profile of Different Pseudocereals

Amino acid (%)	Pseudo cereal				
	Amaranth	**Buckwheat**	**Quinoa**	**Album**	**Chia**
Cysteine	3.1	1.4	1.29	1.27	4.2
Aspartic acid	10.0	8.7	9.05	9.10	12.8
Methionine	2.0	0.4	2.18	2.10	6.7
Threonine	4.9	4.5	4.09	4.0	5.4
Serine	8.8	6.2	5.45	5.37	9.4
Glutamic acid	15.5	16.3	15.22	15.12	28.7
Proline	4.6	4.4	3.22	3.18	12.8
Glycine	14.3	9.9	6.40	6.30	9.1
Alanine	6.2	6.3	4.40	4.32	9.4
Valine	4.8	5.8	4.19	4.10	7.9
Isoleucine	3.6	4.3	3.22	3.18	7.4
Leucine	6.2	6.9	6.70	6.72	14.2
Histidine	2.0	2.5	3.15	3.10	6.1
Lysine	8.0	4.8	5.95	5.90	9.3
Arginine	12.7	5.9	8.75	8.70	20.0
Phenylalanine	N.D	4.2	5.79	5.73	11.6
Tryptophan	N.D	N.D	0.95	0.92	N.D
Tyrosine	N.D	1.6	3.62	3.52	6.1

N.D = not detected.

(Adapted and modified from: Mir, N. A., Riar, C. S., & Singh, S. (2019). Effect of pH and holding time on the characteristics of protein isolates from Chenopodium seeds and study of their amino acid profile and scoring. *Food Chemistry, 272*, 165–173).

12.15 CONCLUSION

Cereals and pseudocereals are thus gaining importance due to the above-mentioned nutritional properties and their possible impacts on human health. Western diet is usually composed of cereal grains and due to this reason, the global status of bakery industry has increased dramatically from the past years. The bakery products developed from cereal grains are gaining popularity among all cross-sections of population irrespective of age group and economic conditions and urban areas share a major portion this is possibly due to rising disposable incomes and time constraints in the working population residing in these areas. In order to meet the demands of burgeoning population we have to produce good amount of cereal grains with excellent nutritional profile and increased yield and for this reason, new technologies and hybrid varieties of cereal grains are continuously tested. In addition to this, pseudocereals are also gaining importance as their nutritional and functional profile is far better than traditional cereal crops. Nowadays, most of the westernized countries are switching towards pseudocereals like quinoa, buckwheat, and chia due to their increased demand for the development of novelty foods. However, some anti-nutritional factors like saponins reduce their take off in large segment of populations, but these anti-nutritional factors have also got some agro pharmacological applications. So if better agricultural practices and innovative methods of processing are placed in line, then conditions would be more conducive for pseudocereal crops to act as a substitute for many traditional cereal crops.

KEYWORDS

- **cereals**
- **health benefits**
- **nutritional properties**
- **pseudocereals**

REFERENCES

Abdelrahman, A. A., & Hoseney, R. C., (1984). Basis for hardness in pearl millet, grain sorghum, and corn. *Cereal Chemistry, 61*(3), 232–235.

Abeshu, Y., & Abrha, E., (2017). Evaluation of proximate and mineral composition profile for different food barley varieties grown in central highlands of Ethiopia. *World Journal of Food Science and Technology, 1*(3), 97–100.

Abugoch, J. L. E., (2009). Quinoa (Chenopodium quinoa Willd.): Composition, chemistry, nutritional, and functional properties. *Advances in Food and Nutrition Research, 58*, 1–31. http://dx.doi.org/10.1016/S1043-4526(09)58001-1 (Accessed on 14 June 2019).

Andrews, D. J., & Kumar, K. A., (1992). Pearl millet for food, feed, and forage. In: *Advances in Agronomy* (Vol. 48, pp. 89–139). Academic Press.

Anukam, K. C., & Reid, G., (2009). African traditional fermented foods and probiotics. *Journal of Medicinal Food, 12*(6), 1177–1184.

Awika, J. M., Rooney, L. W., & Waniska, R. D., (2004). Anthocyanins from black sorghum and their antioxidant properties. *Food Chem., 90*, 239–304.

Ayerza, R., & Coates, W., (2005). Ground chia seed and chia oil effects on plasma lipids and fatty acids in the rat. *Nutrition Research, 25*, 995–1003.

Azhakanandam, K., Power, J. B., Lowe, K. C., Cocking, E. C., Tongdang, T., Jumel, K., & Davey, M. R., (2000). Qualitative assessment of aromatic Indica rice (Oryza sativa L.): Proteins, lipids and starch in grain from somatic embryo-and seed-derived plants. *Journal of Plant Physiology, 156*(5/6), 783–789.

Badr, A., Sch, R., El Rabey, H., Effgen, S., Ibrahim, H., Pozzi, C., Rohde, W., & Salamini, F., (2000). On the origin and domestication history of barley (Hordeum vulgare). *Mol. Biol. Evol., 17*, 499–510.

Baik, B. K., & Ullrich, S. E., (2008). Barley for food: Characteristics, improvement, and renewed interest. *Journal of Cereal Science, 48*, 233–242.

Belton, P. S., Delgadillo, I., Halford, N. G., & Shewry, P. R., (2006). Kafirin structure and functionality. *Journal of Cereal Science, 44*(3), 272–286.

Berghofer, E., & Schoenlechner, R., (2002). Grain amaranth. In: Belton, P. S., & Taylor, J. R. N., (eds.), *Pseudocereals and Less Common Cereals: Grain Properties and Utilization Potential* (pp. 219–260). Berlin, Springer, Verlag.

Berghofer, E., & Schoenlechner, R., (2010). *Pseudocereals–An Overview*. Department of Food Science and Technology, University of Natural Resources and Applied Life Sciences, Vienna-Austria.

Bhargava, A., Shukla, S., & Ohri, D., (2003). Genetic variability and heritability of selected traits during different cuttings of vegetable Chenopodium. *Indian J. Genet. Plant Breed., 63*, 359–360.

Bonafaccia, G., Marocchini, M., & Kreft, I., (2003). Composition and technological properties of the flour and bran from common and tartary buckwheat. *Food Chemistry, 80*, 9–15.

Boyer, C. D., & Hannah, L. C., (2001). Kernel mutants of corn. In: Hallauer, A. R., (ed.), *Specialty Corns*, (pp. 1–31.).

Brennan, C. S., (2005). Dietary fiber, glycaemic response and diabetes. *Molecular Nutrition & Food Research, 49*, 560–570.

Bressani, R., & Amaranth, (2003). In: Caballero, B., (ed.), *Encyclopedia of Food Sciences and Nutrition* (pp. 166–173). Oxford: Academic Press.

Brinegar, C., & Goundan, S., (1993). Isolation and characterization of chenopodin, the 11S seed storage protein of quinoa (*Chenopodium quinoa*). *J. Agric. Food Chem., 41*, 182–185.

Brinegar, C., Sine, B., & Nwokocha, L., (1996). High-cysteine 2S seed storage proteins from quinoa (*Chenopodium quinoa*). *J. Agric. Food Chem., 44*, 1621–1623.

Burger, H. M., & Rheeder, P., (2017). Important mycotoxins relevant to maize. http://www.grainsa.co.za/important-mycotoxins-relevant-to-maize (Accessed on 14 June 2019).

Caselato-Souza, V. M., & Amaya-Farfán, J., (2012). State of knowledge on amaranth grain: A comprehensive review. *Journal of Food Science, 77*, 93–104.

Chan, K. W., Khong, M. H., Iqbal, S., & Ismail, M., (2013). Isolation and antioxidative properties of phenolics-saponins rich fraction from defatted rice bran. *Journal of Cereal Science, 57*, 480–485.

Chauhan, A., Saxena, D. C., & Singh, S., (2015). Total dietary fiber and antioxidant activity of gluten-free cookies made from raw and germinated amaranth (Amaranthus spp.) flour. *LWT-Food Science and Technology, 63*(2), 939–945.

Choudhury, N. H., & Juliano, B. O., (1980). Lipids in developing and mature rice grain. *Phytochemistry, 19*, 1063–1069.

Chung, O. K., & Pomeranz, V., (1985). Amino acids in cereal proteins fractions. In: Finely, J. W., & Hopkins, D. T., (eds.), *Digestibility and Amino Acid Availability in Cereals and Oil Seeds* (pp. 65–707). St. Paul, MN: American Associations of Cereal Chemists.

Cicero, A. F. G., & Derosa, G., (2005). Rice bran and its main components: Potential role in the management of coronary risk factors. *Current Topics in Nutraceutical Research, 3*(1), 29–46.

Clotault, J., Thuillet, A. C., Buiron, M., De Mita, S., Couderc, M., Haussmann, B. I., & Vigouroux, Y., (2011). Evolutionary history of pearl millet (*Pennisetumglaucum* [L.] R. Br.) and selection on flowering genes since its domestication. *Molecular Biology and Evolution, 29*(4), 1199–1212.

Collar, C., & Angioloni, A., (2014). Pseudocereals and teff in complex bread making matrices. Impact on lipid dynamics. *Journal of Cereal Science, 59*(2), 145–154.

Comai, S., Bertazzo, A., Bailoni, L., Zancato, M., Costa, C. V. L., & Allegri, G., (2007). The content of proteic and nonproteic (free and protein-bound) tryptophan in quinoa and cereal flours. *Food Chem., 100*, 1350–1355.

Da Silva, M. R., Moraes, É. A., Lenquiste, S. A., Godoy, A. T., Eberlin, M. N., & Maróstica, M. R., Jr., (2014). Chemical characterization and antioxidant potential of Chilean chia seeds and oil (Salvia hispanica L.). *LWT-Food Science and Technology*, 59, 1304–1310.

Dar, R. A., Dar, E. A., Kaur, A., & Phutela, U. G., (2017). Sweet sorghum-a promising alternative feedstock for biofuel production. *Renewable and Sustainable Energy Reviews.*

Dass, S., (2013). Maize and its diversified uses. In: Kumar, A., Jat, S. L., Kumar, R., & Yadav, O. P., (eds.), *Maize Production Systems for Improving Resource-Use Efficiency and Livelihood Security* ((pp. 1–3)). DMR, New Delhi, India.

Dykes, L., & Rooney, L. W., (2006). Sorghum and millet phenols and antioxidants. *J. Cereal Sci., 44*, 236–251.

EFSA, (2011). Panel on dietetic products, nutrition and allergies (NDA), scientific opinion on the substantiation of health claims related to beta-glucans from oats and barley and maintenance of normal blood LDL-cholesterol concentrations (ID 1236, 1299), increase in satiety leading to a reduction in energy intake (ID 851, 852), reduction of post-prandial glycaemic responses (ID 821, 824), and "digestive function" (ID 850) pursuant to Article 13(1) of Regulation (EC) No 1924/2006. *EFSA Journal, 9*, 2207. http://dx.doi.org/10.2903/j.efsa.2011.2207 http://www.efsa.europa.eu/en/efsajournal/pub/2207 (Accessed on 14 June 2019).

Eggum, B. O., Juliano, B. O., & Maniñgat, C. C., (1982). Protein and energy utilization of rice milling fractions by rats. Qual. Plant. *Plant Foods for Human Nutrition, 31*, 371–376.

Elyas, S. H., El Tinay, A. H., Yousif, N. E., & Elsheikh, E. A., (2002). Effect of natural fermentation on nutritive value and *in vitro* protein digestibility of pearl millet. *Food Chemistry, 78*(1), 75–79.

FAOSTAT, (2014). *FAO Statistics Division.* Available from: http://www.faostat.fao.org (Accessed on 14 June 2019).

FAOSTAT, (2016). *Database of Food and Agricultural Organization*, Rome, Italy. Available from: http://www.faostat.fao.org (Accessed on 14 June 2019).

Fletcher, R. J., (2004). *Pseudocereals: Overview* (pp. 488–493). Encyclopedia of Grain Science.

Föste, M., Elgeti, D., Brunner, A. K., Jekle, M., & Becker, T., (2015). Isolation of quinoa protein by milling fractionation and solvent extraction. *Food Bioprod. Process, 96*, 20–26.

Frei, M., Siddhuraju, P., & Becker, K., (2003). Studies on the *in vitro* starch digestibility and the glycemic index of six different indigenous rice cultivars from the Philippines. *Food Chemistry, 83*, 395–402.

Fujino, Y., & Mano, Y., (1972). Classification of lipids and composition of fatty acids in brown rice. *Eiyo to Shokuryo, 25*, 472–474.

Gorinstein, S., Pawelzik, E., Delgado-Licon, E., Haruenkit, R., Weisz, M., & Trakhtenberg, S., (2002). Characterization of pseudocereal and cereal proteins by protein and amino acid analyses. *J. Sci. Food Agric., 82*, 886–891.

Gul, K., Singh, A. K., & Jabeen, R., (2016b). Nutraceuticals and functional foods: The foods for the future world. *Critical Reviews in Food Science and Nutrition*, *56*(16), 2617–2627.
Gul, K., Singh, A. K., & Sonkawade, R. G., (2016a). Physicochemical, thermal and pasting characteristics of gamma-irradiated rice starches. *International Journal of Biological Macromolecules*, *85*, 460–466.
Gul, K., Yousuf, B., Singh, A. K., Singh, P., & Wani, A. A., (2015). Rice bran: Nutritional values and its emerging potential for development of functional food—A review. *Bioactive Carbohydrates and Dietary Fiber*, *6*(1), 24–30.
ICRISAT, (2002). International crops research institute for semi-arid tropics. *Ann. Rep., 10*, 102–114.
Izydorczyk, M. S., & Dexter, J. E., (2008). Barley b-glucans and arabinoxylans: Molecular structure, physicochemical properties, and uses in food products–A review. *Food Research International*, *41*, 850–868.
Jamriška, P., (1990). The effect of the stand organization on the yield of amaranth (*Amaranthus hypochondriacus*). *Rost. Vyroba*., *36*, 889–896.
Jan, R., Saxena, D. C., & Singh, S., (2016). Physico-chemical, textural, sensory and antioxidant characteristics of gluten-free cookies made from raw and germinated Chenopodium (Chenopodium album) flour. *LWT-Food Science and Technology*, *71*, 281–287.
Jialal, I., & Devaraj, S., (2005). Scientific evidence to support a vitamin E and heart disease health claim: Research needs. *The Journal of Nutrition*, *135*, 348–353.
Jones, R. W., & Beckwith, A. C., (1970). Proximate composition and proteins of three-grain sorghum hybrids and their dry-mill fractions. *Journal of Agricultural and Food Chemistry*, *18*(1), 33–36.
Ju, Z. Y., Hettiarachchy, N. S., & Rath, N., (2001). Extraction, denaturation and hydrophobic properties of rice flour proteins. *Journal of Food Science*, *66*(2), 229–232.
Juliano, B. O., (1985b). *Rice: Chemistry and Technology* (2nd edn., p. 774). St Paul, MN, USA, Am. Assoc. Cereal Chem.
Kahlon, T. S., & Smith, G. E., (2004). Rice bran, a health-promoting ingredient. *Cereal Foods World, 49,* 188–194.
Kato, N., Kayashita, J., & Tomotake, H., (2001). Nutritional and physiological functions of buckwheat protein. *Recent Research Development Nutrition*, *4*, 113–119.
Kaul, H. E., Aufhammer, W., Laible, B., Nalborczyk, E., Pirog, S., & Wasiak, K., (1996). The suitability of amaranth genotypes for grain and fodder use in Central Europe. *Die Bodenkultur*., *47*, 173–181.
Kent, N. L., (1978). Technology of cereals, with special reference to wheat (No. 664.729 K4 1978).
Kent, N. L., (1994). *Kent's Technology of Cereals: An Introduction for Students of Food Science and Agriculture*. Elsevier publishers.
Kocková, M., Dilongová, M., & Hybenová, E., (2013). Evaluation of cereals and pseudocereals suitability for the development of new probiotic foods. *Journal of Chemistry*.
Kreft, I., Plestenjak, A., Golob, T., Skrabanja, V., Rudolf, M., & Draslar, K., (1999a). Functional value of buckwheat as affected by the content of inositol, phosphate, minerals, dietary fiber and protein. In: Sanberg, A. S., Andersson, H., Amado, R., Schlemmer, H., & Serra, F., (eds.), *Bioactive Inositol Phosphates and Phytosterols in Food* (pp. 69–72). Cost 916. Office for Official Publications of the European Communities, Luxemburg.
Kreft, I., Srabanja, V., Ikeda, S., Ikeda, K., & Bonafaccia, G., (1996). Dietary value of buckwheat. *Research Reports Biotechnical Faculty of the University of Lubljana, 67*, 73–78.
Krkošková, B., & Mrázová, Z., (2005). Prophylactic components of buckwheat. *Food Research International, 38*, 561–568.

Léder, I., (2004). Sorghum and millets. *Cultivated Plants, Primarily as Food Sources, 1*, 66–84.

Li, S., & Zhang, Q. H., (2001). Advances in the development of functional foods from buckwheat. *Critical Reviews in Food Science and Nutrition, 41*, 451–464.

Liu, Z., Ishikawa, W., Huang, X., Tomotake, H., Watanabe, H., & Kato, N., (2001). Buckwheat protein product suppresses 1,2-dimethylohydrazine-induced colon carcinogenesis in rats by reducing cell proliferation. *Nutrition and Cancer–Research Communication*, 1850–1853.

Mano, Y., Kawaminami, K., Kojima, M., Ohnishi, M., & Ito, S., (1999). Comparative composition of brown rice lipids (lipid fractions) of indica and japonica rice. *Bioscience, Biotechnology, and Biochemistry, 63*(4), 619–626.

Marshall, W. G., & Wordsworth, J. I., (1994). *Rice Science and Technology* (pp. 237–259). New York: Marcel Dekker, Inc.

Milán-Carrillo, J., Montoya-Rodríguez, A., Gutiérrez-Dorado, R., Perales-Sánchez, X., & Reyes-Moreno, C., (2012). Optimization of extrusion process for producing high antioxidant instant amaranth (*Amaranthus hypochondriacus L.*) flour using response surface methodology. *Applied Mathematics,* 3.

Mir, N. A., Riar, C. S., & Singh, S., (2018). Nutritional constituents of pseudocereals and their potential use in food systems: A review. *Trends in Food Science & Technology, 75*, 170–180.

Mlakar, S. G., Turinek, M., Jakop, M., Bavec, M., & Bavec, F., (2009). Nutrition value and use of grain amaranth: Potential future application in bread making. *Agricultura., 6*, 43–53.

Mokrane, H., Amoura, H., Belhaneche-Bensemra, N., Courtin, C. M., Delcour, J. A., & Nadjemi, B., (2010). Assessment of Algerian sorghum protein quality [Sorghum bicolor (L.) Moench] using amino acid analysis and in vitro pepsin digestibility. *Food Chemistry, 121*, 719–723.

Nitrayová, S., Brestenský, M., Heger, J., Patráš, P., Rafay, J., & Sirotkin, A., (2014). Amino acids and fatty acids profile of chia (Salvia hispanica L.) and flax (Linumusitatissimum L.) seed. *Potravinarstvo Slovak Journal of Food Sciences, 8*(1), 72–76.

Nongonierma, A. B., Le Maux, S., Dubrulle, C., Barre, C., & Fitz, G. R. J., (2015). Quinoa (*Chenopodium quinoa* Willd.) protein hydrolysates with in vitro dipeptidyl peptidase IV (DPP-IV) inhibitory and antioxidant properties. *J. Cereal Sci., 65*, 112–118.

Obilana, A. B., & Manyasa, E., (2002). Millets. In: Belton, P. S., & Taylor, J. R. N., (eds.), *Pseudocereals and Less Common Cereals: Grain Properties and Utilization Potential* (pp. 177–217). Springer-Verlag: New York.

Onipe, O. O., Jideani, A. I., & Beswa, D., (2015). Composition and functionality of wheat bran and its application in some cereal food products. *International Journal of Food Science & Technology, 50*(12), 2509–2518.

Pachauri, T., Satsangi, A., Lakhani, A., & Kumari, K. M., (2013). Chemical characterization of nutrients in seeds of underutilized grain: Chenopodium album. *Res. Environ. Life Sci., 6*, 43–46.

Paiva, C. L., Queiroz, V. A. V., Simeone, M. L. F., Schaffert, R. E., Oliveira, A. C., Silva, D., & Da, C. S., (2017). Mineral content of sorghum genotypes and the influence of water stress. *Food Chemistry, 214*, 400–405.

Pal, V., Pandey, J. P., & Sah, P. C., (1999). Effect of degree of polish on proximate composition of milled rice. *Journal of Food Science and Technology, 36*, 160–162.

Pedersen, B., & Eggum, B. O., (1983). The influence of milling on the nutritive value of flour from cereal grains. *IV. Rice. Qual. Plant. Plant Foods for Human Nutrition, 33*, 267–278.

Písaříková, B., Kráčmar, S., & Herzig, I., (2005). Amino acid contents and biological value of protein in various amaranth species. *Czech Journal of Animal Science, 50*(4), 169–174.

Popkin, B. M., Keyou, G., Zhai, F., Guo, X., Ma, H., & Zohoori, N., (1993). The nutrition transition in China: A cross-sectional analysis. *European Journal of Clinical Nutrition, 47*(5), 333–346.

Prakash, D., Narain, P., & Misra, P. S., (1987). Protein and amino acid composition of Fagopyrum (buckwheat). *Plant Foods for Human Nutrition, 36*(4), 341–344.

Repo-Carrasco, R., Espinoza, C., & Jacobsen, S. E., (2003). Nutritionalvalue and use of the Andean crops Quinoa (*Chenopodiumquinoa*) and Ka˜niwa (*Chenopodium pallidicaule*). *Food Rev. Int., 19*, 179–189.

Ryynanen, M., Lampi, A.-M., Salo-Vaananen, P., Ollilainen, V., & Piironen, V., (2004). A small-scale sample preparation method with HPLC analysis for determination of tocopherols and tocotrienols in cereals. *Journal of Food Composition and Analysis, 17*, 749–765.

Sapna, C. D. P., & Srivastava, P., (2013). Qualitative dynamics of maize for enhanced livelihood security. In: Kumar, A., Jat, S. L., Kumar, R., & Yadav, O. P., (eds.), *Maize Production Systems for Improving Resource-Use Efficiency and Livelihood Security* (pp. 1–3). DMR, New Delhi, India.

Stibilj, V., Kreft, I., Smrkolj, P., & Osvald, J., (2004). Enhanced selenium content in buckwheat (Fagopyrum esculentumMoench) and pumpkin (Cucurbita pepo L.) seeds by foliar fertilization. *European Food Research and Technology, 219*, 142–144.

Tiwari, U., & Cummins, E., (2009). Nutritional importance and effect of processing on tocols in cereals. *Trends in Food Science & Technology, 20*, 511–520.

USDA, (1989). *United States Standards for Rice* (Rev. edn.). Federal Grain Inspection Service, Washington DC, USDA.

USDA, (2009). *USDA National Nutrient Database for Standard Reference*. https://ndb.nal.usda.gov/ndb/ (Accessed on 14 June 2019).

Valenzuela, C., Abugoch, L., Tapia, C., & Gamboa, A., (2013). Effect of alkaline extraction on the structure of the protein of quinoa (*Chenopodium quinoa* Willd.) and its influence on film formation. *Int. J. Food Sci. Technol., 48*, 843–849.

Watson, S. A., (1987). Structure and composition. In: Watson, S. A., & Ramstad, P. E., (eds.), *Corn: Chemistry and Technology*, (pp. 53–82). St Paul, Minn., USA, American Association of Cereal Chemist.

Wei, Y., Hu, X., Zhang, G., & Ouyang, S., (2003). Studies on the amino acid and mineral content of buckwheat protein fractions. *Nahrung/Food, 47*, 114–116.

Williams, J. T., & Brenner, D., (1995). Grain amaranth (*Amaranthus* species). In: Williams, J. T., (ed.), *Cereals and Pseudocereals* (pp. 129–186). London: Chapman Hall. Chapter 3.

Wright, K. H., Pike, O. A., Fairbanks, D. J., & Huber, C. S., (2002). Composition of *Atriplexhortensis*, sweet and bitter *Chenopodium quinoa* seeds. *J. Food Sci., 67*, 1383–1385.

Wu, Y., Li, X., Xiang, W., Zhu, C., Lin, Z., Wu, Y., Li, J., et al. (2012). Presence of tannins in sorghum grains is conditioned by different natural alleles of Tannin. *Proc. Natl. Acad. Sci., 109*(26), 10281–10286.

Xiao, Y., Liu, H., Wei, T., Shen, J., & Wang, M., (2017). Differences in physicochemical properties and in vitro digestibility between tartary buckwheat flour and starch modified by heat-moisture treatment. *LWT-Food Science and Technology, 86*, 285–292.

Yu, L., Perret, J., Harris, M., Wilson, J., & Haley, S., (2003). Antioxidant properties of bran extracts from "Akron" wheat grown at different locations. *Journal of Agricultural and Food Chemistry, 51*(6), 1566–1570.

CHAPTER 13

Indigenous Fermented Beverages of the Indian Subcontinent: Processing Methods, Nutritional, and Nutraceutical Potential

PIYUSH K. JHA,[1] ALAIN LE-BAIL,[1] ASHISH RAWSON,[2] SASWATI PARUA,[3] PRADEEP K. DAS MOHAPATRA,[4] and KESHAB C. MONDAL[4]

[1]*ONIRIS-GEPEA (UMR CNRS 6144), Site de la Géraudière CS 82225, 44322 Nantes Cedex 3, France*

[2]*IIFPT, MOFPI, Pudukkottai Road, Thanjavur – 613005, Tamil Nadu, India*

[3]*Department of Physiology, Bajkul Milani Mahavidyalaya, Purba Medinipur, West Bengal, India*

[4]*Department of Microbiology, Vidyasagar University, Midnapore–721102, West Bengal, India*

ABSTRACT

Indigenous fermented foods share an integral part of the diet in the Indian subcontinent. Amongst the various methods of food preservation such as drying or salting, the problem arises when the need is to preserve water. And drinking from an unsafe water source is a life or death risk. Here fermentation plays a key role in acting as a method of preservation of food as well as water. The fermentation method is the oldest and economical method for preservation, development of aromas, flavors, textures, and over the years, it has improved. Fermented foods and beverages are region specific and have their own substrates and processing or preparation methods. They can be made from a variety of sources such as fruits, cereals, and milk. They also contribute to the health of the consumer, as they may contain a healthy mix of prebiotics, probiotics, and enzymes, and increase the digestibility of the starter material. Different alcoholic and non-alcoholic fermented beverages are produced in India, plays an important role in the socio-cultural lives of the natives, which are unique in preparation. Detailed studies and information on the preparation of raw materials, processing methods/preparation methods, nutritive, and medicinal value (composition) of each beverage can provide valuable information and would be beneficial for wider application and development of large-scale production. Further, it helps in conservation of the knowledge and proper documentation of the processes for sustaining the livelihood of local communities of India. This chapter would deal with the biochemical aspect of fermentation of beverages, its processing, and its properties. Along with that, it would also introduce to the reader, some uncommon, but equally interesting fermented beverages from the Indian subcontinent. These include but are not limited to buttermilk, lassi, feni, toddy, and many more.

13.1 INTRODUCTION

"Fermentation and civilization are inseparable."—John Ciardi

The relationship between man and fermentation is probably older than many civilizations, dating to the Neolithic age, with the earliest instance reported of a vessel used for fermentation being 7000 BC in China (McGovern et al., 2004). Salting, drying, and fermentation is the earliest methods of preservation.

It was probably initially a serendipity which eventually became a more streamlined process, where the hunter-gatherer lifestyle might investigate to check if the fermented fruits can be safe to consume. With the advent of agriculture and coalescing of humans into urban centers, availability of constant food supplies, fermentation may have been experimented and eventually perfected. Although fruits would have been the ideal choice due to the availability of free sugars along with the surface flora of fermenting species of microorganisms. Cereals posed a unique problem that they did not have sugars in the form that is readily broken down by microbes (starch is a polysaccharide, as opposed mono/disaccharides that the microbes use). Also, they do not have any natural microflora that might accelerate fermentation if it were to happen.

The process of fermentation is so intertwined with our history that we have unknowingly contributed to the genetic drift of our prime fermenting agent, *Saccharomyces cerevisiae.* As individual strains are not isolated from plants, it was believed by some that *S. cerevisiae* has been domesticated, rather than being a chance compatriot in our food fermentation process (Martini, 1993). As a result, it was seen that the genetic diversity in *S. cerevisiae* is very low, which supports the hypothesis of the species being domesticated (Fay and

Benavides, 2005). All the major strains were seen to have originated from Lebanon, and the migration is believed to have occurred near the last ice age, around 10,000 BC (Legrase et al., 2007).

The key requirements for fermentation are sugars, water, and the right starter culture. These features are not limited to just fruits. Milk, a very common food that humans adapted to, can also be fermented to prepare fermented foods. The difference arises in the sugar being fermented and the fermenting organism. Different species produce a different by-product, like ethanol or lactic acid, and thus, a very different food product.

Fermentation is a low-input enterprise and provides individuals with limited purchasing power, access to safe, inexpensive, and nutritious foods (Marshall and Mejia, FAO, 2011). Fermented foods make up a third of the worlds food consumption (Campbell-Platt, 1994). Fermentation of foods is still practiced on a domestic level or small scale level in many parts of the world. This confers to it an immense diversity, as locally available foods are used to produce the products and at its core, it remains largely a food preservation technique, helping people tide over times of scarcity. Since fermented foods are largely unaffected by weather for their production and safekeeping, it is practiced by numerous people groups, throughout the globe.

Another advantage of fermented foods is that the nutritional components of the product are very different from its original raw material. Lactic acid foods are traditionally used to wean off infants, since the fermentation causes the long chain starches to be broken down into shorted portions which are easy for digestion. Anti-nutritive factors (ANF) that prevent the absorption of magnesium, zinc, calcium, and iron are eliminated. ANF for essential amino acids, needed for building protein is also eliminated (Holzapfel, 2002).

13.2 INDIAN SCENARIO

In India, the word *madhu* is believed to be the earliest written reference of any mead, given in the *Rigveda* around 1700–1100 BC, where it refers to both honey and mead (Sage, 1975). For a country as diverse as India, food shortages in the past had to be tided through using fermentation as a method of preservation. Along with that, fermentation increases the digestibility in the case of many traditional foods. In the case of fermented beverages, they are as diverse as the raw material they use. Rice being high in calories and a staple, was the primary raw material used in fermentation, especially in the hills of northeast India. Other major ingredients which are fermented include milk, bamboo shoots, fruits, and other sugary foods. The flower of *mahua*, which is rich in sugars, is used in some parts of northern and central India to make a fermented beverage, which is later distilled.

13.2.1 FERMENTED NON-ALCOHOLIC BEVERAGES

Fermented non-alcoholic beverages (FNAB) from milk and cereals are very popular in India. FNAB products are known for their taste, nutritive value, and therapeutic properties (Sankaran, 1998). These are prepared by controlled fermentation of milk/cereals to produce acidity and flavor to a desired level. Fermented milk/cereals non-alcoholic beverages are more stable than fresh product because they are more acidic and/or contain less moisture (Fox et al., 2015; Panesar, 2011; Goswami, 2017; Widyastuti and Febrisiantosa, 2014). Moreover, they also increase the digestibility of food through metabolic activity. For instance, fermentation process of milk reduces the lactose content by converting it into another product like lactic acid, acetic acid, ethanol, and CO_2 (Widyastuti and Febrisiantosa, 2014), and hence, decreases the problems associated with lactose intolerance (Panesar, 2011). In addition to that, it leads to production of substantial quantities of ß-D-galactosidase which in turn improves lactose digestibility and absorption in lactose intolerant people (Tamang, Shin, Jung, and Chae, 2016; Vonk, Reckman, Harmsen, and Priebe, 2012). FNAB products from milk and cereals are laden with probiotics; the intake of probiotics replenishes the intestinal microflora and provides certain health benefits. Some of the well-known benefits of probiotics in food are: (i) it enhances antibody responses (Haghighi et al., 2005); (ii) prevents the recurrence of superficial bladder cancer (Aso and Akazan, 1992); (iii) prevents traveler's diarrhea (Oksanen et al., 1990), antibiotic-associated diarrhea (Siitonen et al., 1990) and shortens the course of acute diarrhea (Isolauri et al., 1991); (iv) lowers blood cholesterol level (Fuller, 1989); (v) provides relief from constipation (Fuller, 1989); (vi) alleviates lactose intolerance (Panesar, 2011; Tamang et al., 2016); and (vii) antiallergenic (Panesar, 2011; Tamang et al., 2016).

13.2.1.1 CLASSIFICATION

On the basis of the substrate used, the FNAB can be divided into:

- milk-based FNAB;
- cereals-based FNAB;
- fruits and vegetables-based FNAB.

1. Milk-Based FNAB: These are derived from fermented milk product such as curd/*dahi* or fermented cream. They are known for their taste, nutritive value, and medicinal properties. In India, two types of milk based FNAB are widely consumed, one is *lassi,* and other is buttermilk.

a. Lassi: This is one of the indigenous FNAB of the Indian subcontinent, usually consumed in summer as a cold, refreshing therapeutic beverage (George et al., 2012; Joshi, 2016). It is considered as digestive, nutritive, and useful in gastrointestinal ailments (Padghan et al., 2015).

Lassi has a creamy consistency, sweetish rich aroma, and mild to acidic flavor, which makes the product refreshingly palatable (Behare et al., 2010; George et al., 2012). The main base material for *lassi* is *dahi* (Indian yogurt), which is prepared milk with 1.5–4.5% fat content. *Lassi* is a good source of probiotic due to the presence of *Streptococcus*, *Lactobacillus*, and *Enterococcus* genera as active cultures: *Streptococcus thermophiles* is the predominant bacterium present in lassi (Behare et al., 2010; Joshi, 2016; Sarkar et al., 2015; Sathe and Mandal, 2016).

For the preparation of *lassi,* firstly, *dahi* is stirred vigorously to break the well-set curd into fine particles, and then it is blended with water, sugar, salt, and spices such as cumin seeds and coriander leaves until frothy; Figure 13.1. (George et al., 2012; Joshi, 2016; Sankaran, 1998; Sarkar et al., 2015; Sathe and Mandal, 2016; Tamang, 2016). The presence of spices decreases thirst (Sarkar et al., 2015). Sometimes, a little milk is used to reduce the acid tinge and is topped with a thin layer of *malai* or clotted cream. There are two main variants of *lassi*, one is *sweetened lassi,* and another is salted *lassi. Sweetened lassi* is prepared by breaking up *the dahi* into fine particles by agitation and addition of sugar syrup and flavor (Shuwu et al., 2011). Some of the types of *sweetened lassi* available in the Indian market are *fruit lassi* (e.g., *mango lassi, pineapple lassi, banana lassi,* etc.), *honey lassi, bhang lassi,* etc. *Fruit lassi* is prepared by blending of fruit pulp such as pineapple, mango, and banana pulp with *dahi*/curd. The *fruit lassi* contains 7–8% of fruit pulp (Puniya, 2016; Shuwu et al., 2011). *Pineapple* lassi is mildly more acidic as compared to *mango* and *banana lassi*. *Honey lassi* is prepared by adding honey syrup to the stirred *dahi.* The combined optimum sweetness, honey flavor, and mild acidic taste contribute to the flavor of *honey lassi* (Shuwu et al., 2011). *Bhang lassi* is a special type of *lassi* which are prepared on special occasion such as Holi and Shivarathri festival season by incorporating the extract of cannabis plant leaves in the *sweetened lassi. Bhang lassi* provides health benefits due to the presence of cannabis plant extracts, which help in relieving digestive problems, skin-related problems, fever, and sunstroke (Backes, 2014; Sarkar et al., 2015; Sathe and Mandal, 2016).

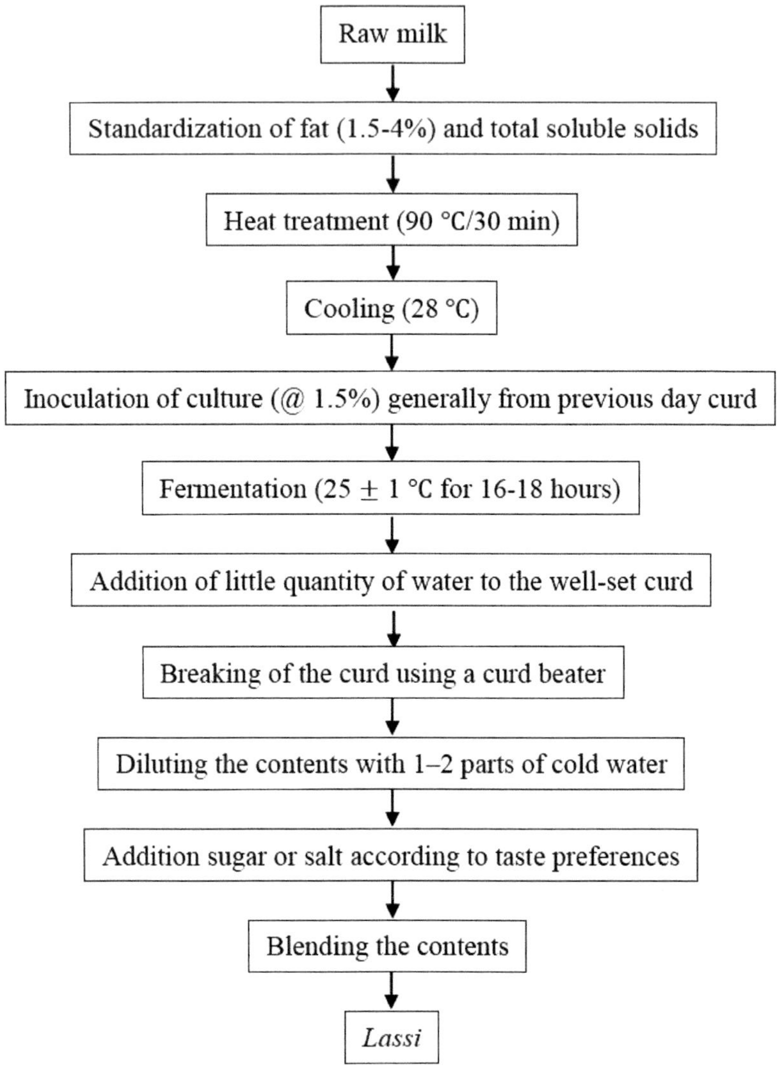

FIGURE 13.1 Steps of *lassi* manufacturing. Inspired from George et al. (2012) and Joshi (2016).

Salted lassi is mainly popular in south India, while its sweetened version is widely relished in the northern parts of the country (Kirstbergsson and Oliveira, 2016). *Salted lassi* also known buttermilk or *chaas, taak* or *mahi* in India is consumed mostly after meals.

The average composition of *lassi* stands as: water 96.2%, fat 0.8%, protein1.29%, lactose 1.2%, lactic acid 0.44%, ash 0.4%, calcium 0.6%, and phosphorus 0.04% (De, 2004; Joshi, 2016; Padghan et al., 2015; Rangappa and Achaya, 1974; Tamang et al., 2016). The factors that affect its composition are the type of milk used, extent of dilution during churning, and the efficiency of fat removal (Joshi, 2016; Padghan et al., 2015). Curd to water ratio in *lassi* determines its overall acceptability. The lassi is more acceptable at an optimum thickness and ideal flavor balance. The *lassi* blend of 3:2 curd to water ratio has the highest overall acceptability among people as it provides ideal consistency and a good flavor balance (Shuwu et al., 2011). The health benefits of *lassi* are the result of biologically active components that are present in native milk and also, due to their suitably modulated activities produced through the action of lactic acid bacteria (LAB), recognition of the immense therapeutic and nutritional value and used for the treatment for diarrhea, dysentery, chronic specific and non-specific colitis, piles, and jaundice (Padghan et al., 2015). According to *Ayurveda,* buttermilk warmed with curry and/or coriander leaves, turmeric, ginger, and salt is a therapy for obesity and indigestion (Kirstbergsson and Oliveira, 2016).

b. Buttermilk: Buttermilk (*chaach*): It is a medium-acid, slightly viscous cow's or buffalo's milk beverage with a buttery flavor (Sankaran, 1998). The piquant taste comes from LAB, which remains as an integral part of buttermilk even after fermentation (Puniya, 2016). Buttermilk is the liquid that is left over when butter is churned out of fermented or non-fermented cream (Joshi, 2016; Puniya, 2016; Sarkar et al., 2015; Sathe and Mandal, 2016). It is also produced by fermentation of fluid skim milk, either by spontaneous souring by the action of lactic acid-forming or aroma-forming bacteria, or by inoculation of milk with pure bacterial cultures (cultured buttermilk); Figure 13.2 (Joshi, 2016; Puniya, 2016; Sarkar et al., 2015; Sathe and Mandal, 2016). It has less fat content and fewer calories compared with regular milk or *dahi* (Puniya, 2016; Sarkar et al., 2015; Sathe and Mandal, 2016). Buttermilk is a rich source calcium and protein as milk, at the same time, presence of live cultures keeps the intestine healthy (Sarkar et al., 2015; Sathe and Mandal, 2016). Large portion of these active cultures (e.g., *Lactococcus, Lactobacillus, Streptococcus,* and *Leuconostocs*) belong to LAB (Puniya, 2016).

In general, there are three types of buttermilk, namely sweet cream buttermilk (obtained by churning fresh/pasteurized cream; develops little or no acidity and taste similar to regular skim milk), sour buttermilk (traditional buttermilk) (obtained by churning sour cream or milk: the milk is allowed to sour naturally), and cultured buttermilk (obtained by churning of curd produced by culturing skim milk). Since the present chapter focuses on the indigenous FNABs, the topic sweet cream buttermilk is not discussed here.

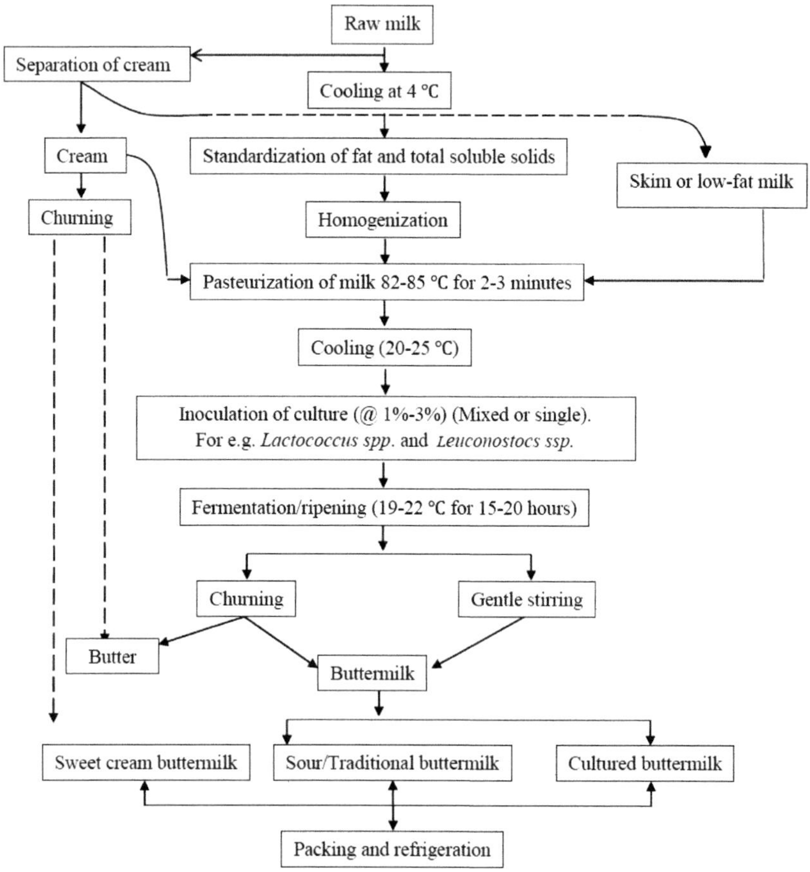

FIGURE 13.2 Different methods of buttermilk manufacturing (Chandan & Arun Kilara, 2011; Goswami, Singh, & Sharma, 2011; Joshi, 2016; Puniya, 2016).

Sour or traditional buttermilk has taste, similar to yogurt or sour cream. The fat content is very low (since most of the fat goes to the butter part). Sour buttermilk has wider variations in its composition as it varies with milk quality used for the preparation of curd and the amount of water added in between the churning. Yet, on the average, it consists of total solids (4%), lactose (3%–4%), lactic acid (1.2%), protein (1.3%), and fat (0.8%) (Puniya, 2016).

Cultured buttermilk is obtained by fermenting pasteurized skim milk with a mixed culture of *Lactococcus* spp. (*Lc. Lactis* subsp. *lactis* and *Lc. Lactis* subsp. *cremoris*), which are the main acid producers, and *Lc. Lactis* subsp. *Lactis* biovar *diacetylactis* and *Leu. mesenteroides,* the latter of which is responsible for aroma and flavor production (Chandan and Arun Kilara, 2011). Upon completion of the fermentation (15–20 hours), the buttermilk attains pH of 4.4–4.6 and acidity 0.7–0.9% as lactic acid. It possesses a mild acid flavor with a diacetyl overtone and a smooth texture. Cultured buttermilk has a soft white color and may contain added butter flakes, fruit condiments, or flavorings. The approximate composition of cultured buttermilk is presented in Table 13.1 ("Fermented Milk Products," 2011).

TABLE 13.1 Composition Cultured Buttermilk ("Fermented Milk Products," 2011).

Composition (Base on 100 g)	Cultured Buttermilk	
	Low Fat	Fat-Free
Fat (g)	2.0	0.88
Protein (g)	4.1	3.3
Carbohydrates (g)	5.3	4.8
Calcium (mg)	143	116

Various types of starter culture can be used to produce cultured buttermilk. Depending on the type of culture used, the flavor, texture, and consistency can vary widely. They are broadly classified into two categories, namely mesophilic lactic starter cultures (20–30°C) and thermophilic lactic starter cultures (40–45°C). The cultured buttermilk made by use of mesophilic lactic starter cultures may use one of the following starter culture types (Bylund, 2003; Puniya, 2016):

i) **O-Type:** This type of starter culture mainly comprises of lactic acid producing homofermentative bacterial species such as *Lactococcus lactis* subsp. *lactis* and *Lactococcus lactis* subsp. *cremoris.*

ii) **D-Type:** These starter cultures are a combination of a flavor producing lactic bacteria known as *Streptococcus lactis* subsp. *lactis var. Diacetylactis* and the O-type bacteria. In addition to lactic acid, this type of starter culture produces diacetyl and CO_2.

iii) **L-Type:** This type of starter culture contains, in addition to the O-type bacteria, also *Leuconostoc mesenteroides* subsp. *mesenteroides* as the main flavor compound producing bacteria. It produces diacetyl, acetic acid, acetaldehyde, and other flavor compounds but less carbon dioxide than the D-type.

iv) LD-Type: These contain a combination of *Str. Lactis* subsp. *Lactis var diacetylactis* and *Leuconostoc mesenteroides* subsp. *mesenteroides* to give a fine blend of dedicated flavor and aroma.

While, thermophilic lactic starter culture comprising of *Streptococcus salivaricus* subsp. *thermophilus* and *Lactobacillus delbrueckii* subsp. *bulgaricus* can also be used to prepare cultured buttermilk (Bylund, 2003).

The buttermilk has several medicinal value, its consumption on a regular basis imparts following health benefits: decrease the chances of cancer (Larsson et al., 2008); reduces the cholesterol levels by inhibiting the intestinal absorption of cholesterol (Conway et al., 2013); removes undesired stomach acid and reduces the chances of indigestion and heartburn (Puniya, 2016). It is generally consumed with or without added salt and spices (Sankaran, 1998; Sarkar et al., 2015).

Buttermilk in Ladakh region of India is known as Tara. Milk (*oma*) from milching animals reared in the region such as sheep, goat, cow, *dzomo* (female of a crossbreed of cow and yak) and *dri* (female yak) is boiled, allowed to cool, and then, inoculated with the previous batch of buttermilk and incubated overnight in a warm place to form a curd (known as "*jho*" in Leh). In the Changthang and Zanskar areas of Ladakh, the curd (*jho*) is shaken vigorously in a bag made of goatskin to separate the butter from the buttermilk. In other parts of the region, however, buttermilk is made by churning curd in a special wooden vessel (*zem*) made of juniper wood. In Kargil (India), the nomenclature for these products is different again: they call milk "*orjen*," curd "*oma*," and buttermilk "*derba*" (Joshi, 2016; Raj and Sharma, 2015).

c. Pheuja* or *Suja: It is a refreshing beverages made by adding fermented yak butter and salt to tea. It tastes salty and has a buttery flavor. It is very popular in north-eastern states of India (Arunachal Pradesh, Ladakh, and Sikkim), Nepal, and Bhutan. The microorganism associated to fermentation of yak butter is unknown to date (Tamang et al., 2012; Tamang et al., 2010; Tamang and Kailasapathy, 2010).

2. Cereal-Based FNAB: Whole grains are rich in vitamins, especially B vitamins, and good sources of minerals, particularly trace minerals. They are also important sources of many phytochemicals, including phytoestrogens, phenolic compounds, antioxidants, phytic acid, and sterols (Katina et al., 2007). Moreover, they are considered as effective substrates for the production of probiotic-incorporated functional food, as they can be used as a source of non-digestible carbohydrates, which stimulate the growth of

Lactobacilli and Bifidobacteria (Enujiugha and Badejo, 2017). Thus, cereal-based FNAB offer opportunities to include probiotics, prebiotics, vitamins, minerals, and fiber in human diet.

a. Millet-Based Kambarq Coozhi/Ambali/Ambli: *Kambarq Coozhi/ Ambali* is a liquid product (beverages) prepared by fermented millet or *ragi (Eleusine coracana*) and pearl millets (*Pennisetum glaucum*) (Anon, 2017). It is quite popular in the rural area of Maharashtra (India) (Ramakrishnan, 1979), as well as in the small villages and towns of Tamil Nadu (India) and is especially consumed in summer (sold by the roadside during the summer months). In earlier times, it was used as a breakfast food, and is generally regarded as providing energy and being good for one's health overall. For its preparation, firstly, coarse broken grain or flour of pearl millet and finger millet flour is mixed with water thoroughly, and the resultant slurry/batter is fermented overnight in an earthen pot. The fermented grain slurry is added to the boiling water and cooked. In case the porridge becomes too thick, additional water or sometimes buttermilk is added to dilute it. Salt is added to taste. It is drunk alongside some side dishes, such as raw mango, chili, fried vegetables, etc. Alternatively, *ambali* can also be prepared by cooking the fermented batter (millet flour batter) along with partially cooked rice. Once cooking is done, sour milk is added to the porridge, and is ready to consumption. *L. mesenteroides* (1.6×10^9 per g), *L. fermentum* (1.6×10^9 per g) and *S. faecalis* (8×10^8 per g) have been isolated from the fermented *ragi*. During the fermentation process, the pH decreases from 6.4 to 4.0 and the volume increases by about 20%, indicating some CO_2 production (Das, Raychaudhuri, and Chakraborty, 2012; Ramakrishnan, 1979; Steinkraus, 1995). Moreover, there is an increase in thiamine, riboflavin, and niacin contents during its fermentation (Aliya and Geervani, 1981; Rajyalakshmi and Geervani, 1990). *Ambali* is a part of daily diet of rural and agricultural workers and is claimed to be a nourishing health food (Antony and Chandra, 1997). It is also given to children at weaning age (Mbithi-Mwikya et al., 2002).

b. Fermented Rice-Based Drink (Pazhaiyasoru): Fermented rice drink is one of the indigenous cereals based FNAB products of India. It is made by fermenting overnight the leftover cooked rice along with some water. The water is then drained off and mixed with buttermilk and salt and consumed directly. *S. faecalis* (2.7×10^7 per g), *Pediococcus acidilactici* (2.7×10^7 per g), *Bacillus* sp. (1.6×10^8 per g), and *Microbacterium flavum* (1.1×10^8 per g) have been isolated from fermented rice. The pH decreases from 6.1 to 5.7 in 16 h. There is no change in volume, amino nitrogen, or free sugar

(Ramakrishnan, 1979; Steinkraus, 1995). It is believed that consumption of fermented rice-based drink cools the body and is very good for one's health.

- ***Kanjika:*** It is a lactic fermented rice product. It also contains raw material of plant origin, but is devoid of dairy products (Madhu et al., 2010; Reddy et al., 2007). According to Reddy et al. (2007), for the preparation of *Kanjika,* firstly, broken rice is cooked in 1 L of water till the rice becomes soft. The resulting starchy liquid after straining is made up to 1 L, and cooled to room temperature. Then raw materials of plant origin (50 g) along with ground mustard (25 g), rock salt (25 g) and mustard oil (12 g) is added to the 1 L of starchy liquid and mixed well. The mixture is fermented in a closed vessel at 37°C ± 2°C up to 7 days. It has been prescribed for a number of chronic diseases by Indian *ayurvedic* practitioners (Reddy et al., 2007). *Kanjika* is a very good source of probiotic LAB as it has the advantage of being cooked first followed by fermentation (Madhu et al., 2010). The LAB isolates from *kanjika* exhibit probiotic properties such as acid tolerance, bile salt tolerance, antimicrobial activity against foodborne pathogens, β-galactosidase activity, antibiotic susceptibility and cholesterol assimilation (Reddy et al., 2007). *L. plantarum* isolated from *kanjika* is a potential source of Vitamin B12 (Madhu et al., 2010).

3. **Fruits and Vegetables-Based FNAB:** Intake of vegetables rich in antioxidants and other nutritional components is suggested to provide various health benefits to humans. The lactic acid fermentation of these vegetables, applied as a preservation method for the production of finished and half-finished products, is considered as an important technology because of its capability to improve the nutritive value, palatability, acceptability, microbial quality and shelf life of the fermented produce.

a. Ready-to-Serve (RTS) Drinks: Most recently, different RTS drinks have been formulated using lactic acid-fermented radish, carrot, and cucumber blended with different pulps of fruits. On basis of the physicochemical characteristics and sensory evaluation scores of different attributes, the carrot and mango RTS with 40% fermented carrot + 60% mango was ranked the best, and thus, was considered the most appropriate blend for fermented RTS drinks. Similarly, among the radish-based RTS drinks, those containing 30% fermented radish pulp + 70% mango were pronounced the best. Likely, among the cucumber-based products, the RTS drink prepared with 20% fermented cucumber + 80% mango pulp was found to be the best (Joshi, 2016). Figure 13.3 presents the processing steps of ready-to-serve (RTS) drinks made from fermented vegetables and fruit pulp (Joshi, 2016).

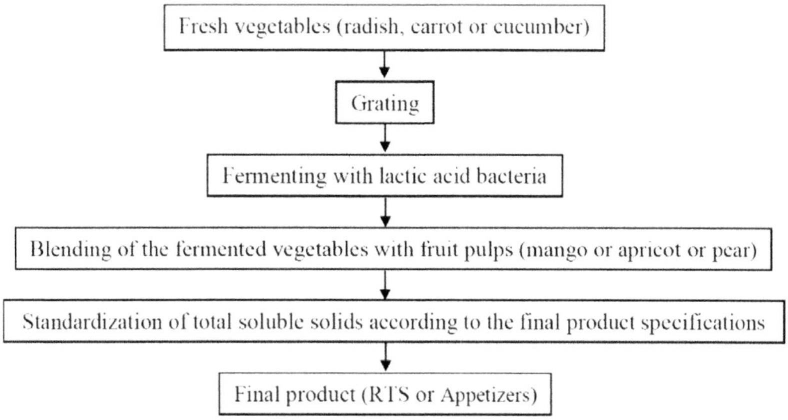

FIGURE 13.3 Processing steps of fruits and vegetables based lactic acid fermented drinks. Adapted from Joshi (2016) and Joshi et al. (2011).

b. Lactic Acid-Fermented Appetizers: Lactic acid fermented appetizers can be prepared using a blend of fermented vegetables (radish, carrot, and cucumber) and fruit pulps (mango, apricot, and pear). The vegetables are fermented using LAB viz., *Lactobacillus plantarum* (NCDC, 020), *Pediococcus cerevisiae* (NCDC 038) and *Streptococcus lactis* var *diacetylactis* (NCDC 061). Based on the sensory evaluation, it was found that the appetizer made with 20% fermented radish + 10% apricot had the highest overall acceptability (Joshi, 2016; Joshi et al., 2011). The procedure to prepare lactic acid-fermented appetizers is shown in Figure 13.3. Other traditional lactic acid-fermented appetizers are discussed below.

- ***Sinki:*** It is a non-salted fermented radish tap root product, traditionally consumed as a base for soup and as a pickle in some north-eastern states of India, in Nepal and a few places of Bhutan (Tamang and Sarkar, 1993; Sekar and Mariappan, 2007). Figure 13.4 demonstrates the steps followed to prepare hot beverages from *sinki (sinki* soup). *Sinki* has 14.5% of proteins, 2.5% fat and 11.3% ash on dry weight basis. The sun-dried market sinki has mean pH and titratable acidity (% of lactic acid) of 4.4 and 0.72%, respectively. The fermentation is initiated by *L. fermentum*, and is successively followed by *L. brevis* and *L. plantarum*. During fermentation process of *sinki, L. plantarum* utilizes mannitol and removes the bitter flavor of mannitol produced by the gas-forming lactobacilli (Tamang and Sarkar, 1993). It is said to be a good appetizer, and people use it for remedies for indigestion (Tamang and Sarkar, 1993; Sekar and Mariappan, 2007).

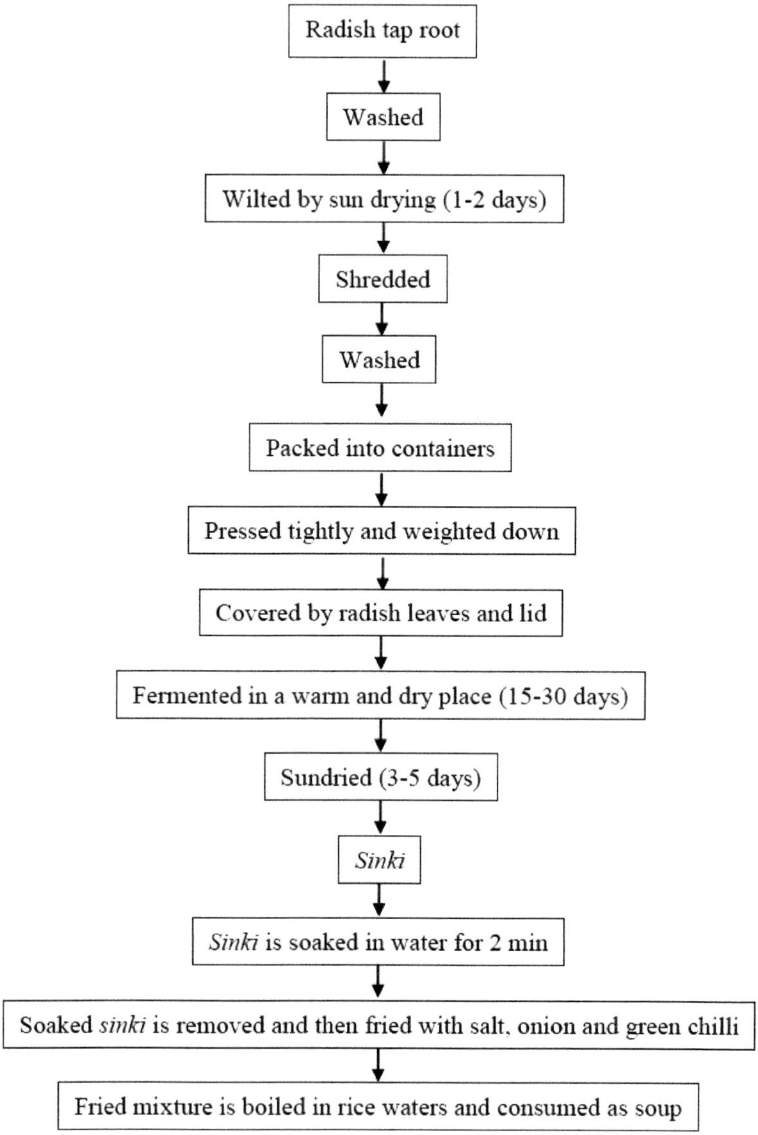

FIGURE 13.4 Flow sheet for the production of sinki soup (Tamang & Sarkar, 1993; Sekar & Mariappan, 2007).

- ***Kanji:*** It is a very popular drink of North India. It is made by fermenting vegetables such as carrot or beetroot with crushed mustard seed, hot chili powder, and salt for 7–10 days (Figure 13.5). Carrot Kanji is considered to have high nutritional value and cooling and soothing properties (Sura, Garg, and Garg, 2001). Beetroot kanji is considered to have potential to prevent infection and malignant disease (Kingston et al., 2010; Winkler et al., 2005). The LAB associated with fermentation of carrot/beetroot during *kanji* preparation are *L. mesentroides*, *Pediococcus* species, *L. dextranicum,* and *L. plantarum* group (includes *L. paraplantarum* and *L. pentosus*) (Kingston et al., 2010; Sura et al., 2001).

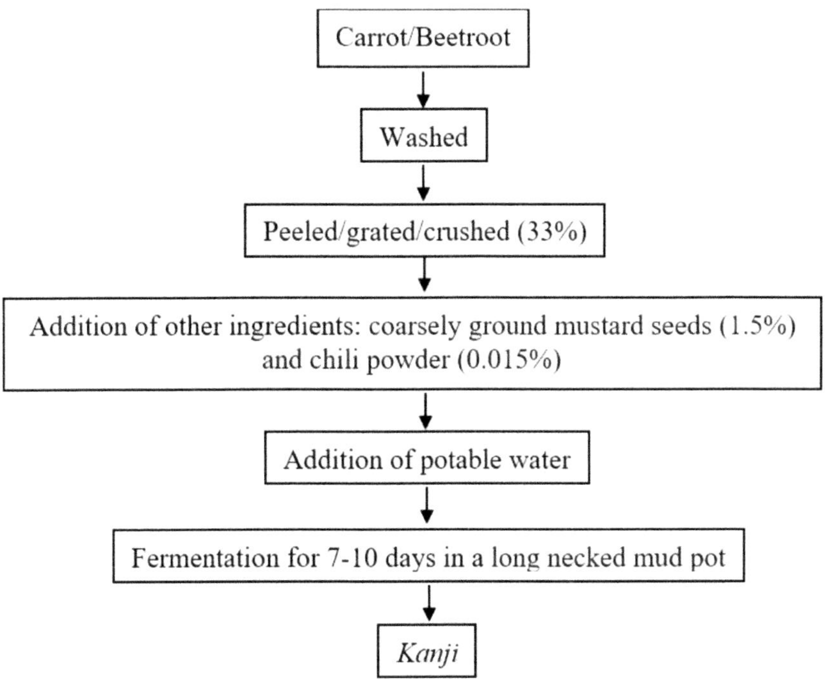

FIGURE 13.5 Flowchart for kanji preparation (Kingston et al., 2010).

Intake of vegetables rich in antioxidants and other nutritional components is suggested to provide various health benefits to humans. The lactic acid fermentation of these vegetables, applied as a preservation method for the production of finished and half-finished products, is considered as an important technology because of its capability to improve the nutritive value, palatability, acceptability, microbial quality and shelf life of the fermented produce.

13.3 ALCOHOLIC BEVERAGES

13.3.1 HISTORICAL EVIDENCES

Fermented food preparation is the inherited culture of age-old wisdom of early civilization depends upon availability of food resources (Ray et al., 2016). Grain, fruit juice, and honey have been used to make alcoholic beverage for thousands of years. The early Vedic literature suggest the use of alcoholic drink in the Indus Valley Civilization around 3000–2000 B.C. The consumption of alcoholic beverages also mentioned in the two great Hindu epics, Ramayana, and Mahabharata (Singh et al., 2010). The process of fermentation was probably discovered by observing the changes in the juices of several fruits and other substances that had been kept for a day or more. It appears that fermentation technology started simultaneously with settled agriculture during the Neolithic period. The two primitive alcoholic drinks that are quoted in the Vedic texts are *Soma* and *Sura* (Oort, 2002). The *Rgveda* describes *Soma* as a godly concoction, while *Sura* is described as mainly a human drink. However, the *Yajurveda* deifies *Sura* and uplifts it to the status of *Soma*. The Rgveda (c.1500 BC) shows that fermentation technology took its first step in connection with the preparation of Soma juice. The *Sukla Yajurveda (*Vaj. XIX. 13–15; 82–83) describes that *Sura* was supposed to be prepared from germinated paddy, germinated barley, and parched rice with the help of ferment. Yeast was used most often as the fermenting agent. The *Katyayana Srauta sutra* (XV, 9.28–30; XIX, 1–2) also gives a complete description of the preparation of *sura* where either boiled rice or boiled barley was mixed with the ferment and also with *masara* and the mixture was filled in a jar. The jar was then kept in a pit for three nights (Oort, 2002; Singh et al., 2010).

Kautilyya, the prime minister of India's great emperor, Chandragupta Maurya, in his treaties (Arthashastra, circa 4th Century BC) mentioned the instruction for the preparation of at least 12 different varieties of alcoholic beverage. The text also codifies the rules and regulation pertaining to the production, sales, and taxation of alcoholic beverage. At that time alcoholic beverages were prepared from fermentation of grain (*masara* from barley gruel; *kashaya, prasanna* from fermented rice flour and *svetasura*); flowers (*parisruta* and *varuni* from the mahua flower and *jathi* from the jasmine); and fruits (*mahasara* from mango, *khajurasara* from dates; *kadambari* from kadamba fruit; *madhu* and *mridvika* from grapes). The sweet exudates from the Palmyra or coconut palm were fermented into *thalakka* (thari or toddy), and it could be distilled to yield arrack, and both these practices have survived till today (http://nimhans.ac.in/cam/sites/default/files/Publications/25.1.pdf).

13.3.2 USES AND PROCESS OF PREPARATION

Traditionally, alcohol has been recognized one of the most popular chemical substances used for intoxication by man since time immemorial. It has extensively used for medicinal, social, religious, and recreational purposes across different cultures. Both soma and *sura* have been used in different medicinal preparations, surgical procedures, and in many chemical and alchemical operations (Oort, 2002). The Ayurvedic texts describe both the beneficent uses of alcoholic beverages and the consequences of intoxication and alcohol-related diseases (Sekar, 2007). Distillation was known in the ancient Indian subcontinent, evident from baked clay retorts and receivers found at Indus valley civilization. In his excavations at Taxila, now in Pakistan, Sir John Marshall (1951) discovered, a group of vessels which he reconstructed as apparatus for condensing water. Allchin (1975) in his report mentioned that 'India appears on present evidence to have been the first culture to exploit widespread distillation of alcohol for human consumption; and it may well be that the art of distillation was India's gift to the world!' Till now tribal people of India are using traditional distillation system with a group of earthen pots for distilling homemade alcoholic beverage. In Southern and Northern India, closed process is employed where the apparatus is commonly consisting of a globular pot covered by an inverted bowl with a hole. A third pot (collector) immersed in cold water and a bamboo pipe is connecting in between second and third. In Eastern and Northeast part, tribal people used a vertical series of three pots or tube process. Where the second pot connected with the first through a hole and the top (3rd) containing cold water. A round shaped wooden collector is remained in the second pot and from this, a bamboo or metal tube is connected with the external collector. In both the cases, joints were sealed with clay and wet cloth. Fermented materials kept in the first large pot that heated and maintained a certain temperature. In India, the true distillation process of alcohol was initiated during the period of Delhi Sultanate by the 14th century, but the first distilleries on European lines were established in around I835 (Allchin, 1979; Singh et al., 2010).

13.3.3 EXPLOITATION OF HERBAL STARTER AND MICROBIAL INTERPLAY

The alcohol making process from cereal proceed via two separate biochemical steps: saccharification and fermentation. Saccharification leads to hydrolysis of the starch in the cereal into fermentable sugars, and fermentation is the process of converting sugars to ethyl alcohol and carbon dioxide by yeast (Mondal et al., 2016). The participating microbes in these multi-phase processes arise from the nature (food itself or environment) or from addition of starter culture (Ray et al., 2016, 2017). These microbes must have innate ability to grow anaerobically, produce amylolytic enzyme to hydrolyze starchy raw ingredients and alcohol fermenting capacity. Yeast is considered as a useful and effecting microbes for both the processes from ancient time. People also invented two different methods for saccharification: they use sprouted grains and wet processing (soaking) of grains for prolonged duration for induction of amylase activity that fastened the saccharification process. The concept of starter development is also unique in Indian sub-continent as most of the starter for alcoholic beverage preparation are made up of plant residues (Ray et al., 2016). The wild microbes resides in plant parts (endophytic organisms) and participate in the multi-stage and multi-species fermentation process, and the phytochemicals also act as a biopreservative and improve the nutraceutical and therapeutic potentialities of the beverage (Ray et al., 2016). *Haria* or *Handia* is very popular rice-based alcoholic beverage in Central and Eastern India and its starter *Bakhar* or *Ranu* tablet that consist of 3–7 plant parts (most are tubers) without addition of old ferment. In this starter tablet, a group of microbial consortia are present, which sequentially participate in the fermentation process of rice. The initial stage of fermentation is facilitated by the molds, which saccharify the rice and decompose it, and therefore, created an anaerobic condition. In this environment, LAB, and yeasts are begins to multiply, which liberated acid and alcohol. These reactive metabolites and anaerobic condition fasten the destruction of molds community within three days of fermentation and there is no existence of molds in the final ferment (Ghosh et al., 2015a,b; Ray et al., 2017). The microbial composition and alcohol content in the popular indigenous alcoholic drinks are depicted in Table 13.2. Now scientific exploration evident that this type of unique fermentation and microbial dynamics cannot be initiated by the addition of old ferments. The tribal people by their experience and mastery in the art, selected few plant and their specific parts as starter materials only considering the improvement of the sensory quality of the final product. *Bakhar* is also used for preparation of Mahuwa,

TABLE 13.2 Some Popular Alcoholic Drinks and Their Associated Microbes

Name of beverage	Substrate	Fermentable microbes	Alcohol content	References
Bhaati jaanr	Rice	*Mucor circinelloides, Rhizopus chinensis, Rhizopus stolonifer, Saccharomycopsis fibuligera, Pichia anomala, Saccharomyces cerevisiae, Pediococcus pentosaceus, L. bifermentans*	5.9%	Tamang and Thapa, 2006; Tamang et al., 2012
Chhang	Barley	*Saccharomyces cerevisiae, Saccharomyces fibuligera, Lactobacillus plantarum, Pediococcus pentosaceus,* and *Enterococcus pentosaceiis,*	8.6%	Thakur et al., 2004; Kanwar et al., 2011
Haria	Rice	*Saccharomyces cerevisiae,* lactic acid bacteria like *Lactobacillus fermentum, Bifidobacterium* sp, mold	6.0–9.0%	Ghosh et al., 2014; Ghosh et al., 2015a; Ghosh et al., 2015b
Judima	Rice	*P. pentosaceous, Bacillus circulans, B. catarosporous, B. pumilus, B. firmus, Debaryomyces hansenii, Sacharomyces cerevisiae*	16%	Arjun et al., 2014; Chakrabarty et al., 2010
Kodo ko jaanr	seeds of finger millets	*Muco rcircinelloides, Rhizopus chinensis; Saccharomycopsis fibuligera, Pichia anomala, S. cerevisiae, Candida glabrata; Pediococcus pentosaceus, Lb. bifermentans6*	4.8%	Dutta et al., 2012; Thapa and Tamang, 2004
Jann	Different cereals and fruits	A total number of 32 microbial cultures were isolated among them two species of Gram-positive spore-forming bacteria (belonging to genus Bacillus) and three of yeasts (*Saccharmycopsis fibuligera, Kluyveromycesmaxianus, and Sacharomyces sp.*) are dominant in balam	~10% depends upon the food items	Roy et al., 2004; Das and Pandey, 2007.

TABLE 13.2 *(Continued)*

Name of beverage	Substrate	Fermentable microbes	Alcohol content	References
Po:ro Apong	Rice	*S. cerevisiae, Hanseniaspora sp., Kloeckera sp., Pischia sp., and Candida sp.*	7.5–18%	Kardong et al., 2012
Zutho	Rice	*Saccharomyces cerevisiae*, *Rhizopus* sp.	5.0%	Das and Deka, 2012; Tamang et al., 2012; Teramoto et al., 2002
Mahua	Dried flower bud (corollas) of *Madhucaindica*	*Saccharomyces cerevisiae, Lactic acid bacteria*	8.5%	Singh et al., 2013
Daru/Chakti	Jaggery	*S. cerevisiae, Candida famata, C. valida, Kluveromyces thermotolerance*	5.0%	Thakur et al., 2004
Angoori/ Kinnauri	Grapes	*S. cerevisiae*	Red grapes Angoori–5.1% Green grapes *Angoori-3.44%*	Thakur et al., 2004
Sura	Millet flour	Natural fermentation with predominant *Saccharomyces cerevisiae Zygosaccharomyces bisporus*	15.28	Thakur et al., 2004

a popular alcoholic drink from the flower of *Madhuca longifolia*. This type of herb-based starter is also used for preparation of many alcoholic beverages in the North-East states of India (Das et al., 2012).

In spite of economic and social development, ethnic people in India still consume local made alcoholic beverage. These beverages can be broadly categorized into two groups, one is un-distilled raw alcoholic beverage and, another is partially distilled alcoholic drink (spirit). Both are prepared from available food resources like rice, barley, ragi, beat roots, flower, and fruits (grapes, apricots, apple), etc., with or without starter (Thakur et al., 2004; Mondal et al., 2016). The native skill for preparation of this type of alcoholic beverages propagated from one generation to another and, this ancestral knowledge and skill is modified by the availability of raw materials, environmental, and social changes. But till day, traditional foods and beverages are very popular in the villages, and people followed the traditional methods for their preparation. Most of the traditional beverages are scientifically un-touched and unexplored their hidden beauty. The evolving of bioactive compounds during traditional fermentation and the age-old claims of health beneficial impacts of some popular alcoholic drinks are represented in Table 13.3. The wealth of microbial resources, bioactive compounds, nutraceuticals, minerals, etc. in these beverages can serve the nation for different curative purposes as these are well tested by the ethnic people from generation to generation. Few of them are now scientifically enlightened, their production process and compositions are now optimized and exploited.

13.4 PROCESSING METHODS

13.4.1 CHHANG

In Ladakh, *chhang* is the general word for alcohol. It relief thirst, gives energy and provides nutrition. This alcoholic beverage is mostly prepared from barley grains at Ladakh, and rice dust at Lahaul and Spiti districts of Himachal Pradesh and other parts of the Himalaya. *Chhang* become an integral part of sociocultural life of the people and serving of this beverage is common from birth ceremony to death feast. In Ladakh, steps involve in *chhang* preparation from barley grain includes cleaning and boiling of barley grains, cooling, addition of starter culture, fermentation, filtration, and blending. It also involves multi-stage fermentation process, initially solid state (in poly bags) and followed by submerged (in earthen pot) fermentation (Bhardwaj et al., 2016; Targais et al., 2012; Thakur, 2013).

TABLE 13.3 Nutraceuticals and Functional Properties of Popular Alcoholic Beverages

Alcoholic drink	Nutri–chemical composition	Claimed 54health benefits	References
Haria	pH–3.61, Acidity (ml) 5.0, Reducing sugar (mg/ml) 6.5, Total sugar (mg/ml) 51, Protein content (mg/ml) 0.63 Maltooligosaccharides, pyranose sugar derivatives, vitamins, minerals, and phenolic-rich beverage	Protects from gastrointestinal ailments like dysentery, diarrhea, amebiosis, acidity, vomiting. It exerted significant level of antioxidant activity. It works effective against insomnia, headache, body ache, inflammation of body parts, diarrhea, and urinary problems, expelling worms and as a treatment of cholera.	Roy et al., 2012; Ghosh et al., 2014; Ghosh et al., 2015a, 2015b
Chhang	pH 4.1, protein 13%, L–lactate 2.0 g/l Phenolics 0.40 g/l Furfurals 0.03 g/l	It quenches thirst, gives energy and provides nutrition. It is considered to provide protection against cold during winter months	Bhardwaj et al., 2016; Targais et al., 2012; Thakur, 2013
	Reducing sugars 7.65 g/l; enhanced the B vitamin levels especially thiamine (B_1), pyridoxine (B_6), folic acid (B_9), and amino acids		
Apong	pH–4.06, lactic acid–0.5%, carbohydrate–46 mg/ml, reducing sugar 3 mg/ml, protein–1.05 mg/ml, free amino acids–2.43mg/ml, ethanol–7.52%–18.5%, amylase–2.4 U/ml	It is nutricious, energy-rich refreshing drink having antimicrobial, antioxidant, and other age preventing effects. Apong is also helpful in preventing formation of kidney stones.	Das et al., 2012; Kardong et al., 2012
Judima	pH–4.4, titratable acidity–0.45%, carbohydrate–32 mg/ml, protein–0.97 mg/ml, free amino acids–3.21 mg/ml. ethanol–16% (v/v), trace elements like Cu, Cr, Mn, Fe, K, Na, Se	*Judima*has anti-inflammatory, anti-allergic, anti-oxidant, anti-bacterial, anti-fungal, anti-spasmodic, hepatoprotective, hypolimidemic, neuroprotective, hypotensive, anti-aging, and anti- diabetic potentialities.	Chakrabarty et al., 2014; Arjun et al., 2014
Zutho	pH–3.6, acidity–5.1 ml, reducing sugar–6.3 mg/ml, total sugar–39.7 mg/ml, ethanol–5%	Zutho boosts immune system, lowers insulin level of blood, prevents loss of appetite, lowers bad cholesterol, assists in healing wound and prevents infection.	Das et al., 2012; Teramoto et al., 2002

TABLE 13.3 *(Continued)*

Alcoholic drink	Nutri–chemical composition	Claimed 54health benefits	References
Mahua	pH–3.8–4.8; titratable acidity–0.4–0.8; ethanol–8.5%, TSS–6.0, ascorbic acid (mg/100 ml)–2.11, reducing sugar (%) – 0.1, Phenolics like gallic acid, chlorogenic acid, catechin, epicatechin, caffeic acid, 4–hydroxybenzaldehyde, ascorbic acid and tannic acid in mahua wine	It is sweet, strength promoting, *stambhana* (which increases the power of retention), cooling, and heavy	Jadav et al., 2009; Vora et al., 2016; Singh et al., 2013
Daru (un-distilled)/ Chakti	Not available	It is said to promote the power of digestion, complexation, and strength. Gauda eases passage of flatus and feces and thus, improves appetite	Sekar, 2007
Bhaati Jaanr	pH–3.5, ash–1.7%, protein–9.5%, fat–2.0%, crude fiber–1.5%, carbohydrate–86.9%, food value (per 100 g)–404.1kcal, trace elements like Ca, K, P, Fe, Mg, Mn, Zn are also present.	It is consumed as a staple food or mild alcoholic sweet beverage. It is recommended for ailing persons and post-natal women to regain their strength.	Tamang and Thapa, 2006
Apricot wine	Total soluble solids 8.57, Reducing sugars 0.35 0 (%), Total sugars 1.28 (%), Titratable acidity 0.80%, Ethanol, 11.70 (%v/v), Total esters 124.7 (mg/l), Total phenols 264.0 (mg/l), Total carotenoids 0.78 (mg/l)	The fruits are anti-diarrheal, anti-pyretic, emetic, allaying thirst whereas; the seeds are tonic used in liver troubles, piles, earache, and deafness.	Sharma et al., 2014
Jann	Carbohydrate, amino acid, vitamins, reduction in phytic acid content While increase the availability of minerals increases.	Nutritious food items that can cope up the inhospitable climatic conditions at high altitude areas	Roy et al., 2004; Das and Pandey, 2007.
Kodo ko jaanr	Moisture: 69.7%, pH: 4.1, ash:5.1% DM, protein: 9.3% DM, fat: 2.0% DM, crude fiber: 4.7% DM, carbohydrate: 83.7% DM, food value: 389.6 kcal/100 gm, Ca: 281.0 mg/100 gm, K: 398.0	Because of high calorie, ailing persons and post-natal women consume the extract of kodoko jaanr to regain the strength.	Thapa and Tamang, 2004; Tamang et al., 2012

TABLE 13.3 *(Continued)*

Alcoholic drink	Nutri–chemical composition	Claimed 54health benefits	References
	mg/100 gm, P: 326.0 mg/100 gm, Fe: 24.0 mg/100 gm, Mg: 118.0 mg/100 gm, Mn: 9.0 mg/100 gm and Zn: 1.2 mg/100 gm; high level of valine, threonine, leucine, and isoleucine, improved level of cyanocobalmin, which is deficient in millet		
Toddy	pH–5.5, sucrose–12–15%, protein–0.23%, fat–0.02%, mineral matter and ascorbic acid 5.7 mg/100 ml). There is very little reducing sugar, although glucose, fructose, maltose, and raffinose are present. Alcohol may 5–6% after 36–48h of fermentation. Thirty volatile compounds are present, among them, lupeol and squalane are remain in higher quantity.	Toddy is believed to be good for the health, particularly for eyesight and also serves as a sedative. It is also a mild laxative relieving constipation. It is prescribed as a tonic for those recovering from chicken pox.	Sekar and Mariappan, 2007; Karthikeyan et al., 2014

Phab is the common and traditional starter/inoculum for preparation of alcoholic beverage in Ladakh, Himachal Pradesh and Uttaranchal states of India. The ingredients of phab include ground roasted barley, powered black pepper, dried ginger powder, crushed paddy, and wild herbs/medicinal plants (local availability). All these ingredients are kneaded and make a small ball shaped structure, allowed to ferment for 2–3 days by covering the branches of wild shrub, *burtse* (*Artemisia sp.*). Then the balls are air dried and stored, and it remain active for about a year (Thakur et al., 2004; Kanwar et al., 2011).

The conventional starter, *phabs* is essential for *chhang* preparation that facilitate the initial fermentation. Generally, three tablets of *phab* (each around 1.3–1.7g) is added to 5 kg of grain for optimum fermentation (Targais et al., 2012). For its preparation tradition utensils are used that includes *bho*, a wooden container used for measuring grains; *khol-char*, canvas made of Yak's fur used for air drying of boiled grains; *rZa-ma*, an earthen pot used for fermentation; *zem*, wooden drum for extraction and filtration; and *chapskan*, a decorative brass vessel for serving *chhang*.

13.4.2 HARIA OR HANDIA

This is a most popular rice-based alcoholic beverage in India. Tribal people of any age from Central to Eastern and North Eastern India consumed this beverage regularly. The consumption of this beverage is the integral part of their social and ritual life. The process of this ethnic beverage is more or less common in India. Initially, rice is boiled vigorously up to charring, then air dried under shade. Traditional starter for this beverage is *bakhar* or *ranu* tablet. This tablet is made up of 3–7 plant residue (according to the availability in the locality) that blended with rice dust and make a ball-shaped structure (Ghosh et al., 2014). The major plants used for *bakhar* preparation are *Asparagus racemosus* roots, *Elephantopus scaber* L. root, *Cissampelos pareira* L. roots, *Diospyros melanoxylon* Roxb. Bark & leaves, *Lygodium smithianum* C. Presl., whole plants, *Orthosiphon rubicundus* (D. Don) Benth. tubers, *Ruellia macrantha* L. tubers, and bark of *Terminalia alata* Roth. The old ferment is not used for preparation of this starter (Kumar and Rao, 2007; Ghosh et al., 2014).

The dust of *bakhar* tablet mixed properly with air-dried boil rice (about 2–3 g of starter added to 200 g of boiled rice) and packed in an earthen pot and that sealed with an earthen lid. Fermentation is carried out 3 days during summer and 5 days during winter (Ghosh et al., 2014). The glutinous ferment is diluted with drinking water and sieve with a fine cloth, and the filtrate is drink. Apart from its positive nutritional values, tribal people apply it as a remedy for many degenerative and infectious diseases, particularly against gastrointestinal ailments.

13.4.3 BHAATI JAANR

It is a rice based mild-alcoholic and paste like soft product very popular in the Himalayan region. *Bhaati jaanr* is a Nepali word for fermented rice beverage. Its local name is different among the different community of the ethnic people such as *tak thee* (Limboo), *kokumaak* (Rai), *kaiyanpaa* (Gurung), *kaan chi* (Tamang), *kameshyaabu* (Sunwar), *chhohaan* (Magar), *ja thon* (Newar), *dacchhang* (Sherpa), *laayakaachhyaang* (Bhutia), and *jo chee* (Lepcha). During traditional preparation, boiled rice is spread on a mat and mixed well with powdered traditional starter marcha (1 to 2%), and kept in a vessel or an earthen pot for 1–2 days at room temperature for saccharification. *Marcha* (*Murcha*) is prepared by blending Rice/millet/wheat dust, roots of *Plumbagozeylanica*, leaves of *Buddlejaasiatica*, flowers of *Vernonia cinerea*, ginger, red dry chili, other available spices, and powdered old marcha (Tamang and Thapa, 2006; Tamang et al., 2012). *Marcha* like starters in other regions of NE such as *hamei* of Manipur, *pham*, ipoh, and phab of Arunachal Pradesh, *humao* of Assam, and *thiat* of Meghalaya are used for the preparation of alcoholic beverage (Das et al., 2012).

During the preparation of *bhaati jaanr*, presence of sweet aroma in the ferment indicated the end of saccharification. After that, the vessel is made airtight and fermented for 2–3 days in summer and 7–8 days in winter. *bhaati jaanr* is made into a thick paste by using a traditional hand-driven wooden or bamboo-made stirrer. It is consumed directly. Only 5% of people of the Sikkim state consume *bhaati jaanr* daily. Per capita, daily consumption of *bhaati jaanr* in the Darjeeling hills and Sikkim is 252.5 g and 323.5 g, respectively (Tamang and Thapa, 2006; Tamang et al., 2012).

13.4.4 DARU (UN-DISTILLED)/*CHAKTI*

Daru is a jaggery-based popular alcoholic beverage prepared in the rural areas of Shimla, Kullu and other regions of Himachal Pradesh (Takhur, 2004). Daruis also called Chakti in Kullu valley. This is served during local festivals and marriage ceremonies. Babool wood (*Acacia nilotica*) is added to give taste and aroma. For its preparation, jaggery (*gur*) mixed with water and local starter phab. A small portion of babool wood is also added. The mixer is kept in the simmering heat for a few hours. Then the mixture put into a traditional vessel (earthen pot or wooden chamber) and kept for 4–5 days in static condition. Mixture is then filtered by using a fine cloth or poly mosquito net. The filtrate is daru and consumed with spicy vegetables (Takhur, 2004).

13.4.5 CHULLI

Dried wild apricots, locally called *chulli* are used for making the alcoholic beverage. It is a traditional fermented beverage of the tribal district of Kinnaur of Himachal Pradesh and Ladakh region of J&K state. Dried apricots are initially boiled and when it is still warm then *phab* is added. Then mixer is transferred into a traditional container and fermented for 2–3 days. The filtrate is consumed (Takhur, 2004).

13.4.6 JANN

Jann is a traditional low alcoholic drink in Uttaranchal state and very common drink among the Bhotiya community. A good number of substrates of both cereals and fruits like rice (*Oryzasativa*), wheat (*Triticum aestivum*), jau (*Hordeum vulgare*), koni (*Setaria italica*), china (*Panicum miliaceum*), oowa (*Hordeum himalayens),* and chuwa (*Amaranthus paniculatus*) are using for jann preparation. Similarly, amongst the fruits, apple is most desired and is also very delicious. But *jann* prepared from *koni* is considered to be the best in quality in respect to its taste (sweetness), smell, and strength. In the making process of rice *jaan,* first rice is boiled and cooked rice is thoroughly mixed with powdered *balam*, a traditional starter in this region. The quantity of the *balam* powder required is proportionate to the quantity of rice (usually 40 gm *balam* in 5 kg of rice) to be fermented (Roy et al., 2004).

Balam is made up of wheat flour by mixing a number of herbs and spices. First, the raw wheat flour is roasted over fire till it becomes brown in Color. The roasted flour is then mixed with spices like long *(Syzygium aromaticum),* kali elachi (*Amomum subulatum),* kalimirch (*Piper nigrum),* leaves of mirchi-ghash (wild chilies), and seeds of pipal (*Ficus religiosa*). In this mixture, powder of old *balam* is also added. A thick paste of this mixture is then pressed between palms to make *balam* balls of the required size. These balls are then dried in shade and stored for future use for an indefinite period of time (Das and Pandey, 2007; Roy et al., 2004).

The mixture rice and *balam* is then kept in a narrow mouth earthen pot or traditional container. The container is made airtight by sealing with a piece of cloth and is kept in a dark and warm place for fermentation. The extended as long as prolonged fermentation period is essential for a better quality of *jann*. Bhotiyas prepare *jaan* and leave it for fermentation, and thereafter they migrate to the lower valleys for winter settlement. After their return (~6 months later) they consume this beverage and *jann* produced in this way is considered to be the best in quality (Das and Pandey, 2007). After the completion of fermentation, the *jann* so produced is filtered with the help of a sieve. The filtrate is a whitish liquid, which is consumed. Jann is also prepared from fruits like pumpkin, apple, banana, and orange. Apples are first cut into pieces and then are mixed with *balam* powder for fermentation. Other process is as similar as rice *jann*. After one or two days of fermentation, the fermented liquor is called *sez*, which is also consumed by people.

13.4.7 KODOKO JAANR

It is a mild alcoholic beverage prepared from finger millets. *Kodoko jaanr* or *chyang* or *chee* is a very popular refreshing drink with sweet-sour taste among the ethnic people of north-eastern Himalaya (Thapa and Tamang, 2004; Dutta et al., 2012). Seeds of finger millet (*Eleusine coracana*), locally called kodo, are cooked and spread on a bamboo mat and mixed with marcha (1–2%). The mixture is generally packed in a bamboo basket lined with fresh fern, locally called *thadreuneu* (*Thelypteris erubescens*) and kept at room temperature. After 2–4 days of fermentation, the mixture in then transferred into an earthen pot and made air-tight. This phase of fermentation is continued for 3–4 days during summer and 5–7 days during winter. During consumption, fermented millet is packed into a 'U' shaped container locally called *toongbaa* and a bamboo straw is also pushed into the bottom of the container. Then mild warm water is added onto its top and the filtrate is consumed by using the straw. Water can be added twice or thrice after sipping of the extract.

13.4.8 ZUTHO

Zutho is a popular rice beer in Nagaland with characteristics of fruity aroma and sour taste. This is a naturally fermented alcoholic beverage. The sprouted rice grain with notable hull are taken crushed in a mortar with a wooden pestle, form a grist. About 3 kg of rice dust boiled in about 5 L of water and prepared a porridge. Then one to two handful of grist is added to the porridge and mixed properly. The mixture poured in a wooden vessel and kept for fermentation for 3 (during summer) to 7 (during winter) days. The viscous ferment can drink directly as strong beverage, this is called as *Zutshe* or diluted with water and strained through a bamboo vessel then it is called as *Zutho*. *Zutho* served in a cup made of a wide bamboo stalk or in a cow's horn (Das and Deka, 2012; Tamang et al., 2012).

13.4.9 ZUDIMA/ZU

It is a mild alcoholic rice beverage. It has different names among the tribes of North East. For its preparation, a starter cake called as *umhu* or *humao* is used, it is a mixture of rice and bark of *thempra* (*Acacia pennata*) plant, for variation in taste and flavor they use to mix leaves powder of *Piper betle*, *Buddleja asiatica*, and *Hedyotis scandens*. For preparing Judima, boiled rice is mixed with powdered *humao* (1–5%) and kept in a large container. After 5–7 days, slightly yellowish juices come out from the fermented mass, which indicates the completion of fermentation process. The mixture is transferred to *khulu* (a triangle shaped bamboo cone) or bottom hole earthen pot, and the pale to dark yellow colored leachate *judima* is collected. *Judima/Zu* is drunk directly with or without water. *Judima* is served during celebration of the traditional rituals and festivals (Chakrabarty et al., 2014; Arjun et al., 2014).

13.4.10 MAHUA

Mahua distilled beverages are prepared from dried flower bud (corollas) of *Madhuca indica/longifolia* plants (Kumar and Rao, 2007). It is a very popular drink among the tribal community in states of Andhra Pradesh, Bihar, Gujarat, Karnataka, Madhya Pradesh, Orissa, Rajasthan, Uttar Pradesh and West Bengal (Jadav et al., 2009). The flowers are small in size and musky-scented appear from March to April. The tree blooms at night and each short-lived flower falls to the ground at dawn. *Mahua* flowers are in dense fascicles near end of the branches having 1.5 cm long fleshy cream colored corolla tube and are scented. *Mahua* flowers are rich in sugar (68–72%), in addition to a number of minerals and one of the most important raw materials for alcohol fermentation. For preparation of *mahua* wine, dried flower firstly cleaned and washed in water, and again dried. *Bakhar* tablet is commonly used as starter along with the juice of *Buchanania lanzan*. Spring leaves used as flavoring agent. The mixture is kept in an earthen vessel with adequate of water (usually 1:4 ratio). The mouth of the pot is sealed tightly using cloth/straw and that further layered with mud. The mixture is fermented for 3–5 days. Household distillation process viz. closed process or tube process is employed for distillation of wine. Normally yield of alcohol is about 300–400 ml per kg of dried flower. Sometimes, jaggery is added to the flower (half of weight of flower) for higher-yielding of alcohol. During distillation, the alcohol content decrease with respect to time (Jadav et al., 2009; Vora et al., 2016; Singh et al., 2013).

13.4.11 ANGOORI

This is one of the very popular traditional alcoholic beverages consumed by the tribal people of Kinnaur, Himachal Pradesh during local festivals and marriage ceremonies. For this region its local name is *Kinnauri*. In this area, red, and white variety of grapes is widely cultivated. For the preparation of this beverage, grapes are cleaned, washed, and then crushed. Sometime jaggary is added to increase the alcohol production rate. A local starter *phab* is mixed with the crushed grapes. The mixture is poured in an earthen pot or aluminum container then made airtight. The fermentation is allowed for at least 15 days. Fermented material is then filtered and distilled (Thakur et al., 2004).

13.4.12 DARU

Daru is a distilled alcoholic beverage. Rice and jaggery is used as a starchy based substrate for alcohol production (Roy et al., 2004). This type of alcoholic beverage preparation is very popular among the different tribal and low-income group of people in around the country. In many places, it is also called as country liquor. Only difference is the use of regional starter but substrates are more or less common. In Himachal Pradesh *balam*, in central India *bahka*r and in NE *marcha* is used to prepare this alcoholic beverage. Apart from rice, the cereals like *Setaria italica* (koni), *Amaranthus paniculatus* (chuwa), *Hordeum himalayens* (oowa) and wheat are used in the preparation of daru in Himachal Pradesh. The taste (i.e., quality) of the daru does not vary according to the type of substrate used (Roy et al., 2004).

For *daru* preparation, cooked rice is mixed with powdered *balam* (higher quantity of *balam* is added for *daru* preparation than *jann*), and the mixture is kept in an airtight container and in preferably warmer place. After about a week of fermentation, when a sweet smell is released, it is distilled in a distillation vessel. The distillate substance is the daru collected in bottles.

The undigested residue is called *chak*, this can be recycled for preparation of daru by addition with fresh jaggary and balam powder (Sekar and Mariappan, 2007). *Chak* cannot be reused more than three times then it is used as animal feed. Like other traditional distilled alcohol, daru is also graded into three categories, the initial few bottles are of high strength of alcohol, and the middle few bottles containing moderate level of alcohol, which are rated as good for consumption. Final few bottles containing very low contents of alcohol. For making the daru attractive in appearance, a small quantity of turmeric is hanged right at the mouth of the collecting tube. This makes the liquid a light but brilliantly shining yellow in Color (Roy et al., 2004).

13.4.13 FENI/FENNY

It is a symbolic alcoholic beverage of Goa. In India, for the first time, this drink was registered as a geographical indication (GI) on 27th February 2009 (Dwijen, 2009). *Feni* is a distilled alcoholic beverage of coconut toddy or fermented cashew (*Anacardium occidentale* L.) apple. The word *feni* is originated from Konkani, where 'Fennõ' means froth. This is related to the bubbles formed when *feni* is poured, or froth formed during fermentation. The pungent cashew apples are crushed in a large rock, juice is extracted (about 1L of juice is produced from 3.5 kg of fruit) and then juice is fermented naturally. After three days of fermentation, the product is sequentially distilled for three times. The *feni* prepared from the third distillation containing 30–40% alcohol and has a long shelf-life (Sekar and Mariappan, 2007).

13.4.14 PALM WINE OR TODDY

The toddy is prepared from the sap of varieties of palm trees and very popular in Southern India from coconut palm (*Cocos nucifera* L.) and from Asian palmyra palm (*Borassus flabellifer*) in the Eastern India (Karthikeyan et al., 2014). The fresh sap containing sucrose (12–15% by weight) and other monosaccharides that favors the growth of natural microflora. Fresh palm sap (*neera*) is generally a dirty brown, but due to microbial growth it becomes pale and eventually milky white, this beverage is called toddy. Thus, palm wine is generally sweetish with vigorously effervescent alcoholic beverage. It is a mild alcoholic (1.5–2%) beverage but during its storage alcohol content may increase (5–6%).

KEYWORDS

- **alcoholic beverages**
- **Chhang**
- **fermented non-alcoholic beverages**
- ***Madhuca longifolia***

REFERENCES

Allchin, F. R., (1979). India: The ancient home of distillation? *Man, New Series, 14*(1), 55–63.

Anon, (2017). https://millets.wordpress.com/recipes/pearl-millet-bajra-kambu-recipes/kambu-ragi-koozh-fermented/ (Accessed on 14 June 2019).

Arjun, J., Verma, A., & Prasad, S., (2014). Method of preparation and biochemical analysis of local tribal wine *Judima*: An indigenous alcohol used by Dimasa tribe of North Cachar Hills district of Assam, India. *Int. Food Res. J., 21*, 463–470.

Aso, Y., & Akazan, H., (1992). Prophylactic effect of a *Lactobacillus casei* preparation on the recurrence of superficial bladder cancer. *Urologia Internationalis, 49*(3), 125–129.

Backes, M., (2014). *Cannabis Pharmacy: The Practical Guide to Medical Marijuana.* New York: Black Dog & Leventhal, pp. 10–245.

Behare, P. V., Singh, R., Tomar, S. K., Nagpal, R., Kumar, M., & Mohania, D., (2010). Effect of exopolysaccharide producing strains of *Streptococcus thermophilus* on technological attributes of fat-free lassi. *Journal of Dairy Science, 93*(7), 2874–2879. http://doi.org/10.3168/jds.2009–2300 (Accessed on 14 June 2019).

Bhardwaj, K. N., Jain, K. K., Kumar, S., & Kuhad, R. C., (2016). Microbiological analyses of traditional alcoholic beverage (*Chhang*) and its starter (*Balma*) prepared by *Bhotiya* tribe of Uttarakhand, India. *Indian J. Microbiol., 56*(1), 28–34.

Bylund, G., (2003). *Dairy Processing Handbook* (2nd rev. edn.). Tetra Pak Processing Systems AB, Lund, pp. 233–262.

Chakrabarty, J., Sharma, G. D., & Tamang, J. P., (2014). Traditional technology and product characterization of some lesser-known ethnic fermented foods and beverages of North Cachar Hills district of Assam. *Indian J. Tradit. Know., 13*(4), 706–715.

Chakrabarty, J., Sharma, G., & Tamang, J. P. T., (2010). Substrate utilization in traditional fermentation technology practiced by tribes of North Cachar Hills district of Assam. *Assam University Journal of Sci. & Technol., 4*, 66–72.

Chandan, R. C., & Arun, K., (2011). *Dairy Ingredients for Food Processing.* Blackwell Publishing Ltd. http://doi.org/10.1002/9780470959169.ch5 (Accessed on 14 June 2019).

Conway, V., Couture, P., Richard, C., Gauthier, S. F., Pouliot, Y., & Lamarche, B., (2013). Impact of buttermilk consumption on plasma lipids and surrogate markers of cholesterol homeostasis in men and women. *Nutrition, Metabolism and Cardiovascular Diseases, 23*(12), 1255–1262. http://doi.org/10.1016/j.numecd.2013.03.003 (Accessed on 14 June 2019).

Das, A. J., Deka, S. C., & Miyaji, T., (2012). Methodology of rice beer preparation and various plant materials used in starter culture preparation by some tribal communities of North-East India: A survey. *Int. Food Res. J., 19*(1), 101–107.

Das, A., & Deka, S., (2012). Fermented foods and beverages of the North-East India. *Int. Food Res. J., 19*, 377–392.

Das, A., Raychaudhuri, U., & Chakraborty, R., (2012). Cereal-based functional food of Indian subcontinent: A review. *Journal of Food Science and Technology, 49*(6), 665–672. http://doi.org/10.1007/s13197-011-0474-1 (Accessed on 14 June 2019).

Das, C. P., & Pandey, A., (2007). Fermentation of traditional beverages prepared by Bhotiya community of Uttaranchal Himalaya. *Indian J. Tradit. Know., 6*(1), 36–140.

De, S., (2004). Indian dairy products. In: *Outlines of Dairy Technology* (pp. 463–464). New Delhi: Oxford University Press.

Dutta, A., Das, D., & Goyal, A., (2012). Purification and characterization of fructan and fructan sucrase from *Lactobacillus fermentum* AKJ15 isolated from *Kodoko jaanr*, a fermented beverage from north-eastern Himalayas. *Int. J. Food Sci. Nutr., 63*, 216–224.

Dwijen, R., (2009). *Geographical Indications and Localization: A Case Study of Feni.* ESRC Report, Available at SSRN: https://ssrn.com/abstract=1564624 (Accessed on 14 June 2019).

Elaine, M., & Danilo, M., (2011). *Traditional Fermented Foods and Beverages for Improved Livelihood (FAO).* Diversification Booklet 21.

Fay, J. C., & Benavides, J. A., (2005). Evidence for domesticated and wild populations of *Saccharomyces cerevisiae*. *PLoS Genet, 1*(1), e5. https://doi.org/10.1371/journal.pgen.0010005 (Accessed on 14 June 2019).

Fermented Milk Products, (2011). Retrieved from: http://www.milkingredients.ca/index-eng.php?id=180 (Accessed on 14 June 2019).

Fox, P. F., Uniacke-Lowe, T., McSweeney, P. L. H., & O'Mahony, J. A., (2015). Water in Milk and Dairy Products. In: *Dairy Chemistry and Biochemistry* (2nd edn., pp. 299–320). Switzerland: Springer International Publishing Switzerland. http://doi.org/10.1007/978-3-319-14892-2 (Accessed on 14 June 2019).

Fuller, R., (1989). Probiotics in man and animals. *Journal of Applied Bacteriology, 66*(5), 365–378. http://doi.org/10.1111/j.1365–2672.1989.tb05105.x (Accessed on 14 June 2019).

Geoffrey Campbell-Platt, (1994). *Fermented Foods–A World Perspective, Food Research International* (Vol. 27, No. 3, pp. 253–257).

George, V., Arora, S., Sharma, V., Wadhwa, B. K., & Singh, A. K., (2012). Stability, physicochemical, microbial and sensory properties of sweetener/sweetener blends in lassi during storage. *Food and Bioprocess Technology, 5*(1), 323–330. http://doi.org/10.1007/s11947-009-0315-7 (Accessed on 14 June 2019).

Ghosh, K., Maity, C., Adak, A., Halder, S. K., Jana, A., Das, A., Parua, S., Das, M. P. K., Pati, B. R., & Mondal, K. C., (2014). Ethnic preparation of Haria, a rice-based fermented beverage, in the province of lateritic West Bengal, India. *Ethnobot. Res. and Appl., 12*, 39–49.

Ghosh, K., Ray, M., Adak, A., Dey, P., Halder, S. K., Das, A., Jana, A., Parua, S., Das, M. P. K., Pati, B. R., & Mondal, K. C., (2015a). Microbial, saccharifying and antioxidant properties of an Indian rice-based fermented beverage. *Food Chem., 168*, 196–202.

Ghosh, K., Ray, M., Adak, A., Halder, S. K., Das, A., Jana, A., Parua, S., Vágvölgyi, C., Das, M. P. K., Pati, B. R., & Mondal, K. C., (2015b). Role of probiotic *Lactobacillus fermentum* KKL1 in the preparation of a rice-based fermented beverage. *Bioresour Technol., 188*, 161–168.

Goswami, P., Singh, S., & Sharma, K. P., (2011). *Fermented Milk Products—Nature's Blessings*. Retrieved from: http://www.pfonline.com/index.php/columns/value-addition/146-fermenenter-milkproducts (Accessed on 14 June 2019).

Haghighi, H. R., Gong, J., Gyles, C. L., Hayes, M. A., Sanei, B., Parvizi, P., & Sharif, S., (2005). Modulation of antibody-mediated immune response by probiotics in chickens.

Clinical and Diagnostic Laboratory Immunology, *12*(12), 1387–1392. http://doi.org/10.1128/CDLI.12.12.1387–1392.2005 (Accessed on 14 June 2019).

Holzapfel, W. H., (2002). Appropriate starter culture technologies for small-scale fermentation in developing countries. *International Journal of Food Microbiology*, *75*, pp. 197–212.

Jean-Luc, L., Didier, M., Jean-Marie, C., & Francis, K., (2007). Bread, beer and wine: *Saccharomyces Cerevisiae* diversity reflects human history. *Molecular Ecology, 16*, 2091–2102.

Joshi, V. K., (2016). *Indigenous Fermented Foods of Southeast Asia*. CRC Press, Taylor & Francis Group. Retrieved from https://books.google.com/books?id=EWVYBQAAQBAJ & pgis=1 (Accessed on 14 June 2019).

Joshi, V. K., Sharma, S., & Rana, N., (2011). Preparation and evaluation of appetizers from lactic acid fermented vegetables. *Journal of Hill Agriculture*, *2*(1), 20–27.

Kanwar, S. S., Gupta, M. K., Katoch, C., & Kanwar, P., (2011). Cereal-based traditional alcoholic beverages of Lahaul and Spiti area of Himachal Pradesh. *Indian J. Tradit. Know.*, *10*, 251–257.

Kardong, D., Deori, K., Sood, K., Yadav, R., Bora, T., & Gogoi, B., (2012). Evaluation of nutritional and biochemical aspects of *Po:ro apong* (Saimod): A homemade alcoholic rice beverage of Missing tribe of Assam, India. *Indian J. Tradit. Know., 11*, 499–504.

Karthikeyan, R., Suresh, K. K., Singaravadivel, K., & Alagusundaram, K., (2014). Volatile elements of coconut toddy (*Cocos Nucifera*) by gas chromatography-mass spectrometry. *J. Chromatograph Separat Techniq., 5*, 1.

Kirstbergsson, K., & Oliveira, J., (2016). *Traditional Foods: General and Consumer Aspects*. New York: Springer Science+Business Media. http://doi.org/10.1007/978-1-4899-7648-2 (Accessed on 14 June 2019).

Kumar, V., & Rao, R. R., (2007). Some interesting indigenous beverages among the tribals of Central India. *Indian J. Tradit. Know., 6*(1), 141–143.

Larsson, S. C., Andersson, S., Johansson, J., & Wolk, A., (2008). Cultured milk, yogurt, and dairy intake in relation to bladder cancer risk in a prospective study of Swedish women and men. *American Society for Nutrition*, *88*, 1083–1087.

Le Sage, D. E., (1975). The language of honey. In: Crane, E., (ed.), Honey. Crane, Russak & Company, Inc., New York.

Mondal, K. C., Ghosh, K., Mitra, B., Parua, S., & Das, M. P. K., (2016). Rice-based fermented foods and beverages: Functional and nutraceutical properties. In: Ramesh, C., Ray, R. C., & Montet, D., (eds.), *Fermented Foods, Part II: Technological Intervention* (pp. 150–176). New York, CRC Press.

Oksanen, P. J., Salminen, S., Saxelin, M., Hämäläinen, P., Ihantola-Vormisto, A., Muurasniemi-Isoviita, L., & Salminen, E., (1990). Prevention of travelers' diarrhea by Lactobacillus GG. *Annals of Medicine*, *22*(1), 53–56. http://doi.org/10.3109/07853899009147242 (Accessed on 14 June 2019).

Oort, M. S., (2002). Surā in the PaippalādaSaṃhitā of the Atharvaveda. *J. Am. Orient. Soc., 122*(2), 355–360.

Padghan, P. V., Mann, B., Rajeshkumar, S. R., & Kumar, A., (2015). Studies on bio-functional activity of traditional Lassi. *Indian Journal of Traditional Knowledge*, *14*(1), 124–131.

Panesar, P. S., (2011). Fermented dairy products: Starter cultures and potential nutritional benefits. *Food and Nutrition Sciences*, *2*(1), 47–51. http://doi.org/10.4236/fns.2011.21006 (Accessed on 14 June 2019).

Patrick, E. M., Juzhong, Z., Jigen, T., Zhiqing, Z., Gretchen, R. H., Robert, A. M., et al., (2004). *Fermented Beverages of Pre and Proto-Historic China*. PNAS vol. 101 no. 51, 17593–17598.

Puniya, A. K., (2016). *Fermented Milk and Dairy Products* (1st edn.). CRC Press, Taylor & Francis Group, Boca Raton.

Ramakrishnan, C. V., (1979). The studies on Indian fermented food. *Baroda Journal of Nutrition, 6*, 1–57.

Rangappa, K. S., & Achaya, K. T., (1974). *Indian Dairy Products* (2nd edn.). Bombay, New Delhi: Asia Publishing House.

Ray, M., Ghosh, K., Har, P. K., Singh, S. N., & Mondal, K. C., (2017). Screening of health beneficial microbes with potential probiotic characteristics from the traditional rice-based alcoholic beverage, *haria*. *Acta Biol. Szeged., 61*, 51–58.

Ray, M., Ghosh, K., Singh, S., & Mondal, K. C., (2016). Folk to functional: An explorative overview of rice-based fermented foods and beverages in India. *Journal of Ethnic Foods, 3*, 5–18.

Roy, A., Khanra, K., Bhattacharya, C., Mishra, A., & Bhattacharyya, N., (2012). *Bakhar-Handia* fermentation: General analysis and a correlation between traditional claims and scientific evidences. *Adv. Biores., 3*, 28–32.

Roy, B., Prakash, K. C., Farooquee, N. A., & Majila, B. S., (2004). Indigenous fermented food and beverages: A potential for economic development of the high altitude societies in Uttaranchal. *J. Hum. Ecol., 15*(1), 45–49.

Sankaran R. (1998) Fermented foods of the Indian subcontinent. In: Wood B. J. B. (eds.) *Microbiology of Fermented Foods* (Vol. 1, 2nd edn., pp. 753–789). Springer, Boston, MA.

Sarkar, P., Lohith, K. D. H., Dhumal, C., Panigrahi, S. S., & Choudhary, R., (2015). Traditional and ayurvedic foods of Indian origin. *Journal of Ethnic Foods, 2*(3), 97–109. http://doi.org/10.1016/j.jef.2015.08.003 (Accessed on 14 June 2019).

Sathe, G. B., & Mandal, S., (2016). Fermented products of India and its implication: A review. *Asian Journal of Dairy and Food Research, 35*(1), 1–9. http://doi.org/10.18805/ajdfr.v35i1.9244 (Accessed on 14 June 2019).

Sekar, S., & Mariappan, S., (2007). Usage of traditional fermented products by Indian rural folks and IPR. *Indian J. Tradit. Know., 6*(1), 111–120.

Sekar, S., (2007). Traditional alcoholic beverages from Ayurveda and their role on human health. *Indian J. Tradit. Know., 6*(1), 144–149.

Sharma, R., Gupta, A., Abrol, G. S., & Joshi, V. K., (2014). Value addition of wild apricot fruits grown in North-West Himalayan regions: a review. *J. Food Sci. Technol., 51*(11), 2917–2924.

Shuwu, M. P., Ranganna, B., Suresha, K. B., & Veena, R., (2011). Development of value added lassi using honey. *The Mysore Journal of Agricultural Sciences, 45*(4), 757–763.

Siitonen, S., Vapaatalo, H., Salminen, S., Gordin, A., Saxelin, M., Wikberg, R., & Kirkkola, A. L., (1990). Effect of Lactobacillus GG yogurt in prevention of antibiotic-associated diarrhea. *Annals of Medicine, 22*(1), 57–59. http://doi.org/10.3109/07853899009147243 (Accessed on 14 June 2019).

Singh, N. L., Ramprasad, M. P. K., Shukla, S. K., Kumar, J., & Singh, R., (2010). Alcoholic fermentation techniques in early Indian tradition. *Indian J. Hist. Sci., 45*(2), 163–173.

Singh, R., Mishra, B. K., Shukla, K. B., Jain, N. K., Sharma, K. C., Kumar, S., Kant, K., & Ranjan, J. K., (2013). Fermentation process for alcoholic beverage production from mahua (*Madhucaindica* J. F. Mel.) flowers. *Afr. J. Biotechnol., 12*, 5771–5777.

Steinkraus, K. H., (1995). *Handbook of Indigenous Fermented Foods, Revised and Expanded* (2nd edn.), Marcel Dekker, Inc., New York, pp. 363–508.

Tamang, J. P., Shin, D. H., Jung, S. J., & Chae, S. W., (2016). Functional properties of microorganisms in fermented foods. *Frontiers in Microbiology, 7*, 1–13. http://doi.org/10.3389/fmicb.2016.00578 (Accessed on 14 June 2019).

Tamang, J. P., Tamang, N., Thapa, S., Dewan, S., Tamang, B. M., Yonzan, H., Rai, A. K., Chettri, R., Chakrabarty, J., & Kharel, N., (2012). Microorganisms and nutritional value of ethnic fermented foods and alcoholic beverages of North East India. *Indian J. Tradit. Know., 11*, 7–25.

Targais, K., Stobdan, T., Mundra, S., Ali, Z., Yadav, A., Korekar, G., & Singh, S. B., (2012). *Chhang*–A barley based alcoholic beverage of Ladakh, India. *Indian J. Tradit. Know., 11*, 190–193.

Teramoto, Y., Yoshida, S., & Ueda, S., (2002). Characteristics of a rice beer (*Zutho*) and a yeast isolated from the fermented product in Nagaland, India. *World J. Microbiol. Biotechnol., 18*, 813–816.

Thakur, N., & Savitri, B. T. C., (2004). Characterization of traditional fermented foods and beverages of Himachal Pradesh. *Indian J. Tradit. Know., 3*, 325–335.

Thakur, N., (2013). *Characterization of Traditional Fermentation Processes Used for the Production of Some Alcoholic Beverages (Chhang, Suraand Jau Chhang) in Himachal Pradesh* (pp. 120–123). PhD. Thesis, Himachal Pradesh University, Summer Hill, Shimla.

Thapa, S., & Tamang, J. P., (2004). Product characterization of *kodoko jaanr*: Fermented finger millet beverage of the Himalayas. *Food Microbiol., 21,* 617–622.

Vonk, R. J., Reckman, G. A. R., Harmsen, H. J. M., & Priebe, M. G., (2012). Probiotics and lactose intolerance. In: *Probiotics* (pp. 149–160). In Tech. http://doi.org/10.5772/51424 (Accessed on 14 June 2019).

Vora, J. D., Srinivasan, P., & Singnurkar, T., (2016). Biochemical and organoleptic study of the mahua flower and mahua flower wine. *IOSR J. Biotechnol. Biochem., 2*, 01–06.

Widyastuti, Y., & Febrisiantosa, A., (2014). The role of lactic acid bacteria in milk fermentation. *Food and Nutrition Sciences*, *5*, 435–442. http://doi.org/10.4236/fns.2014.54051 (Accessed on 14 June 2019).

Yadav, P., Garg, N., & Diwedi, D. H., (2009). Effect of location of cultivar, fermentation temperature and additives on the physico-chemical and sensory qualities on Mahua (*Madhucaindica* J. F. Gmel.) wine preparation. *Nat. Prod. Rad., 8*(4), 406–418.

CHAPTER 14

Food Industry Byproducts: Resource for Nutraceuticals and Biomedical Applications

WINNY ROUTRAY

Marine Bioprocessing Facility, Center for Aquaculture and Seafood Development, Fisheries and Marine Institute, Memorial University of Newfoundland, P.O. Box 4920, St. John's, NL, A1C 5R3, Canada,
E-mail: routrayw@yahoo.com

ABSTRACT

Byproducts and wastes from agricultural and food industries can be valorized through biorefinery approach to obtain bioactive compounds and biopolymers applicable as biomedical compounds and development of medicinal concoctions and devices. This chapter briefly summarizes the different sources, the respective bioactive compounds and biopolymers and their possible use as nutraceuticals, pharmaceuticals, and in biomedical applications. The major benefits of these compounds and their prospect as a valuable resource for maximum utilization in current and future scenarios have also been discussed.

14.1 INTRODUCTION

Agricultural and food industry wastes contribute significantly to the worldwide solid and liquid pollution. Increasing world population with essential nutritional requirements has augmented agricultural production and subsequent agricultural and food industry processing and unit-operations. This has led to further increased accumulation of byproducts and waste production. These biomaterials are rich sources of phenolics, proteins, protein hydrolysates, amino acids, dietary fibers, fats, and waxes and other components including pigments and saponins, which have several nutritional, health beneficial and medicinal components. Other components such as chitosan (CS) and collagen can be utilized or combined with other components for pharmaceuticals and biomedical applications. In most parts of the world, agricultural, and food industry wastes are disposed in the landfills, composted, applied for energy generation through biogas production (El-Mashad and Zhang, 2010; Zhang et al., 2014) or for biochar production, which is utilized in various applications (Kwapinski et al., 2010; Opatokun et al., 2015; Zhao et al., 2013). However, systematic biorefinery approach with maximum stepwise separation and valorization of the extracted biomolecules can lead to the production of high-quality value-added products, followed by composting and conversion to biochar of the residual biomass with low-value applications.

Also, changing food diversity and cultivation of cash crops and traditional crops with high demand have accelerated the profit-oriented farming approach, simultaneously affecting the biodiversity, which can also lead to endangering or extinction of certain indigenous food crop species. *Perilla frutescens* is a local crop mostly cultivated in central Himalaya in India, where different parts of the plant are utilized by the indigenous population as an ingredient in various recipes, as oil source and as a traditional medicine (Negi et al., 2011). Hence, crops like this can be exploited as an evolved resource of basic ingredients such as oil, along with the application of their byproduct as a source of medicine.

This chapter has summarized some of the agricultural and food industry byproducts, potential bioactive nutraceutical components, and other components with potential biomedical applications. Process technologies of production and current and future aspects of byproducts application have also been briefly discussed.

14.2 FOOD INDUSTRY BYPRODUCTS

Food industry wastes can be solid and/or liquid, and can be derived from both plant and animal based food-processing unit-operations. Food industry byproducts derived from plants can be categorized as wastes and byproducts generated during unit operations applied for (a) agricultural processes (twigs or parts produced during trimming and maintenance of the crops), (b) harvest activities (twigs, leaves, and parts discarded during harvest), (c) post-harvest processes (discarded or degraded parts and samples, husk),

(d) ready to cook food commodities (wastewater, discarded parts during the processing of vegetables or fruits), and (e) ready to eat food preparations (wastewater, discarded oils, spoiled, and discarded parts).

Animal-based wastes include byproducts obtained through (a) initial processing during harvesting (blood water, hair, and skin), (b) size reduction processes for market-friendly commodity production (waste produced during cleaning and cutting, water used for cleaning of cut flesh), (c) ready to cook food commodity (flesh pieces produced during size reduction), (d) ready to serve food product (waste produced during culinary unit-operations including oils and water).

Wastewater is a major effluent of food industry, which is rich in biomolecules with high content of carbon, nitrogen, and other nutritional components. Currently, wastewater has been increasingly applied for irrigation of crops, which is an efficient way of utilization of this nutrient source for nutrient recycling in the ecosystem; however, it is a low-value application. Furthermore, extraction of biomolecules from the wastewater will valorize the resource, which can also lead to high-value nutraceuticals production. Astaxanthin and bioactive peptides were extracted by Amado et al. (2015) from shrimp cooking water. Hydroxytyrosol is a potent phenolic compound, which was obtained from olive mill wastewater and displayed high antioxidant capacity of rat plasma (Visioli et al., 2001). Also, wastewater can be utilized as growth medium for microorganisms, intended for bioactive compounds production and accumulation, which can be applied for nutraceuticals, pharmaceuticals, and biomedical purposes. Olive wastewater was utilized by different strains of *Yarrowia lipolytica* for production of enzymes and other bioactive compounds, such as lipase and citric acid (Lanciotti et al., 2005). Similar observations have been obtained in other studies.

14.3 HEALTH BENEFICIAL AND MEDICINAL COMPONENTS

There are several food constituents, which trigger and lead to physiological processes for prevention of diseases, lowering of the symptoms or decreasing the severity of health ailments. For example, there are a number of legumes which can lead to hypocholesterolemic and hypoglycemic effect that are often identified as anti-diabetic effect (Kaushik et al., 2010). One of the bioactive properties, which has been widely identified and correlated with other health beneficial properties, is antioxidant properties. According to the extensive studies conducted in past few decades, parts of the plants and animals other than the sections traditionally consumed as food, also contain the bio-compounds or their precursors, which make them highly valuable sources for these compounds. Some of the bioactive compounds have been summarized in Table 14.1 with their respective applications.

The mechanisms of bioactivity and bioavailability of several bioactive compounds are not yet fully documented and still require comprehensive study. In case of protein hydrolysates, mechanism was reported to be dependent on composition, configuration of proteins, the oxidation system (emulsions, other additives, etc.) involved (Slizyte et al., 2016). Factors affecting the degree of bioactivities of the different biomolecules and their mechanisms vary with the physio-chemical properties of biomaterials. Concentration of bioactive compounds in plants vary with species, cultivars, agricultural practices, along with supplied nutritional components and corresponding concentrations, harvest practices and harvest season, and post-harvest practices. Furthermore, the processing methods applied for commodity production can affect the composition of the waste/byproduct generated during the process. In animals, the production and/or accumulation of the biomolecules in different parts is determined by the environmental factors employed during the growth of these organisms, the supplied nutrient source (plant and/or animals), harvesting methods, and processing methods. In addition to these factors, processing methods applied to the byproducts and unit-operations employed for the production of end-products also affect the produced bioactive commodity and their bioactivities.

Phenolics are aromatic secondary metabolites present in various parts of the plants including parts consumed as food and agricultural and food industry byproducts such as twigs, leaves, branches, peels, and seeds (Table 14.2). Phenolics are major constituents of various popular food commodities, including various beverages such as tea and wine, comfort foods such as chocolate and other common food components. However, in some cases, the quantity of phenolics in different byproducts have been reported to have equal or comparatively higher content than the parts consumed as food (Routray and Orsat, 2014). Phenols can be polar as well as non-polar, can be mildly or strongly acidic. The degree of polarity and acidity affects the choice of solvents and efficiency of different extraction methods (Routray and Orsat, 2013). Several authors have discussed in detail the different properties and classifications of phenolics, which can be referred to consider different extraction methods and their applications (Harborne, 1989; Orsat and Routray, 2017; Vermerris and Nicholson, 2006).

TABLE 14.1 Bioactive Compounds from Various Food Industry Byproducts and Their Effects and Applications

Components	Effects/Applications	References
Phenolics	Antioxidant effect, anticarcinogenic, antidiabetic, antibacterial along with positive ocular, and gastric effects. Protective effects against cardiovascular disorders, hepatic damage, neurodegenerative disorders and cytotoxic effects of oxidized low-density lipoprotein and consequently atherosclerosis	Routray and Orsat (2013); Ali Asgar (2013); Croft (1998)
Poly-unsaturated fatty acids (EPA, DHA, DPA)	Effective against cardiovascular disease, inflammatory disease. Helpful for proper brain function and mental health	Ruxton et al. (2004)
Proteins, protein hydrolysates, amino-acids	Antioxidant, antihypertensive, immunomodulatory, and antimicrobial peptides. Inhibit lipid oxidation, Angiotensin converting enzyme inhibition activity, Anti-proliferative effect	Chalamaiah et al. (2012); Slizyte et al. (2016); Villamil et al. (2017); Halim et al. (2016)
Polysaccharides	Dietary fibers: Applied as fat replacer and texturizer	Anderson et al. (2009)
	Antioxidant activity, effectiveness against coronary heart disease, coronary stroke, diabetes, hypertension, obesity, and certain gastrointestinal diseases	
	Chitosan: Encapsulating agents for nutraceuticals, which can be fortified in other food commodities.	Gültekin-Özgüven et al. (2016); Anraku et al. (2009); Anraku et al. (2012)
	A significant decrease in levels of plasma glucose, atherogenic index and led to increase in high-density lipoprotein cholesterol; Lowered the ratio of oxidized to reduced albumin and increased total plasma antioxidant activities; Effective reno-protective activity	
Plant wax	Policosanols: Protective effect against atherosclerotic plaques on aortas in monkeys; Effective against hypertension and type II hypercholesterolemia; Antiplatelet effect	Noa and Mas (2005); Castaño et al. (2002); Arruzazabala et al. (2002)
Carotenoids (astaxanthin, lycopene, carotene)	Coloring agent; Antioxidant, prevention of chronic diseases	Rao and Rao (2007); Hussein et al. (2006)

TABLE 14.1 *(Continued)*

Components	Effects/Applications	References
Saponins	Inhibitory effects against the infectivity of the AIDS virus (HIV) and the activation of Epstein-Barr virus early antigen; Protective effect on PC12 cells; Neuroprotective effect	Tsukamoto et al. (1994); Zhang et al. (2012); Liao et al. (2002)
Enzymes (Pepsin, trypsin, chymotrypsin, collagenase, elastase)	Applied in extraction of other bioactive and medicinal complexes and components required for preparation of complexes in medical aids	Castillo-Yánez et al. (2005)

TABLE 14.2 Functional and Medicinal Benefits of Food Industry Waste

Source/Byproducts	Byproduct Components	Application/ Beneficial Effect	Reference
Okara (Soybean byproduct)	Polysaccharide fractions	Antioxidant activity	Mateos-Aparicio et al. (2010)
–	Protein, lipids, dietary fiber, isoflavones	A significant decrease in weight gain and total cholesterol. Increase in antioxidant activity. Hypoglycemic effect	Jiménez-Escrig et al. (2008); Lu et al. (2013); Muliterno et al. (2017)
Asparagus officinalis byproduct	Steroidal saponins (Diosgenin and Sarsasapogenin)	Chinese medicine: antifungal, antiviral, and antitumoral activities	Wang et al. (2011a)
Pomegranate	–	Antimicrobial, antioxidant, anti-inflammatory, anticancer, and immune-suppressive activities	Miguel et al. (2010)
Pumpkin seed	Minerals, vitamins, essential fatty acids, phytosterols, proteins, and dietary fibers, Cucurbitacin	Cucurbitacin is a deworming agent applied in case of domestic livestock species	Maheshwari et al. (2015)
Saccharum officinarum (Desugared sugar cane extract)	Neolignan glucosides (Saccharnan A and B), phenolics	Antioxidant and tyrosinase inhibitor activity	Chung et al. (2011)
Onion byproducts extract	–	Lowers plasma lipids	Roldán-Marín et al. (2010)
Wheat bran	Protein derived hydrolysates and peptides	ACE-inhibitory activities and antihypertensive effect	Nogata et al. (2009)
Peach (*Prunus persica* (L.) Batsch) byproduct	peptides	Angiotensin I converting enzyme inhibition activity	Vásquez-Villanueva et al. (2015)
Tea seed pomace (*Camellia oleifera* Abel)	Saponin	Protective effect on PC12 cells against H_2O_2-induced cell death	Zhang et al. (2012)
Asparagus officinalis byproduct	Saponin	Antifungal, antiviral, and antitumoral activities	Wang et al. (2011a)
Olea europaea L. leaves	Oleuropein, luteolin-7-O-glucoside and verbascoside	Radical scavengers	Wang et al. (2011b)

TABLE 14.2 *(Continued)*

Source/Byproducts	Byproduct Components	Application/ Beneficial Effect	Reference
Drumstick leaves	–	Antioxidant and free radical scavenging activities	Ekaluo et al. (2015b)
Whey protein concentrate	Antibodies against potential pathogenic bacteria and their toxins	Potential therapeutic use for autoimmune disorders such as rheumatoid arthritis, in which bacterial toxins and an abnormal balance of intestinal bacterial flora are major factors.	Kijima et al. (2009)
Olive leaves	Hydroxytyrosol and oleuropein	Antidiabetic and antioxidant effect in alloxan-diabetic rats	Jemai et al. (2009)
Iraqi olive leaves	–	Protective properties against oxidation and pathogenic bacteria in food applications	Altemimi (2017)
Capsicum frutescens (L.) Var. Longa (Solanaceae) leaves	–	Antibacterial and anthelmintic activity	Vinayaka et al. (2010)
Cauliflower, broccoli, and okara byproducts	–	Antimicrobial potential against gram-negative and gram-positive bacteria	Sanz-Puig et al. (2015)
Guava leaves	Total phenolic content	In vitro antioxidant and free radical activity	Ekaluo et al. (2015a)
Camellia seed cake	Polysaccharides	Antioxidant activity	Shen et al. (2014)
Korean *Citrus hallabong* peels	Pectic polysaccharides	Antitumor metastasis activity	Lee et al. (2014)
Ipomoea batatas leaf	Flavone	Antidiabetic activity in non-insulin dependent diabetic rats	Zhao et al. (2007)
Grape skin	Polyphenols	Antitumor and antimetastatic activities in a murine model of breast cancer	Sun et al. (2012)
Grape seed	Oil	Attenuates oxidative and inflammatory responses in human primary monocytes	del Carmen Millán-Linares et al. (2018)

TABLE 14.2 *(Continued)*

Source/Byproducts	Byproduct Components	Application/ Beneficial Effect	Reference
Bambangan (*Mangifera pajang* Kort.) peels	Phenolics	Radical scavenging activity	Hassan et al. (2011)
Rice Bran	Oil, methanolic extracts	Antioxidative effects	Mariod et al. (2010)
–	Policosanols (wax)	–	Cravotto et al. (2004)
Sugarcane peel	Wax (policosanols, octacosanol)	–	Inarkar and Lele (2012); Gnanaraj (2012)
Ash gourd peel	Wax	An edible coat for fruits to enhance shelf life	(Sreenivas et al., 2011)
Almond (*Prunus dulcis*) skin	Proanthocyanidins and chlorogenic acid	–	Ma et al. (2014)
Olive leaves	–	Antioxidative effects	Papoti and Tsimidou (2009)
Sugarcane molasses	–	Antioxidants	Guan et al. (2014)
Onion (*Allium cepa*) solid waste	Phenolics; Quercetin and biosugar	Antioxidant properties and several other associated health beneficial effects	Kiassos et al. (2009); Choi et al. (2015)
Citrus bergamia Risso peel	Flavonoids	Antimicrobial activity	Mandalari et al. (2007)
Cornsilk (*Zea mays* hairs)	Polyphenol content	Antioxidant activity	Nurhanan and WI (2012)
Food industry byproducts	Dietary fibers	–	Nawirska and Kwaśniewska (2005); Elleuch et al. (2011); O'Shea et al. (2012)
Orange byproducts	Dietary fibers	Fat replacer	(de Moraes Crizel et al., 2013)
Tomato seed and skin waste	Amino acid and mineral salt content	–	Tsatsaronis and Boskou (1975)
Tomato paste waste	Lycopene	–	Jun (2006)
Olive mill wastewater	Hydroxytyrosol	Increases the antioxidant capacity of rat plasma	Visioli et al. (2001)

TABLE 14.2 *(Continued)*

Source/Byproducts	Byproduct Components	Application/ Beneficial Effect	Reference
Flax processing waste (rich in cuticles)	Plant wax, policosanol	Effective against hypertension and type II hypercholesterolemia	Castaño et al. (2002); Morrison III et al. (2006)
Eggplant peels	Anthocyanins, other phenolic compounds	–	Ferarsa et al. (2018)
Shrimp shell waste	Chitin, chitosan, and its oligomers	Antibacterial activity	Benhabiles et al. (2012)
Crawfish shell waste	Chitin	–	No et al. (1989)
Herring (*Clupea harengus*) Byproduct	Protein hydrolysates	Antioxidant activity	Sathivel et al. (2003)
Fish waste	Unsaturated fatty acids, Omega-3/6/9 fatty acids	Reduction in atopy risk leading to reduced allergies	Nges et al. (2012); Kremmyda et al. (2011)
Shrimp waste	Astaxanthin	Antioxidant	López et al. (2004)
Blue crab shell waste	Astaxanthin	–	Felix-Valenzuela et al. (2001)
Shrimp waste	Carotenoprotein complex	–	Armenta-López et al. (2002)
Carrot pulp	Carotenoid	–	Chen and Tang (1998)
Tomato waste	Carotenoid	–	Strati and Oreopoulou (2011)
Date seed oil	Carotenoid	–	Habib et al. (2013)
Fish waste (skin, head, viscera, trimmings, frames, bones, and roes)	Fish protein hydrolysate	Antioxidant, antihypertensive, immunomodulatory, and antimicrobial peptides	Chalamaiah et al. (2012)
Salmon backbones	Protein hydrolysates	Angiotensin converting enzyme inhibition activity, antioxidant activity	Slizyte et al. (2016)

Saponins are another group of biochemicals, which are mainly amphipathic glycosides. Aqueous solutions of these compounds lead to soap-like foaming. The beneficial bioactivity of several popular nutraceutical-rich food including *Asparagus officinalis* L. and soybean have been correlated with high concentration of saponins, which include antifungal, antiviral, and antitumoral activities (Wang et al., 2011a). Other bioactive effects include protective effect on PC12 cells, neuroprotective effect on spinal cord neurons *in vitro* (Liao et al., 2002), inhibitory effects against the infectivity of the AIDS virus (HIV) and the activation of Epstein-Barr virus early antigen (Tsukamoto et al., 1994; Zhang et al., 2012). Saponins are mainly known for their beneficial effects; however, many saponins have highly undesirable taste. Hence, in novel nutraceutical formulations, addition of food byproducts and corresponding saponin extracts should be supplemented with flavor additives for acceptability and marketability of developed products.

Carotenoids are mainly isoprenoid metabolites, which contribute significantly to the color quality of different food commodities. Studies on extraction and application of carotenoids from food industry waste and byproducts have concentrated on lycopene, carotene, astaxanthin, and corresponding complexes and derivatives (Table 14.2). These pigments contribute significantly to the photoprotective effect against photo-oxidative damage in plants, and in communication of plants and animals, including attraction of pollinators and seed dispersing by animals. These functions are also supported by the carotenoid cleavage derivatives and corresponding volatile aromas (Concepcion et al., 2018). Beneficial effects for humans include antioxidant effect (Naguib, 2000), anticancer, antiobesity, antidiabetic, anti-inflammatory, and cardioprotective activity (Chuyen and Eun, 2017), which are being increasingly investigated in case of development of carotenoid rich nutraceuticals and/or food commodities. Carotenoids from different byproducts can be significant sources of nutritional pigments, which add aroma, color, and nutritional properties to the developed commodities.

Dietary fibers can be considered as a broad category of polysaccharide which includes both soluble and insoluble fibers. It is an integral part of the cell wall of biomolecules. Cereal processing waste such as oat and barley bran (which also contain oils) and pectin extracted from several fruit residues have been applied as dietary fibers. Vegetable wastes have also been applied as the dietary fiber additives. These components can be applied as texturizers for both solid foods, emulsions, and beverages. These can also be applied as fat and flour replacers in several products.

Dough prepared with dietary fibers have been observed to have increased water absorption, decreased dough stability, decreased extensibility, and reduced peak viscosity (PV) of the dough along with reduced bulk volume. Apple pomace (AP) and tomato waste have been applied in several cereal-based products, where they provided these properties due to the higher insoluble dietary fibers (O'Shea et al., 2012; Routray and Orsat, 2017). In commercial products such as pork based products, cookies, ice cream and muffins, dietary fibers from various byproducts including apricot kernel fiber (Seker et al., 2010), rice Bran fiber (Choi et al., 2010), orange byproducts (de Moraes Crizel et al., 2013) and cocoa fiber (Martínez-Cervera et al., 2011), were applied as fat replacers. Sensory and physiochemical properties of the products have also been explored in some products with fortified food byproducts, where percentage of dietary fiber fortification can be optimized or proportion can be controlled to have minimum effect on the desirable properties (Vasantha Rupasinghe et al., 2009). As observed in case of fortification of orange fiber in yogurt, gel structure and viscosity can be modulated, and targeted gelling and thickening effect can be obtained through application and subsequent pasteurization, with desirable sensory properties (Sendra et al., 2010). In case of beverages, dietary fiber with gritty texture are avoided, as it affects the mouthfeel. Hence along with the enhancement of the nutritional properties, these components can increase the overall quality of products.

Pectin and CS are mainly categorized as prebiotic dietary fibers, which have been extensively extracted from the plant and crustacean byproducts (Table 14.2). Prebiotics are mainly described as non-digestible food component beneficially affecting host, through selective stimulation of the growth and activity of targeted colon bacteria (Gibson and Roberfroid, 1995). Prebiotics can also affect the activity of antioxidant bioactive compounds and the bioavailability of bioactives, as observed in case of pectin addition in test food matrix, leading to decreased bioavailability of ß-carotene (Palafox-Carlos et al., 2011; Routray and Orsat, 2017). Hence, studies concerning formulation with fortified dietary fibers should optimize the composition of individual dietary fiber and targeted food matrix for obtaining maximum bioactivity.

Plant and animal byproducts digestion can produce *bioactive peptides and protein hydrolysates*. Grains (flax, canola, oat, hemp, wheat), pulses, and animal based food products (milk, fish, crustaceans, meat) wastes are the major sources of protein and associated derivatives (Tables 14.2 and 14.3). Apart from the observed bioactive properties (angiotensin-converting-enzyme inhibitory, hypocholesterolemic, antioxidant effects, immune-modulatory effect, hypotensive effect, cytomodulatory, antimicrobial effects, antithrombotic effect, and anticarcinogenic effect) of proteins and derivatives (peptides and hydrolysates), the bioavailability of amino acids and peptides has been reported to be higher than proteins (Routray and Orsat, 2017; Udenigwe and Aluko, 2012). Also, some amino-acids such as glutamic acid and aspartic acid can be added as flavor components, which can lead to Umami flavor (Zhang et al., 2015). Collagen is another significant protein, which is an integrated constituent of many animal tissues. Gelatin derived from collagen is a major gelling agent utilized in nutraceutical and pharmaceutical formulations.

Enzymes can be extracted from the viscera of various organisms, which can be further applied for enzymatic extraction of other bioactives. Many proteases have been efficiently isolated from fish waste. Trypsin was isolated and characterized from the pyrolic caeca of *Sardinops sagax* caerulea (Castillo-Yánez et al., 2005), the digestive system of carp *Cirrhinus mrigala* (Khangembam and Chakrabarti, 2015), viscera of vermiculated sailfin catfish, *Pterygoplichthys disjunctivus* (Villalba-Villalba et al., 2013) and *Luphiosilurus alexandri* pyloric cecum. New alkaline trypsin was isolated from the viscera of *Liza aurata* and corresponding structural features were studied to explain thermal stability (Bkhairia et al., 2016). In a different study by Balti et al. (2009), heat stable trypsin was extracted from hepatopancreas of the cuttlefish (*Sepia officinalis*). Several other fish waste (viscera) sources have been exploited for protease extraction. Lipase fraction has also been extracted from the fish viscera, as reported in case of grey mullet (*Mugil cephalus*). Enzymes from biological byproducts are preferred and are being increasingly explored for further application in food and nutraceutical industry.

Solid and liquid byproducts can be utilized as the growth medium for probiotic bacteria. Banana peel was utilized as growth medium for *Lactobacillus* species (Farees et al., 2017), and barley spent grain was used as growth media of *Bifidobacterium adolescentis* 94BIM and *Lactobacillus* sp. (Novik et al., 2007). In a different study, increased probiotic growth was reported in yoghurt fortified with pineapple waste powder (Sah et al., 2016), which further supports the byproduct fortification in food. Also, these probiotic bacteria can be identified and isolated from these byproducts. Siderophoregenic *Bacillus* spp. was isolated from dairy waste, which displayed high probiotic characteristics and the corresponding spores demonstrated antimicrobial activity against *Escherichia coli, Micrococcus flavus*, and *Staphylococcus aureus* (Patel et al., 2009).

14.3.1 APPLICATIONS AND TYPES OF CONSUMED COMMODITIES

Bioactive compounds can be produced as fractions in both liquid and solid states. Liquids can be (a) crude extracts dispersed in the water of the biomaterial or added water; (b) extracts in other solvents including organic solvents (alcohols and acids such as ethanol, citric acid and acetic acid), which can be easily added to the targeted food commodity/beverage; (c) extract stored in mediums such as oils, which can be either added in other commodities or can be directly consumed. The final dispersing medium is generally decided based on the resulting stability, expected intake concentration and ease of biosorption. Astaxanthin has been extracted with oil, which has been advocated as an efficient consumer friendly medium. For example, dispersion of astaxanthin in soy oil can add to the benefits of consumption of PUFAs during the frying or formulations with oil application (Chen and Meyers, 1982; Sachindra and Mahendrakar, 2005). Furthermore, traditional food are being fortified with these byproduct extracts leading to culinary items with taste and added health benefits, as observed in case of sausages with grape seed extracts, analyzed by Aminzare et al. (2018).

Many food industry byproducts can be further processed (chopped or ground) and dried to directly consume them as chips with bioactive properties or utilize as ingredient or additive to obtain fortified products. Furthermore, bioactive extracts obtained through extraction can be added as an ingredient of food recipe. Also, bioactive extracts can be encapsulated with appropriate encapsulating agents for efficient bioactivity in different targeted organisms. These encapsulated forms can be applied as food ingredients or directly consumed (Ghorbanzade et al., 2017). Spray dried mulberry waste extract were encapsulated in CS coated liposomes, which was fortified in dark chocolate. This fortified chocolate displayed positive effects, as reported through bioaccessibility studies (Gültekin-Özgüven et al., 2016). Some biomolecules such as fish oil with omega-3/6 fatty acids are also consumed in forms of tablets with biocompatible coatings.

Wax extracted from sources such as ash gourd peel, flax waste and sugarcane peel can be applied as an edible coating material for fruits, which increase the shelf life of these fruits along with the added health benefits of consumption of these components (Gnanaraj, 2012; Sreenivas et al., 2011). Also, polysaccharides such as CS, extracted from various byproducts can be utilized for encapsulating applications for controlled dispersion of different extracted bioactive compounds, which leads to corresponding coagulated beneficial effects (Gültekin-Özgüven et al., 2016). Furthermore, hybrid nanocomposites prepared from combination of CS/gelatin with silver nanoparticles (NPs) can be utilized for active food packaging, which have been observed to increase the shelf life of fruits by 2 weeks (Kumar et al., 2018).

14.3.2 METHODS OF PRETREATMENTS, EXTRACTION, AND PRODUCTION

Valorization of byproducts or waste involves several processing steps which are broadly categorized as (a) pretreatment and pre-processing, (b) extraction and separation, (d) characterization, identification and quantification. Drying is one of the processing methods often applied as pretreatment, which increases the shelf-life and provides convenience of storage at room temperature under unavailability of low temperature storage conditions. Furthermore, freeze-drying is generally applied for processing and preservation of thermo-sensitive compounds in biomaterials. Maceration, grinding methods and centrifugation are some of the other processing methods often applied as pretreatments, to increase the contact surface area and segregation or segmentation. Biomaterials, extracts, intermediates, and final products are generally stored at low temperatures and for long-term storage, samples are stored at –20 or –80°C.

Solvent extraction method is one of the most commonly applied method of extraction, which is often coupled with heat application and mixing through magnetic stirring, aeration or other mixing methods, to increase cellular disruption and increase contact surface area of the targeted compounds and solvents of dispersing medium. Solvents are chosen based on the physiochemical properties of the targeted compounds, the nature of the solvents, solvability, and interaction of these compounds with the dispersing medium and other specific properties, such as polarity of the solvent and compounds. Phenolic compounds are often extracted with polar organic solvents such as ethanol or methanol, organic solvents mixed with strong or weak acids such as HCl, acetic acid or citric acid. Recently, eutectic solvents such as choline chloride derivative-based and glycerol/glycine deep eutectic liquids have been employed for efficient ecofriendly extraction of phenolic compounds from agricultural byproducts, such as *Olea europaea* leaves (Alañón et al., 2018; Athanasiadis et al., 2017). Unexplored and novel sources of fats are exploited at lab-scale using hexane as a solvent with Soxhlet extraction method or extraction at room temperature with magnetic stirring. However, extraction through heat application followed by centrifugation is still a major commercial method of extraction, applied for large-scale production of bioactive fatty acids with omega-3/6 fatty acids. Advanced solvent extraction methods including microwave extraction, radio-frequency assisted extraction and ultrasound assisted extraction can be employed as extraction methods, and also as pretreatment methods, and/or one of the sequential steps of the complete extraction process, leading to efficient separation of the targeted compounds. Extraction efficiency of most compounds generally increase with the application of microwave and ultrasound (Cravotto et al., 2004; Ma et al., 2014). Phenolic compounds were extracted with ultrasound-assisted extraction from orange peel (Khan et al., 2010), wheat bran (Wang et al., 2008), and grape seeds (Ghafoor et al., 2009). Microwave-assisted extraction has been studied and optimized for extraction of phenolic compounds from almond skin (Valdés et al., 2015) peanut skin (Ballard et al., 2010), blueberry leaves (Routray and Orsat, 2014), soybean, and rice Bran oil (Terigar et al., 2011) and also several other bioactive compounds from different biomaterials. Some of the specific benefits of these methods include minimization of solvent volume, energy requirement and time of extraction, along with maximum extracted concentrations of the targeted compounds. Subcritical solvent extraction and high-pressure solvent extraction are some of the other solvent-based extraction methods applied for bioactive compounds extraction (Vergara-Salinas et al., 2013; Wijngaard and Brunton, 2009). All these methods lead to heating of the solvent-biomaterial mixture, which occurs in (a) microwave extraction and radio-frequency, caused by disruption of hydrogen bonds through changing dipole moment and ionic conduction (b) ultrasound extraction, caused by continuous vibration, and (c) pressurized fluid extraction (including accelerated fluid extraction), caused due to associated effect of high pressure leading to high temperature. Furthermore, apart from the disruptive effects of heat, cellular disruption with increased solvent penetration has also been achieved through increased pressure in the biological cells (Naviglio et al., 2008; Plaza et al., 2013; Routray and Orsat, 2012; Shirsath et al., 2012). There are different research groups worldwide which are further improving and augmenting the process parameters (application of low polarity water or controlled pH) to increase extraction efficiency for specific biomaterials (Karacabey et al., 2012; Kim and Mazza, 2009), and/or make the process applicable for multiple biomaterials.

Supercritical fluid (SFC) extraction has been applied for extraction of thermosensitive components and generally non-polar compounds, which can also be efficiently extracted through non-polar solvents, such as hexane. Bioactive fatty acid formulations such as krill oil or salmon and other fish waste oils have also been extracted with supercritical extraction with CO_2 solvent (Ali-Nehari et al., 2012; Rubio-Rodríguez et al., 2012). Fractionation of fish oils to produce omega-3 fatty acids rich fractions has also been efficiently achieved through supercritical CO_2 extraction (Corrêa et al., 2008; Rubio-Rodríguez et al., 2008). Similarly, waxes with bioactive properties have also been extracted using hexane and supercritical extraction with CO_2 from flax processing waste (Morrison III et al., 2006). Efficient extraction of astaxanthin from complex biomaterials such as crab and shrimp waste has also been achieved through supercritical extraction (Felix-Valenzuela et al., 2001; López et al., 2004). Apart from non-polar compounds, extraction of polar compounds has been achieved through utilization of alcohol (especially ethanol) as a co-solvent, as observed in case of extraction of phenolics from grape seed (Murga et al., 2000), and lycopene from tomato processing waste (Baysal et al., 2000). Addition of co-solvents, including ethanol or oil modifies the density of supercritical CO_2, which can further increase the extraction efficiency (Rozzi et al., 2002).

Most of these advanced solvent-based extraction methods have been employed at lab-scale. However, for commercial application, installation, and maintenance of the process equipment and subsequent parts for the unit-operations, along with the ease of maintenance of optimized process parameters, are some of the major considerations. The advanced methods mentioned above are still being further explored; however, these methods have been rarely applied commercially. Other than economic considerations, in certain cases, samples need preparation before extraction, such as drying before supercritical extraction. Hence, solvent extraction methods with organic solvents including alcohols, acids, bases are still some of the most widely applied methods.

Enzymatic extraction is another method widely applied for commercial extraction of several compounds. During past decades, enzymes have been extensively employed for extraction or for pretreatment of biomaterials with complex compounds (Villanueva-Suárez et al., 2013), including biofibers (Reddy and Yang, 2005), pectins (Panouillé et al., 2006), phenolics (Li et al., 2006). Protein hydrolysates are preferably prepared through enzymatic extraction for the retention of its effective properties (Chalamaiah et al., 2012), where bioactivities of protein hydrolysates is also affected by the type of enzyme along with the biomaterial. Among several enzymes used for production of fish protein hydrolysates (FPHs), highest angiotensin converting enzyme inhibition activity was observed for trypsin (Slizyte et al., 2016). Further enhancement of the beneficial properties of protein hydrolysates have been achieved through fat removal prior to enzymatic hydrolysis. Measured bioactivities were correlated by Slizyte et al. (2016) to the degree of hydrolysis and molecular weight profiles, which showed that higher bioactivities could be obtained through prolonged hydrolysis. Also, application of combination of multiple enzymes at different proportions led to increased production of targeted compounds in several cases (Panouillé et al., 2006; Reddy and Yang, 2005). Pectin was extracted through cellulase and protease preparations (Panouillé et al., 2006), and biofibers from byproducts have been extracted through combined application of pectinases, hemicellulases, and cellulases along with pre and post chemical applications (Reddy and Yang, 2005).

Furthermore, unconventional approach of extraction such as extraction of pectin from byproducts with acidified fruit juices (Masmoudi et al., 2008) or astaxanthin with bioactive edible oils (Chen and Meyers, 1982) can provide the combined beneficial effects of the solvent and source biomaterials. Acidic fermentation and microbial digestion followed by other separation/extraction techniques can also be applied for bioactive compounds extraction (Khanafari et al., 2007; Kleekayai et al., 2015; Vergara-Salinas et al., 2013). Novel non-thermal extraction method, such as pulse-electric field extraction can also be explored for bioactive compounds extraction from food waste (Gachovska et al., 2013).

Biomaterials are complex molecules, which consist of a combination of various biomolecules. Hence, biorefinery approach and sequential application of different solvents and or extraction methods can lead to efficient and sequential extraction of various compounds. Astaxanthin and bioactive peptides or carotenoprotenoid complex can be extracted through combination of acid extraction or digestion, fermentation, and further solvent extraction or enzymatic extraction (Armenta-López et al., 2002). This can be supported by the fact that biomolecules are often present as complexes, which needs biodegradation before extraction. Fish byproducts can be exploited for the combined extraction of fatty acids and bioactive peptides as demonstrated by Kang et al. (2005). Phenolics and dietary fiber extraction from date seeds was optimized by Al-Farsi and Lee (2008) through solvent extraction. Date seed oil with carotenoids and fat-soluble vitamins can also be extracted (Habib et al., 2013). Hence further studies regarding combined extraction and biorefinery approach for bioactive compounds exploitation should be encouraged.

Post-extraction separation and purification methods have also been employed for extract concentration and preparation of purified extracts. Centrifugation and filtration methods such as ultrafiltration methods have been applied for concentration of the targeted compounds (Amado et al., 2015; Armenta-López et al., 2002; Manoj et al., 2009). Solid phase extraction has been applied for separation of bioactive phenolics (Rodrıguez et al., 2000). Chromatographic techniques are still some of the most applied methods of separation and purification, applied for wide range of bioactive compounds, where high-performance liquid chromatography (HPLC) and thin layer chromatography are some of the most exploited purification techniques for various bioactive biomolecules (Bieleski and Turner, 1966; Cao et al., 2009; Panfili et al., 2004). However, centrifugation, filtration, and osmotic separation can be identified as some of the economically viable methods applicable at larger scale.

Freeze-drying and spray drying are some of the most popular methods for preparation of encapsulations. Encapsulating agents are selected based on (a) compatibility with targeted bioactive compounds, (b) stability of biomolecules in respective encapsulations under appropriate storage conditions, (c) efficiency of biocompatibility of the compounds and the encapsulating agents in the targeted parts of organisms. In case of Whitemouth croaker (*Micropogonias furnieri*) byproducts, encapsulated protein hydrolysates were obtained using phosphatidylcholine as the wall material, where properties evaluated included particle size, polydispersity, encapsulation efficiency, zeta potential, morphology, thermal properties and antioxidant activity (da Rosa Zavareze et al., 2014). In a different study by Mohan et al. (2016), whey peptides were encapsulated within soy lecithin-derived nanoliposomes, where similar properties were studied.

14.4 COMPONENTS FOR OTHER MEDICINAL APPLICATIONS

Biomolecules, biofibers, and derived biochemical and structural complexes can be employed for other biotechnological and biomedical applications. Several components obtained from animal and plant sources along with their components and applications have been summarized in Table 14.3 and briefly discussed in following subsections. Methods for preparation of medicinal components and aids have been briefly mentioned in the subsections; however further detailed account is beyond the scope of this chapter.

TABLE 14.3 Agro and Food Industry Waste with Applications or Potential Applications in Medical and Other Related Fields

Source/ Byproduct	Application	Reference
Crab shells	Chitin and chitosan derived complexes	Jo et al. (2008); Hamdi et al. (2017)
Shrimp shells	Chitin and chitosan derived complexes	Hamdi et al. (2017); Maruthiah and Palavesam (2017)
Fish waste	Edible gelatin films with curcumin	Musso et al. (2017)
Animal processing waste	Gelatins as bioactive compounds carrier (Active packaging)	Etxabide et al. (2017)
Salmon skin	Gelatin with boldine from external source	López et al. (2017)
Salmon skin and heart	Collagen and elastin sponge for biomedical application	Matsumoto et al. (2011)
Fish waste (skin, bone, and fins)	Collagen	Nagai and Suzuki (2000)
Cuttlefish (*Sepia lycidas)* skin waste	Collagen	Nagai et al. (2001)
Sepia pharaonic outer skin	Type I Collagen	Krishnamoorthi et al. (2017)
Catfish skin (*Pangasianodon hypophthalmus*)	Collagen	Singh et al. (2011)
Marine eel-fish (*Evenchelys macrura*) skin	Thermostable collagen	Veeruraj et al. (2013)
Red drum fish (*Sciaenpos ocellatus*) scales	Pepsin-soluble type I collagen	Chen et al. (2016b)
Squid (*Doryteuthis singhalensis*) outer skin	Collagen	Veeruraj et al. (2015)
Tilapia (*Oreochromis* sp.) scale	Collagen	Huang et al. (2016)
Tilapia (*Oreochromis niloticus*) skin and scales	Collagen	Chen et al. (2016a)
Oil palm black liquor waste	Lignin with starch-based biopolymer utilized for production of packaging films with improved thermochemical and barrier properties	Bhat et al. (2013)
Agro-industry waste	Drug carriers	A Joanitti and P Silva (2014)
Pineapple leaves	Cellulose nanocrystals; medical applications	dos Santos et al. (2013); Cherian et al. (2011)

TABLE 14.3 *(Continued)*

Source/ Byproduct	Application	Reference
Corncob	Cellulose nanocrystals	Silvério et al. (2013)
Pandanus tectorius leaves	Cellulose nanocrystals	Sheltami et al. (2012)
Soy hulls	Cellulose nanocrystals	Neto et al. (2013)
Rice Bran	Wax as ointment base	Sabale et al. (2009); Bhalekar et al. (2004)
Citrus sinensis peel extract	Reducing and capping agent in biosynthesized silver nanoparticles with antibacterial activity against *Escherichia coli*, *Pseudomonas aeruginosa*, and *Staphylococcus aureus*	Kaviya et al. (2011)
Olive leaf extract	Silver nanoparticles with antibacterial activity	Khalil et al. (2014)
Mangosteen leaf extract	Silver nanoparticles with antibacterial activity	Veerasamy et al. (2011)
Lemon peels extract	Silver nanoparticles with antidermatophytic activity	Nisha et al. (2014)
Fish scales	Gelatin–pectin coacervate	Huang et al. (2017)
Salmon blood	Purified thrombin and fibrinogen for Hemostatic bandages	Debes et al. (2015)

14.4.1 NANOMOLECULES

Plant extracts have been successfully employed for the synthesis of NPs with several metals including "cobalt, copper, silver, gold, palladium, platinum, zinc oxide and magnetite," which can be efficiently utilized for treatment of various diseases (Kuppusamy et al., 2016). Silver NPs have been produced with green synthesis, where plant byproduct extracts were applied as reducing agents and capping agents. Some of the byproducts utilized for preparation of silver NPs include *Citrus sinensis* peel extract (Kaviya et al., 2011), olive leaf extract (Khalil et al., 2014) and mangosteen leaf extract (Veerasamy et al., 2011). Gold nanoparticles (AuNP) have been prepared from grape waste extracts through green synthesis techniques (Krishnaswamy et al., 2014). These NPs were reported to have potential antibacterial activity (Khalil et al., 2014).

NPs are also useful for development of drug delivery system. NPs with colloidal systems can be utilized for localized or targeted delivery of the drugs to the targeted tissue (Abdelwahed et al., 2006). Some of the other advantages of NPs in case of drug delivery include: improvement of oral bioavailability, sustenance of drug effect in target tissue, improved solubilization of drugs for intravascular delivery, and improved stability against enzymatic degradation (Abdelwahed et al., 2006).

Some of these components can also be applied as the base material or mixed with base material for preparation of ointments. These can also be mixed with oils for efficient skin application and absorption. Silver NPs synthesized with lemon peels extracts can be utilized for the treatment of dermatophytes. Generally, dermatophytes have been reported to develop drug resistance against broad-spectrum antibiotics, which are commonly applied for the treatment (Nisha et al., 2014).

14.4.1.1 PROPERTIES AFFECTING THE FORMATION AND EFFECTIVITY

Conventional methods of NPs synthesis include microwave irradiation, electro-irradiation, and strong chemical reduction. Recently plant extracts have been reported as a potential reductant, as observed in case of silver NPs synthesis using *Atrocarpus altilis* leaf extract and flavonoids fractions of *Psidium guajava* leaves (Ravichandran et al., 2016; Wang et al., 2018). Also, nanoformulations are formed through efficient homogenization and encapsulation methods such as freeze-drying and spray drying followed by size reduction methods. Freeze drying has been applied in case of colloidal NPs, which improves the long-term stability that overcomes the low stability of aqueous medium of these samples, which can form a barrier for the clinical applications of the NPs (Abdelwahed et al., 2006). Furthermore, freeze drying provides the solid form of the dosage, which can be intended for various administration routes and is suitable for different analytical measurements.

Several factors affect the formation and synthesis of these compositions. Temperature is a major factor and temperature levels such as 25 and 60°C have been extensively applied (Kaviya et al., 2011). Other parameters which affect the bioactivity and bioavailability of these NPs, include properties of the colloidal polymeric carriers, pH of the aqueous dispersion and chemical stability of the entrapped drug or biomolecule (Abdelwahed et al., 2006). For preparation of solid lipid NPs and complex protein NPs, phospholipids, and triglycerides along with protein derivatives extracted from various waste sources, can be potential raw materials (Bakshi, 2011; Matsuno and Ishihara, 2009; Young et al., 2004).

Stability of these biosynthesized NPs can be analyzed through the spectroscopic characterization using UV-visible, Fourier transform infrared (FTIR) spectroscopy, energy dispersive X-ray analysis and transmission electron microscopy methods (Kaviya et al., 2011; Nisha et al., 2014). Morphologies and compositions have also been studied through XRD (Basavegowda and Lee, 2013).

14.4.2 SCAFFOLDS AND TISSUE ENGINEERING APPLICATIONS

Three-dimensional scaffolds have been designed using bone tissue engineering and are applied for the production of bone grafts. According to the review by Balagangadharan et al. (2017), major targeted properties considered for these projects include (a) ability to mimic the extracellular matrix, (b) ability to provide mechanical assistance, (c) assistance in formation of new bone, (d) possession of osteo-conductive, osteo-inductive, and osteogenic properties, which promote adhesion along with survival and migration of osteogenic cells, (e) ability to provide the physical and biochemical parameters for inducing osteoblastic lineage, (f) mechanical properties for the determination of site of action, (g) degradation after the formation of natural tissues (Balagangadharan et al., 2017).

Though scaffolds are prepared using various techniques including fiber bonding, melt molding, solvent casting, gas foaming and phase separation, electrospinning is the desired technique for the preparation of nano-scale and microscale fibers. These fibers have the advantage of resemblance to the native components of extracellular matrix (Balagangadharan et al., 2017). Other unique properties of the prepared nanofibers include high surface area to volumeratio, porosity, stability, and permeability. CS is one of the natural polymers, which has been increasing exploited for application in this field, as nanofibers. This compound is abundantly present in crustacean shells, which can be utilized for nanofiber preparation (Balagangadharan et al., 2017). Chitin or CS can also be utilized for preparation of composite scaffolds with nanoceramics, which have been observed as an highly applicable composite for bone tissue engineering (Deepthi et al., 2016).

14.4.3 MEDICAL IMPLANTS

Nanofibers obtained from the food industry byproducts can be utilized for preparation of nanocomposites and applied for production of various biomedical aids. Cherian et al. (2011) isolated nanofibers from pineapple leaf fibers, and utilized them for preparing nanocomposites with polyurethane, through compression molding of stacks of nanocellulose fiber mats betweenpolyurethane films. The developed nanocomposites could be used for fabrication of multipurpose medical implants (Cherian et al., 2011), with enhanced strength. In a different study, where nanocellulose from pineapple leaves were extracted with steam explosion and steam coupled with acid treatment, the prepared nanocellulose was reported to be suitable for wide range of applications including tissue engineering, drug delivery, wound dressing and medical implants (Cherian et al., 2010). Cellulose nanocrystals can also be extracted from coffee husk and rice husk, (Collazo-Bigliardi et al., 2018), corncob (Silvério et al., 2013), mengkuang leaves (Sheltami et al., 2012), and soy hulls (Neto et al., 2013), which can be employed in similar applications.

14.4.4 COLLAGEN AND ITS DIVERSE APPLICATIONS

Collagen is a major biomolecule which can be extracted from several animal sources including fishes and mammals. It is also a component which is widely applied in several drug delivery systems and other biomedical applications, attributed to its physical and structural properties, low immunogenicity/antigenicity, natural turnover, and biodegradability. Detailed description about the interactions of collagen leading to efficient drug design and delivery has been reported by several authors (An et al., 2016; Friess, 1998). Collagen is also a major natural composite suitable for application in bone-tissue engineering and bone regeneration. Composite scaffolds of bioceramics and collagen/gelatin are also being increasingly employed (Kuttappan et al., 2016). Collagen is also useful as "controlling material for transdermal delivery" (Lee et al., 2001). Fish scales as observed in case of red drum fish (*Sciaenopus ocellatus*) (Chen et al., 2016b), tilapia (*Oreochromis* sp.) (Huang et al., 2016), outer skin of *Sepia pharaonis* (Krishnamoorthi et al., 2017), cuttlefish (Nagai et al., 2001), outer skin of squid (*Doryteuthis singhalensis*) (Veeruraj et al., 2013) and salmon skin (Matsumoto et al., 2011) have been proven as potential sources of collagen, where unconventional method such as extrusion hydro extraction (Huang et al., 2016) was explored for obtaining high quality collagen suitable for multiple applications. Collagen with requisite properties can also be isolated depending on the specific sources, such as thermostable collagen from the marine eel-fish (*Evenchelys macrura*) (Veeruraj et al., 2013). In a different study, isolation of collagen from skin, bone, and fins of several varieties of fish has also been reported, which emphasizes fishery byproducts as a major source of this compound. Furthermore, utilization of mammal-derived collagen is limited by the effect of conceivable prion disease in mammals, which doesn't occur in fishes and further emphasizes their usability as artificial dermis and as a scaffold for tissue engineering (Matsumoto et al., 2011).

14.5 FUTURE PROSPECTS OF THE BYPRODUCT BASED INDUSTRY AND CONCLUSION

Development of byproducts based nutraceutical and biomedical applications industries, is still a novel concept. The interest and research in this regard has increased and new techniques are being developed and optimized. However, depending on the market value and demand, purity, nature, and complexity of the source biomatrix and targeted compounds, pilot scale batch and continuous processes can be established. The major components widely explored in current research projects were summarized in this chapter; however, future research can also focus on other medicinal compounds. Application of biological sources for preparation of biomedical devices, scaffolds, and versatile NPs should be further promoted, as these are substantially more biocompatible and biodegradable, as compared to chemical sources. Overall, the worldwide scientific community should encourage further research in the exploitation of these resources and development of state-of-the-art facilities for efficient application of these resources.

KEYWORDS

- **drug delivery**
- **functional composition**
- **functional property**
- **medicinal component**
- **tissue engineering**

REFERENCES

Abdelwahed, W., Degobert, G., Stainmesse, S., & Fessi, H., (2006). Freeze-drying of nanoparticles: Formulation, process and storage considerations. *Advanced Drug Delivery Reviews, 58*(15), 1688–1713.

Alañón, M., Ivanović, M., Gómez-Caravaca, A., Arráez-Román, D., & Segura-Carretero, A., (2018). Choline chloride derivative-based deep eutectic liquids as novel green alternative solvents for extraction of phenolic compounds from olive leaf. *Arabian Journal of Chemistry*. doi.org/10.1016/j.arabjc.2018.01.003.

Al-Farsi, M. A., & Lee, C. Y., (2008). Optimization of phenolics and dietary fiber extraction from date seeds. *Food Chemistry, 108*(3), 977–985.

Ali Asgar, M., (2013). Anti-diabetic potential of phenolic compounds: A review. *International Journal of Food Properties, 16*(1), 91–103.

Ali-Nehari, A., Kim, S. B., Lee, Y. B., Lee, H. Y., & Chun, B. S., (2012). Characterization of oil including astaxanthin extracted from krill (*Euphausia superba*) using supercritical carbon dioxide and organic solvent as comparative method. *Korean Journal of Chemical Engineering, 29*(3), 329–336.

Altemimi, A. B., (2017). A study of the protective properties of Iraqi olive leaves against oxidation and pathogenic bacteria in food applications. *Antioxidants, 6*(2), 34.

Amado, I. R., González, M., Murado, M. A., & Vázquez, J. A., (2015). Shrimp cooking wastewater as a source of astaxanthin and bioactive peptides. *Journal of Chemical Technology and Biotechnology, 91*, 793–805.

Aminzare, M., Tajik, H., Aliakbarlu, J., Hashemi, M., & Raeisi, M., (2018). Effect of cinnamon essential oil and grape seed extract as functional-natural additives in the production of cooked sausage-impact on microbiological, physicochemical, lipid oxidation and sensory aspects, and fate of inoculated *Clostridium perfringens*. *Journal of Food Safety, 38*, e12459

An, B., Lin, Y. S., & Brodsky, B., (2016). Collagen interactions: Drug design and delivery. *Advanced Drug Delivery Reviews, 97*, 69–84.

Anderson, J. W., Baird, P., Davis, J. R. H., Ferreri, S., Knudtson, M., Koraym, A., Waters, V., & Williams, C. L., (2009). Health benefits of dietary fiber. *Nutrition Reviews, 67*(4), 188–205.

Anraku, M., Fujii, T., Furutani, N., Kadowaki, D., Maruyama, T., Otagiri, M., Gebicki, J. M., & Tomida, H., (2009). Antioxidant effects of a dietary supplement: Reduction of indices of oxidative stress in normal subjects by water-soluble chitosan. *Food and Chemical Toxicology*, *47*(1), 104–109.

Anraku, M., Tomida, H., Michihara, A., Tsuchiya, D., Iohara, D., Maezaki, Y., Uekama, K., Maruyama, T., Otagiri, M., & Hirayama, F., (2012). Antioxidant and renoprotective activity of chitosan in nephrectomized rats. *Carbohydrate Polymers*, *89*(1), 302–304.

Armenta-López, R., Guerrero, I., & Huerta, S., (2002). Astaxanthin extraction from shrimp waste by lactic fermentation and enzymatic hydrolysis of the carotenoprotein complex. *Journal of Food Science*, *67*(3), 1002–1006.

Arruzazabala, M. L., Molina, V., Mas, R., Fernández, L., Carbajal, D., Valdés, S., & Castaño, G., (2002). Antiplatelet effects of policosanol (20 and 40 mg/day) in healthy volunteers and dyslipidaemic patients. *Clinical and Experimental Pharmacology and Physiology*, *29*(10), 891–897.

Athanasiadis, V., Grigorakis, S., Lalas, S., & Makris, D. P., (2017). Highly efficient extraction of antioxidant polyphenols from *Olea europaea* leaves using an eco-friendly glycerol/glycine deep eutectic solvent. *Waste and Biomass Valorization*, 1–8.

Bakshi, M. S., (2011). Nanoshape control tendency of phospholipids and proteins: Protein–nanoparticle composites, seeding, self-aggregation, and their applications in bionanotechnology and nanotoxicology. *The Journal of Physical Chemistry C*, *115*(29), 13947–13960.

Balagangadharan, K., Dhivya, S., & Selvamurugan, N., (2017). Chitosan-based nanofibers in bone tissue engineering. *International Journal of Biological Macromolecules*, *104*, 1372–1382.

Ballard, T. S., Mallikarjunan, P., Zhou, K., & O'Keefe, S., (2010). Microwave-assisted extraction of phenolic antioxidant compounds from peanut skins. *Food Chemistry*, *120*(4), 1185–1192.

Balti, R., Barkia, A., Bougatef, A., Ktari, N., & Nasri, M., (2009). A heat-stable trypsin from the hepatopancreas of the cuttlefish (*Sepia officinalis*): Purification and characterization. *Food Chemistry*, *113*(1), 146–154.

Basavegowda, N., & Lee, Y. R., (2013). Synthesis of silver nanoparticles using Satsuma mandarin (*Citrus unshiu*) peel extract: A novel approach towards waste utilization. *Materials Letters*, *109*, 31–33.

Baysal, T., Ersus, S., & Starmans, D., (2000). Supercritical CO_2 extraction of β-carotene and lycopene from tomato paste waste. *Journal of Agricultural and Food Chemistry*, *48*(11), 5507–5511.

Benhabiles, M., Salah, R., Lounici, H., Drouiche, N., Goosen, M., & Mameri, N., (2012). Antibacterial activity of chitin, chitosan and its oligomers prepared from shrimp shell waste. *Food Hydrocolloids*, *29*(1), 48–56.

Bhalekar, M., Manish, L., & Krishna, S., (2004). Formulation and evaluation of rice bran wax as ointment base. *Ancient Science of Life*, *24*(1), 52.

Bhat, R., Abdullah, N., Din, R. H., & Tay, G. S., (2013). Producing novel sago starch based food packaging films by incorporating lignin isolated from oil palm black liquor waste. *Journal of Food Engineering*, *119*(4), 707–713.

Bieleski, R., & Turner, N., (1966). Separation and estimation of amino acids in crude plant extracts by thin-layer electrophoresis and chromatography. *Analytical Biochemistry*, *17*(2), 278–293.

Bkhairia, I., Khaled, H. B., Ktari, N., Miled, N., Nasri, M., & Ghorbel, S., (2016). Biochemical and molecular characterization of a new alkaline trypsin from Liza aurata: Structural features explaining thermal stability. *Food Chemistry*, *196*, 1346–1354.

Cao, X., Wang, C., Pei, H., & Sun, B., (2009). Separation and identification of polyphenols in apple pomace by high-speed counter-current chromatography and high-performance liquid chromatography coupled with mass spectrometry. *Journal of Chromatography A, 1216*(19), 4268–4274.

Castaño, G., Más, R., Fernández, J. C., Fernández, L., Illnait, J., & López, E., (2002). Effects of Policosanol on older patients with hypertension and type II hypercholesterolemia. *Drugs in R&D, 3*(3), 159–172.

Castillo-Yánez, F. J., Pacheco-Aguilar, R., García-Carreño, F. L., & De los Ángeles Navarrete-Del, M., (2005). Isolation and characterization of trypsin from pyloric caeca of Monterey sardine *Sardinops sagax* caerulea. *Comparative Biochemistry and Physiology Part B: Biochemistry and Molecular Biology, 140*(1), 91–98.

Chalamaiah, M., Hemalatha, R., & Jyothirmayi, T., (2012). Fish protein hydrolysates: Proximate composition, amino acid composition, antioxidant activities and applications: a review. *Food Chemistry, 135*(4), 3020–3038.

Chen, B., & Tang, Y., (1998). Processing and stability of carotenoid powder from carrot pulp waste. *Journal of Agricultural and Food Chemistry, 46*(6), 2312–2318.

Chen, H. M., & Meyers, S. P., (1982). Extraction of astaxanthin plgment from crawfish waste using a soy oil process. *Journal of Food Science, 47*(3), 892–896.

Chen, J., Li, L., Yi, R., Xu, N., Gao, R., & Hong, B., (2016a). Extraction and characterization of acid-soluble collagen from scales and skin of tilapia (*Oreochromis niloticus*). *LWT-Food Science and Technology, 66*, 453–459.

Chen, S., Chen, H., Xie, Q., Hong, B., Chen, J., Hua, F., Bai, K., He, J., Yi, R., & Wu, H., (2016b). Rapid isolation of high purity pepsin-soluble type I collagen from scales of red drum fish (*Sciaenops ocellatus*). *Food Hydrocolloids, 52*, 468–477.

Cherian, B. M., Leão, A. L., De Souza, S. F., Costa, L. M. M., De Olyveira, G. M., Kottaisamy, M., Nagarajan, E., & Thomas, S., (2011). Cellulose nanocomposites with nanofibers isolated from pineapple leaf fibers for medical applications. *Carbohydrate Polymers, 86*(4), 1790–1798.

Cherian, B. M., Leão, A. L., De Souza, S. F., Thomas, S., Pothan, L. A., & Kottaisamy, M., (2010). Isolation of nanocellulose from pineapple leaf fibers by steam explosion. *Carbohydrate Polymers, 81*(3), 720–725.

Choi, I. S., Cho, E. J., Moon, J. H., & Bae, H. J., (2015). Onion skin waste as a valorization resource for the by-products quercetin and biosugar. *Food Chemistry, 188*, 537–542.

Choi, Y. S., Choi, J. H., Han, D. J., Kim, H. Y., Lee, M. A., Jeong, J. Y., Chung, H. J., & Kim, C. J., (2010). Effects of replacing pork back fat with vegetable oils and rice bran fiber on the quality of reduced-fat frankfurters. *Meat Science, 84*(3), 557–563.

Chung, Y. M., Wang, H. C., El-Shazly, M., Leu, Y. L., Cheng, M. C., Lee, C. L., Chang, F. R., & Wu, Y.-C., (2011). Antioxidant and tyrosinase inhibitory constituents from a desugared sugar cane extract, a byproduct of sugar production. *Journal of Agricultural and Food Chemistry, 59*(17), 9219–9225.

Chuyen, H. V., & Eun, J. B., (2017). Marine carotenoids: Bioactivities and potential benefits to human health. *Critical Reviews in Food Science and Nutrition, 57*(12), 2600–2610.

Collazo-Bigliardi, S., Ortega-Toro, R., & Boix, A. C., (2018). Isolation and characterization of microcrystalline cellulose and cellulose nanocrystals from coffee husk and comparative study with rice husk. *Carbohydrate Polymers, 191*, 205–215.

Concepcion, M. R., Avalos, J., Bonet, M. L., Boronat, A., Gomez-Gomez, L., Hornero-Mendez, D., Limon, M. C., Meléndez-Martínez, A. J., Olmedilla-Alonso, B., & Palou, A.,

(2018). A global perspective on carotenoids: Metabolism, biotechnology, and benefits for nutrition and health. *Progress in Lipid Research, 70,* 62–93.
Corrêa, A. P. A., Peixoto, C. A., Gonçalves, L. A. G., & Cabral, F. A., (2008). Fractionation of fish oil with supercritical carbon dioxide. *Journal of Food Engineering, 88*(3), 381–387.
Cravotto, G., Binello, A., Merizzi, G., & Avogadro, M., (2004). Improving solvent-free extraction of policosanol from rice bran by high-intensity ultrasound treatment. *European Journal of Lipid Science and Technology, 106*(3), 147–151.
Croft, K. D., (1998). The chemistry and biological effects of flavonoids and phenolic acids. *Annals of the New York Academy of Sciences, 854*(1), 435–442.
Da Rosa Zavareze, E., Telles, A. C., El Halal, S. L. M., Da Rocha, M., Colussi, R., De Assis, L. M., De Castro, L. A. S., Dias, A. R. G., & Prentice-Hernández, C., (2014). Production and characterization of encapsulated antioxidative protein hydrolysates from Whitemouth croaker (*Micropogonias furnieri*) muscle and byproduct. *LWT-Food Science and Technology, 59*(2), 841–848.
De Moraes, C. T., Jablonski, A., De Oliveira Rios, A., Rech, R., & Flôres, S. H., (2013). Dietary fiber from orange byproducts as a potential fat replacer. *LWT-Food Science and Technology, 53*(1), 9–14.
Debes, J., Elmongy, H., Keenan, S., & Kirby, K., (2015). *Purification of Pharmaceutical Grade Salmon-Derived Thrombin and Fibrinogen for Hemostatic Bandages*. Senior design reports (CBE), University of Pennsylvania.
Deepthi, S., Venkatesan, J., Kim, S. K., Bumgardner, J. D., & Jayakumar, R., (2016). An overview of chitin or chitosan/nano ceramic composite scaffolds for bone tissue engineering. *International Journal of Biological Macromolecules, 93*, 1338–1353.
Del Carmen Millán-Linares, M., Bermudez, B., Martín, M. E., Muñoz, E., Abia, R., Rodríguez, F. M., Muriana, F. J., & Montserrat-de la Paz, S., (2018). Unsaponifiable fraction isolated from grape (*Vitis vinifera* L.) seed oil attenuates oxidative and inflammatory responses in human primary monocytes. *Food & Function, 9,* 2517–2523.
Dos Santos, R. M., Neto, W. P. F., Silvério, H. A., Martins, D. F., Dantas, N. O., & Pasquini, D., (2013). Cellulose nanocrystals from pineapple leaf, a new approach for the reuse of this agro-waste. *Industrial Crops and Products, 50*, 707–714.
Ekaluo, U., Ikpeme, E., Ekerette, E., & Chukwu, C., (2015a). *In vitro* antioxidant and free radical activity of some Nigerian medicinal plants: Bitter leaf (*Vernonia amygdalina* L.) and guava (*Psidium guajava* Del.). *Research Journal of Medicinal Plant, 9*(5), 215–226.
Ekaluo, U., Ikpeme, E., Udensi, O., Ekerette, E., Usen, S., & Usoroh, S., (2015b). Comparative *in vitro* assessment of drumstick (*Moringa oleifera*) and neem (*Azadiracta indica*) leaf extracts for antioxidant and free radical scavenging activities. *J. Med. Plant, 9*, 24–33.
Elleuch, M., Bedigian, D., Roiseux, O., Besbes, S., Blecker, C., & Attia, H., (2011). Dietary fiber and fiber-rich by-products of food processing: Characterization, technological functionality and commercial applications: A review. *Food Chemistry, 124*(2), 411–421.
El-Mashad, H. M., & Zhang, R., (2010). Biogas production from co-digestion of dairy manure and food waste. *Bioresource Technology, 101*(11), 4021–4028.
Etxabide, A., Uranga, J., Guerrero, P., & De la Caba, K., (2017). Development of active gelatin films by means of valorization of food processing waste: A review. *Food Hydrocolloids, 68*, 192–198.
Farees, N., Abateneh, D. D., Geneto, M., & Naidu, N., (2017). Evaluation of banana peel waste as growth medium for probiotic Lactobacillus species. *International Journal of Applied Biology and Pharmaceutical Technology, 8*(4), 19–23.

Felix-Valenzuela, L., Higueral-Ciaparal, I., Goycoolea-Valencia, F., & Arguelles-Monal, W., (2001). Supercritical CO_2/ethanol extraction of astaxanthin from blue crab (*Callinectes sapidus*) shell waste. *Journal of Food Process Engineering, 24*(2), 101–112.

Ferarsa, S., Zhang, W., Moulai-Mostefa, N., Ding, L., Jaffrin, M. Y., & Grimi, N., (2018). Recovery of anthocyanins and other phenolic compounds from purple eggplant peels and pulps using ultrasonic-assisted extraction. *Food and Bioproducts Processing, 109,* 19–28.

Friess, W., (1998). Collagen-biomaterial for drug delivery. *European Journal of Pharmaceutics and Biopharmaceutics, 45*(2), 113–136.

Gachovska, T. K., Ngadi, M., Chetti, M., & Raghavan, G. V., (2013). Enhancement of lycopene extraction from tomatoes using pulsed electric field, Pulsed Power Conference (PPC), 19th IEEE. *IEEE*, pp. 1–5.

Ghafoor, K., Choi, Y. H., Jeon, J. Y., & Jo, I. H., (2009). Optimization of ultrasound-assisted extraction of phenolic compounds, antioxidants, and anthocyanins from grape (*Vitis vinifera*) seeds. *Journal of Agricultural and Food Chemistry, 57*(11), 4988–4994.

Ghorbanzade, T., Jafari, S. M., Akhavan, S., & Hadavi, R., (2017). Nano-encapsulation of fish oil in nano-liposomes and its application in fortification of yogurt. *Food Chemistry, 216*, 146–152.

Gibson, G. R., & Roberfroid, M. B., (1995). Dietary modulation of the human colonic microbiota: Introducing the concept of prebiotics. *The Journal of Nutrition, 125*(6), 1401–1412.

Gnanaraj, R. A., (2012). Applications of sugarcane wax and its products: A review. *Int. J. Chemtech. Res., 4*, 705–712.

Guan, Y., Tang, Q., Fu, X., Yu, S., Wu, S., & Chen, M., (2014). Preparation of antioxidants from sugarcane molasses. *Food Chemistry, 152*, 552–557.

Gültekin-Özgüven, M., Karadağ, A., Duman, Ş., Özkal, B., & Özçelik, B., (2016). Fortification of dark chocolate with spray dried black mulberry (*Morus nigra*) waste extract encapsulated in chitosan-coated liposomes and bioaccessability studies. *Food Chemistry, 201*, 205–212.

Habib, H. M., Kamal, H., Ibrahim, W. H., & Al Dhaheri, A. S., (2013). Carotenoids, fat soluble vitamins and fatty acid profiles of 18 varieties of date seed oil. *Industrial Crops and Products, 42*, 567–572.

Halim, N., Yusof, H., & Sarbon, N., (2016). Functional and bioactive properties of fish protein hydolysates and peptides: A comprehensive review. *Trends in Food Science & Technology, 51*, 24–33.

Hamdi, M., Hammami, A., Hajji, S., Jridi, M., Nasri, M., & Nasri, R., (2017). Chitin extraction from blue crab (*Portunus segnis*) and shrimp (*Penaeus kerathurus*) shells using digestive alkaline proteases from *P. segnis* viscera. *International Journal of Biological Macromolecules, 101*, 455–463.

Harborne, J. B., (1989). *Plant Phenolics*. Academic Press, London, San Diego.

Hassan, F. A., Ismail, A., Abdulhamid, A., & Azlan, A., (2011). Identification and quantification of phenolic compounds in bambangan (*Mangifera pajang* Kort.) peels and their free radical scavenging activity. *Journal of Agricultural and Food Chemistry, 59*(17), 9102–9111.

Huang, C. Y., Kuo, J. M., Wu, S. J., & Tsai, H. T., (2016). Isolation and characterization of fish scale collagen from tilapia (*Oreochromis* sp.) by a novel extrusion–hydro-extraction process. *Food Chemistry, 190*, 997–1006.

Huang, T., Tu, Z. C., Shangguan, X., Wang, H., Zhang, L., & Sha, X., (2017). Gelation kinetics and characterization of enzymatically enhanced fish scales gelatin–pectin coacervate. *Journal of the Science of Food and Agriculture.*

Hussein, G., Sankawa, U., Goto, H., Matsumoto, K., & Watanabe, H., (2006). Astaxanthin, a carotenoid with potential in human health and nutrition. *Journal of Natural Products, 69*(3), 443–449.

Inarkar, M. B., & Lele, S., (2012). Extraction and characterization of sugarcane peel wax. *ISRN Agronomy*, http://dx.doi.org/10.5402/2012/340158.

Jemai, H., El Feki, A., & Sayadi, S., (2009). Antidiabetic and antioxidant effects of hydroxytyrosol and oleuropein from olive leaves in alloxan-diabetic rats. *Journal of Agricultural and Food Chemistry, 57*(19), 8798–8804.

Jiménez-Escrig, A., Tenorio, M. D., Espinosa-Martos, I., & Rupérez, P., (2008). Health-promoting effects of a dietary fiber concentrate from the soybean byproduct okara in rats. *Journal of Agricultural and Food Chemistry, 56*(16), 7495–7501.

Jo, G., Jung, W., Kuk, J., Oh, K., Kim, Y., & Park, R., (2008). Screening of protease-producing *Serratia marcescens* FS-3 and its application to deproteinization of crab shell waste for chitin extraction. *Carbohydrate Polymers, 74*(3), 504–508.

Joanitti, A. G., & Silva, P. L., (2014). The emerging potential of by-products as platforms for drug delivery systems. *Current Drug Targets, 15*(5), 478–485.

Jun, X., (2006). Application of high hydrostatic pressure processing of food to extracting lycopene from tomato paste waste. *High Pressure Research, 26*(1), 33–41.

Kang, K. Y., Ahn, D. H., Jung, S. M., Kim, D. H., & Chun, B. S., (2005). Separation of protein and fatty acids from tuna viscera using supercritical carbon dioxide. *Biotechnology and Bioprocess Engineering, 10*(4), 315–321.

Karacabey, E., Mazza, G., Bayındırlı, L., & Artık, N., (2012). Extraction of bioactive compounds from milled grape canes (*Vitis vinifera*) using a pressurized low-polarity water extractor. *Food and Bioprocess Technology, 5*(1), 359–371.

Kaushik, G., Satya, S., Khandelwal, R. K., & Naik, S., (2010). Commonly consumed Indian plant food materials in the management of diabetes mellitus. *Diabetes & Metabolic Syndrome: Clinical Research & Reviews, 4*(1), 21–40.

Kaviya, S., Santhanalakshmi, J., Viswanathan, B., Muthumary, J., & Srinivasan, K., (2011). Biosynthesis of silver nanoparticles using *Citrus sinensis* peel extract and its antibacterial activity. *Spectrochimica Acta Part A: Molecular and Biomolecular Spectroscopy, 79*(3), 594–598.

Khalil, M. M., Ismail, E. H., El-Baghdady, K. Z., & Mohamed, D., (2014). Green synthesis of silver nanoparticles using olive leaf extract and its antibacterial activity. *Arabian Journal of Chemistry, 7*(6), 1131–1139.

Khan, M. K., Abert-Vian, M., Fabiano-Tixier, A. S., Dangles, O., & Chemat, F., (2010). Ultrasound-assisted extraction of polyphenols (flavanone glycosides) from orange (*Citrus sinensis* L.) peel. *Food Chemistry, 119*(2), 851–858.

Khanafari, A., Saberi, A., Azar, M., Vosooghi, G., Jamili, S., & Sabbaghzadeh, B., (2007). Extraction of astaxanthin esters from shrimp waste by chemical and microbial methods. *Iran J. Environ. Health Sci. Eng., 4*(2), 93–98.

Khangembam, B. K., & Chakrabarti, R., (2015). Trypsin from the digestive system of carp *Cirrhinus mrigala*: Purification, characterization and its potential application. *Food Chemistry, 175*, 386–394.

Kiassos, E., Mylonaki, S., Makris, D. P., & Kefalas, P., (2009). Implementation of response surface methodology to optimize extraction of onion (*Allium cepa*) solid waste phenolics. *Innovative Food Science & Emerging Technologies, 10*(2), 246–252.

Kijima, Y., Iwatsuki, S., Akamatsu, H., Terato, K., Kuwabara, Y., Ueda, S., & Shionoya, H., (2009). Natural antibodies to pathogenic bacteria and their toxins in whey protein concentrate. *Nippon Shokuhin Kagaku Kogaku Kaishi. Journal of the Japanese Society for Food Science and Technology*, *56*(9), 475–482.

Kim, J. W., & Mazza, G., (2009). Extraction and separation of carbohydrates and phenolic compounds in flax shives with pH-controlled pressurized low polarity water. *Journal of Agricultural and Food Chemistry*, *57*(5), 1805–1813.

Kleekayai, T., Harnedy, P. A., O'Keeffe, M. B., Poyarkov, A. A., Cunha, N. A., Suntornsuk, W., & Fitz, G. R. J., (2015). Extraction of antioxidant and ACE inhibitory peptides from Thai traditional fermented shrimp pastes. *Food Chemistry*, *176*, 441–447.

Kremmyda, L. S., Vlachava, M., Noakes, P. S., Diaper, N. D., Miles, E. A., & Calder, P. C., (2011). Atopy risk in infants and children in relation to early exposure to fish, oily fish, or long-chain omega-3 fatty acids: A systematic review. *Clinical Reviews in Allergy & Immunology*, *41*(1), 36–66.

Krishnamoorthi, J., Ramasamy, P., Shanmugam, V., & Shanmugam, A., (2017). Isolation and partial characterization of collagen from outer skin of *Sepia pharaonis* (Ehrenberg, 1831) from Puducherry coast. *Biochemistry and Biophysics Reports*, *10*, 39–45.

Krishnaswamy, K., Vali, H., & Orsat, V., (2014). Value-adding to grape waste: Green synthesis of gold nanoparticles. *Journal of Food Engineering*, *142*, 210–220.

Kumar, S., Shukla, A., Baul, P. P., Mitra, A., & Halder, D., (2018). Biodegradable hybrid nanocomposites of chitosan/gelatin and silver nanoparticles for active food packaging applications. *Food Packaging and Shelf Life*, *16*, 178–184.

Kuppusamy, P., Yusoff, M. M., Maniam, G. P., & Govindan, N., (2016). Biosynthesis of metallic nanoparticles using plant derivatives and their new avenues in pharmacological applications–An updated report. *Saudi Pharmaceutical Journal*, *24*(4), 473–484.

Kuttappan, S., Mathew, D., & Nair, M. B., (2016). Biomimetic composite scaffolds containing bioceramics and collagen/gelatin for bone tissue engineering-A mini review. *International Journal of Biological Macromolecules*, *93*, 1390–1401.

Kwapinski, W., Byrne, C. M., Kryachko, E., Wolfram, P., Adley, C., Leahy, J., Novotny, E. H., & Hayes, M. H., (2010). Biochar from biomass and waste. *Waste and Biomass Valorization*, *1*(2), 177–189.

Lanciotti, R., Gianotti, A., Baldi, D., Angrisani, R., Suzzi, G., Mastrocola, D., & Guerzoni, M., (2005). Use of *Yarrowia lipolytica* strains for the treatment of olive mill wastewater. *Bioresource Technology*, *96*(3), 317–322.

Lee, C. H., Singla, A., & Lee, Y., (2001). Biomedical applications of collagen. *International Journal of Pharmaceutics*, *221*(1–2), 1–22.

Lee, E. H., Park, H. R., Shin, M. S., Cho, S. Y., Choi, H. J., & Shin, K. S., (2014). Antitumor metastasis activity of pectic polysaccharide purified from the peels of Korean *Citrus hallabong*. *Carbohydrate Polymers*, *111*, 72–79.

Li, B., Smith, B., & Hossain, M. M., (2006). Extraction of phenolics from citrus peels: II. Enzyme-assisted extraction method. *Separation and Purification Technology*, *48*(2), 189–196.

Liao, B., Newmark, H., & Zhou, R., (2002). Neuroprotective effects of ginseng total saponin and ginsenosides Rb1 and Rg1 on spinal cord neurons *in vitro*. *Experimental Neurology*, *173*(2), 224–234.

López, D., Márquez, A., Gutiérrez-Cutiño, M., Venegas-Yazigi, D., Bustos, R., & Matiacevich, S., (2017). Edible film with antioxidant capacity based on salmon gelatin and boldine. *LWT-Food Science and Technology*, *77*, 160–169.

López, M., Arce, L., Garrido, J., Rıos, A., & Valcárcel, M., (2004). Selective extraction of astaxanthin from crustaceans by use of supercritical carbon dioxide. *Talanta*, *64*(3), 726–731.

Lu, F., Liu, Y., & Li, B., (2013). Okara dietary fiber and hypoglycemic effect of okara foods. *Bioactive Carbohydrates and Dietary Fiber*, *2*(2), 126–132.

Ma, X., Zhou, X. Y., Qiang, Q. Q., & Zhang, Z. Q., (2014). Ultrasound-assisted extraction and preliminary purification of proanthocyanidins and chlorogenic acid from almond (*Prunus dulcis*) skin. *Journal of Separation Science*, *37*(14), 1834–1841.

Maheshwari, P., Prasad, N., & Batra, E. (2015). Papitas – The underutilized byproduct and the future cash crop: a review. *American International Journal of Research in Formal, Applied & Natural Sciences, 15*(432), 31–34.

Mandalari, G., Bennett, R., Bisignano, G., Trombetta, D., Saija, A., Faulds, C., Gasson, M., & Narbad, A., (2007). Antimicrobial activity of flavonoids extracted from bergamot (*Citrus bergamia* Risso) peel, a byproduct of the essential oil industry. *Journal of Applied Microbiology*, *103*(6), 2056–2064.

Manoj, G., Tripathi, S., & Srivastava, A., (2009). Water a new byproduct of sugar industry by vapors condensate pre-filtration and polishing with resins. *Indian Sugar*, *58*(11), 11–18.

Mariod, A. A., Adamu, H. A., Ismail, M., & Ismail, N., (2010). Antioxidative effects of stabilized and unstabilized defatted rice bran methanolic extracts on the stability of rice bran oil under accelerated conditions. *Grasas Y Aceites*, *61*(4), 409–415.

Martínez-Cervera, S., Salvador, A., Muguerza, B., Moulay, L., & Fiszman, S., (2011). Cocoa fiber and its application as a fat replacer in chocolate muffins. *LWT-Food Science and Technology*, *44*(3), 729–736.

Maruthiah, T., & Palavesam, A., (2017). Characterization of haloalkalophilic organic solvent tolerant protease for chitin extraction from shrimp shell waste. *International Journal of Biological Macromolecules*, *97*, 552–560.

Masmoudi, M., Besbes, S., Chaabouni, M., Robert, C., Paquot, M., Blecker, C., & Attia, H., (2008). Optimization of pectin extraction from lemon by-product with acidified date juice using response surface methodology. *Carbohydrate Polymers*, *74*(2), 185–192.

Mateos-Aparicio, I., Mateos-Peinado, C., Jiménez-Escrig, A., & Rupérez, P., (2010). Multi-functional antioxidant activity of polysaccharide fractions from the soybean byproduct okara. *Carbohydrate Polymers*, *82*(2), 245–250.

Matsumoto, Y., Ikeda, K., Yamaya, Y., Yamashita, K., Saito, T., Hoshino, Y., Koga, T., Enari, H., Suto, S., & Yotsuyanagi, T., (2011). The usefulness of the collagen and elastin sponge derived from salmon as an artificial dermis and scaffold for tissue engineering. *Biomedical Research*, *32*(1), 29–36.

Matsuno, R., & Ishihara, K., (2009). *Molecular-Integrated Phospholipid Polymer Nanoparticles with Highly Biofunctionality, Macromolecular Symposia* (pp. 125–131). Wiley Online Library.

Miguel, M. G., Neves, M. A., & Antunes, M. D., (2010). Pomegranate (*Punica granatum* L.): A medicinal plant with myriad biological properties: a short review. *Journal of Medicinal Plants Research*, *4*(25), 2836–2847.

Mohan, A., McClements, D. J., & Udenigwe, C. C., (2016). Encapsulation of bioactive whey peptides in soy lecithin-derived nanoliposomes: Influence of peptide molecular weight. *Food Chemistry*, *213*, 143–148.

Morrison, III, W. H., Holser, R., & Akin, D. E., (2006). Cuticular wax from flax processing waste with hexane and super critical carbon dioxide extractions. *Industrial Crops and Products*, *24*(2), 119–122.

Muliterno, M. M., Rodrigues, D., De Lima, F. S., Ida, E. I., & Kurozawa, L. E., (2017). Conversion/degradation of isoflavones and color alterations during the drying of okara. *LWT-Food Science and Technology*, *75*, 512–519.

Murga, R., Ruiz, R., Beltrán, S., & Cabezas, J. L., (2000). Extraction of natural complex phenols and tannins from grape seeds by using supercritical mixtures of carbon dioxide and alcohol. *Journal of Agricultural and Food Chemistry*, *48*(8), 3408–3412.

Musso, Y. S., Salgado, P. R., & Mauri, A. N., (2017). Smart edible films based on gelatin and curcumin. *Food Hydrocolloids*, *66*, 8–15.

Nagai, T., & Suzuki, N., (2000). Isolation of collagen from fish waste material—skin, bone and fins. *Food Chemistry*, *68*(3), 277–281.

Nagai, T., Yamashita, E., Taniguchi, K., Kanamori, N., & Suzuki, N., (2001). Isolation and characterization of collagen from the outer skin waste material of cuttlefish (*Sepia lycidas*). *Food Chemistry*, *72*(4), 425–429.

Naguib, Y. M., (2000). Antioxidant activities of astaxanthin and related carotenoids. *Journal of Agricultural and Food Chemistry*, *48*(4), 1150–1154.

Naviglio, D., Caruso, T., Iannece, P., Aragòn, A., & Santini, A., (2008). Characterization of high purity lycopene from tomato wastes using a new pressurized extraction approach. *Journal of Agricultural and Food Chemistry*, *56*(15), 6227–6231.

Nawirska, A., & Kwaśniewska, M., (2005). Dietary fiber fractions from fruit and vegetable processing waste. *Food Chemistry*, *91*(2), 221–225.

Negi, V. S., Rawat, L., Phondani, P., & Chandra, A., (2011). Perilla frutescens in transition: A medicinal and oil yielding plant need instant conservation, a case study from Central Himalaya, India. *International Journal of Environmental Science and Technology*, *6*, 193–200.

Neto, W. P. F., Silvério, H. A., Dantas, N. O., & Pasquini, D., (2013). Extraction and characterization of cellulose nanocrystals from agro-industrial residue–Soy hulls. *Industrial Crops and Products*, *42*, 480–488.

Nges, I. A., Mbatia, B., & Björnsson, L., (2012). Improved utilization of fish waste by anaerobic digestion following omega-3 fatty acids extraction. *Journal of Environmental Management*, *110*, 159–165.

Nisha, S. N., Aysha, O., Rahaman, J. S. N., Kumar, P. V., Valli, S., Nirmala, P., & Reena, A., (2014). Lemon peels mediated synthesis of silver nanoparticles and its antidermatophytic activity. *Spectrochimica Acta Part A: Molecular and Biomolecular Spectroscopy*, *124*, 194–198.

No, H. K., Meyers, S. P., & Lee, K. S., (1989). Isolation and characterization of chitin from crawfish shell waste. *Journal of Agricultural and Food Chemistry*, *37*(3), 575–579.

Noa, M., & Mas, R., (2005). Protective effect of policosanol on atherosclerotic plaque on aortas in monkeys. *Archives of Medical Research*, *36*(5), 441–447.

Nogata, Y., Nagamine, T., Yanaka, M., & Ohta, H., (2009). Angiotensin I converting enzyme inhibitory peptides produced by autolysis reactions from wheat bran. *Journal of Agricultural and Food Chemistry*, *57*(15), 6618–6622.

Novik, G. I., Wawrzynczyk, J., Norrlow, O., & Szwajcer-Dey, E., (2007). Fractions of barley spent grain as media for growth of probiotic bacteria. *Microbiology*, *76*(6), 804.

Nurhanan, A., & WI, W. R., (2012). Evaluation of polyphenol content and antioxidant activities of some selected organic and aqueous extracts of cornsilk (*Zea Mays* Hairs). *Journal of Medical and Bioengineering (JOMB)*, *1*(1).

O'Shea, N., Arendt, E. K., & Gallagher, E., (2012). Dietary fiber and phytochemical characteristics of fruit and vegetable by-products and their recent applications as novel ingredients in food products. *Innovative Food Science & Emerging Technologies*, *16*, 1–10.

Opatokun, S. A., Strezov, V., & Kan, T., (2015). Product based evaluation of pyrolysis of food waste and its digestate. *Energy*, *92*, 349–354.

Orsat, V., & Routray, W., (2017). Microwave-assisted extraction of flavonoids. In: Gonzalez, H. D., & Munoz, M. J. G., (eds.), *Water Extraction of Bioactive Compounds from Plants to Drug Development* (pp. 221–244). Elsevier.

Palafox-Carlos, H., Ayala-Zavala, J. F., & González-Aguilar, G. A., (2011). The role of dietary fiber in the bioaccessibility and bioavailability of fruit and vegetable antioxidants. *Journal of Food Science*, *76*(1), R6–R15.

Panfili, G., Fratianni, A., & Irano, M., (2004). Improved normal-phase high-performance liquid chromatography procedure for the determination of carotenoids in cereals. *Journal of Agricultural and Food Chemistry*, *52*(21), 6373–6377.

Panouillé, M., Thibault, J. F., & Bonnin, E., (2006). Cellulase and protease preparations can extract pectins from various plant byproducts. *Journal of Agricultural and Food Chemistry*, *54*(23), 8926–8935.

Papoti, V. T., & Tsimidou, M. Z., (2009). Impact of sampling parameters on the radical scavenging potential of olive (*Olea europaea* L.) leaves. *Journal of Agricultural and Food Chemistry*, *57*(9), 3470–3477.

Patel, A. K., Ahire, J. J., Pawar, S. P., Chaudhari, B. L., Shouche, Y. S., & Chincholkar, S. B., (2009). Evaluation of probiotic characteristics of *Siderophoregenic bacillus* spp. isolated from dairy waste. *Applied Biochemistry and Biotechnology*, *160*(1), 140.

Plaza, M., Abrahamsson, V., & Turner, C., (2013). Extraction and neoformation of antioxidant compounds by pressurized hot water extraction from apple byproducts. *Journal of Agricultural and Food Chemistry*, *61*(23), 5500–5510.

Rao, A. V., & Rao, L. G., (2007). Carotenoids and human health. *Pharmacological Research*, *55*(3), 207–216.

Ravichandran, V., Vasanthi, S., Shalini, S., Shah, S. A. A., & Harish, R., (2016). Green synthesis of silver nanoparticles using *Atrocarpus altilis* leaf extract and the study of their antimicrobial and antioxidant activity. *Materials Letters*, *180*, 264–267.

Reddy, N., & Yang, Y., (2005). Biofibers from agricultural byproducts for industrial applications. *Trends in Biotechnology*, *23*(1), 22–27.

Rodrıguez, I., Llompart, M., & Cela, R., (2000). Solid-phase extraction of phenols. *Journal of Chromatography A*, *885*(1/2), 291–304.

Roldán-Marín, E., Jensen, R. I., Krath, B. N., Kristensen, M., Poulsen, M., Cano, M. P., Sánchez-Moreno, C. n., & Dragsted, L. O., (2010). An onion byproduct affects plasma lipids in healthy rats. *Journal of Agricultural and Food Chemistry*, *58*(9), 5308–5314.

Routray, W., & Orsat, V., (2012). Microwave-assisted extraction of flavonoids: A review. *Food and Bioprocess Technology*, *5*(2), 409–424.

Routray, W., & Orsat, V., (2013). Preparative extraction and separation of phenolic compounds. In: Ramawat, K. G., & Mérillon, J. M., (eds.), *Natural Products* (pp. 2013–2045). Springer Berlin Heidelberg.

Routray, W., & Orsat, V., (2014). MAE of phenolic compounds from blueberry leaves and comparison with other extraction methods. *Industrial Crops and Products*, *58*, 36–45.

Routray, W., & Orsat, V., (2017). Plant by-products and food industry waste: A source of nutraceuticals and biopolymers. In: Grumezescu, A. M., & Holban, A. M., (eds.), *Food Bioconversion* (pp. 279–316). Elsevier.

Rozzi, N., Singh, R., Vierling, R., & Watkins, B., (2002). Supercritical fluid extraction of lycopene from tomato processing byproducts. *Journal of Agricultural and Food Chemistry*, *50*(9), 2638–2643.

Rubio-Rodríguez, N., Sara, M., Beltrán, S., Jaime, I., Sanz, M. T., & Rovira, J., (2008). Supercritical fluid extraction of the omega-3 rich oil contained in hake (*Merluccius*

capensis–Merluccius paradoxus) by-products: study of the influence of process parameters on the extraction yield and oil quality. *The Journal of Supercritical Fluids*, *47*(2), 215–226.
Rubio-Rodríguez, N., Sara, M., Beltrán, S., Jaime, I., Sanz, M. T., & Rovira, J., (2012). Supercritical fluid extraction of fish oil from fish by-products: A comparison with other extraction methods. *Journal of Food Engineering*, *109*(2), 238–248.
Ruxton, C., Reed, S. C., Simpson, M., & Millington, K., (2004). The health benefits of omega-3 polyunsaturated fatty acids: A review of the evidence. *Journal of Human Nutrition and Dietetics*, *17*(5), 449–459.
Sabale, V., Sabale, P., & Lakhotiya, C., (2009). Comparative evaluation of rice bran wax as an ointment base with standard base. *Indian Journal of Pharmaceutical Sciences*, *71*(1), 77.
Sachindra, N., & Mahendrakar, N., (2005). Process optimization for extraction of carotenoids from shrimp waste with vegetable oils. *Bioresource Technology*, *96*(10), 1195–1200.
Sah, B., Vasiljevic, T., McKechnie, S., & Donkor, O., (2016). Effect of pineapple waste powder on probiotic growth, antioxidant and antimutagenic activities of yogurt. *Journal of Food Science and Technology*, *53*(3), 1698–1708.
Sanz-Puig, M., Pina-Pérez, M. C., Criado, M. N., Rodrigo, D., & Martínez-López, A., (2015). Antimicrobial potential of cauliflower, broccoli, and okara byproducts against foodborne bacteria. *Foodborne Pathogens and Disease*, *12*(1), 39–46.
Sathivel, S., Bechtel, P., Babbitt, J., Smiley, S., Crapo, C., Reppond, K., & Prinyawiwatkul, W., (2003). Biochemical and functional properties of herring (*Clupea harengus*) byproduct hydrolysates. *Journal of Food Science*, *68*(7), 2196–2200.
Seker, I., Ozboy-Ozbas, O., Gokbulut, I., Ozturk, S., & Koksel, H., (2010). Utilization of apricot kernel flour as fat replacer in cookies. *Journal of Food Processing and Preservation*, *34*(1), 15–26.
Sendra, E., Kuri, V., Fernández-López, J., Sayas-Barberá, E., Navarro, C., & Pérez-Alvarez, J., (2010). Viscoelastic properties of orange fiber enriched yogurt as a function of fiber dose, size and thermal treatment. *LWT-Food Science and Technology*, *43*(4), 708–714.
Sheltami, R. M., Abdullah, I., Ahmad, I., Dufresne, A., & Kargarzadeh, H., (2012). Extraction of cellulose nanocrystals from mengkuang leaves (*Pandanus tectorius*). *Carbohydrate Polymers*, *88*(2), 772–779.
Shen, S., Cheng, H., Li, X., Li, T., Yuan, M., Zhou, Y., & Ding, C., (2014). Effects of extraction methods on antioxidant activities of polysaccharides from camellia seed cake. *European Food Research and Technology*, *238*(6), 1015–1021.
Shirsath, S., Sonawane, S., & Gogate, P., (2012). Intensification of extraction of natural products using ultrasonic irradiations: a review of current status. *Chemical Engineering and Processing: Process Intensification*, *53*, 10–23.
Silvério, H. A., Neto, W. P. F., Dantas, N. O., & Pasquini, D., (2013). Extraction and characterization of cellulose nanocrystals from corncob for application as reinforcing agent in nanocomposites. *Industrial Crops and Products*, *44*, 427–436.
Singh, P., Benjakul, S., Maqsood, S., & Kishimura, H., (2011). Isolation and characterization of collagen extracted from the skin of striped catfish (*Pangasianodon hypophthalmus*). *Food Chemistry*, *124*(1), 97–105.
Slizyte, R., Rommi, K., Mozuraityte, R., Eck, P., Five, K., & Rustad, T., (2016). Bioactivities of fish protein hydrolysates from defatted salmon backbones. *Biotechnology Reports*, *11*, 99–109.
Sreenivas, K., Chaudhari, K., & Lele, S., (2011). Ash gourd peel wax: Extraction, characterization, and application as an edible coat for fruits. *Food Science and Biotechnology*, *20*(2), 383–387.

Strati, I. F., & Oreopoulou, V., (2011). Effect of extraction parameters on the carotenoid recovery from tomato waste. *International Journal of Food Science & Technology, 46*(1), 23–29.

Sun, T., Chen, Q., Wu, L., Yao, X., & Sun, X., (2012). Antitumor and antimetastatic activities of grape skin polyphenols in a murine model of breast cancer. *Food and Chemical Toxicology, 50*(10), 3462–3467.

Terigar, B., Balasubramanian, S., Sabliov, C., Lima, M., & Boldor, D., (2011). Soybean and rice bran oil extraction in a continuous microwave system: From laboratory-to pilot-scale. *Journal of Food Engineering, 104*(2), 208–217.

Tsatsaronis, G. C., & Boskou, D. G., (1975). Amino acid and mineral salt content of tomato seed and skin waste. *Journal of the Science of Food and Agriculture, 26*(4), 421–423.

Tsukamoto, C., Kikuchi, A., Kudou, S., Harada, K., Iwasaki, T., & Okubo, K., (1994). Genetic improvement of saponin components in soybean. In *ACS Symposium Series* (Vol. 546, pp. 372–381). Washington, DC: American Chemical Society.

Udenigwe, C. C., & Aluko, R. E., (2012). Food protein-derived bioactive peptides: Production, processing, and potential health benefits. *Journal of Food Science, 77*(1), R11–R24.

Valdés, A., Vidal, L., Beltran, A., Canals, A., & Garrigós, M. C., (2015). Microwave-assisted extraction of phenolic compounds from almond skin byproducts (*Prunus amygdalus*): A multivariate analysis approach. *Journal of Agricultural and Food Chemistry, 63*(22), 5395–5402.

Vasantha Rupasinghe, H., Wang, L., Pitts, N. L., & Astatkie, T., (2009). Baking and sensory characteristics of muffins incorporated with apple skin powder. *Journal of Food Quality, 32*(6), 685–694.

Vásquez-Villanueva, R., Marina, M. L., & García, M. C., (2015). Revalorization of a peach (*Prunus persica* (L.) Batsch) byproduct: Extraction and characterization of ACE-inhibitory peptides from peach stones. *Journal of Functional Foods, 18*, 137–146.

Veerasamy, R., Xin, T. Z., Gunasagaran, S., Xiang, T. F. W., Yang, E. F. C., Jeyakumar, N., & Dhanaraj, S. A., (2011). Biosynthesis of silver nanoparticles using mangosteen leaf extract and evaluation of their antimicrobial activities. *Journal of Saudi Chemical Society, 15*(2), 113–120.

Veeruraj, A., Arumugam, M., & Balasubramanian, T., (2013). Isolation and characterization of thermostable collagen from the marine eel-fish (*Evenchelys macrura*). *Process Biochemistry, 48*(10), 1592–1602.

Veeruraj, A., Arumugam, M., Ajithkumar, T., & Balasubramanian, T., (2015). Isolation and characterization of collagen from the outer skin of squid (*Doryteuthis singhalensis*). *Food Hydrocolloids, 43*, 708–716.

Vergara-Salinas, J. R., Bulnes, P., Zúñiga, M. C., Pérez-Jiménez, J., Torres, J. L., Mateos-Martín, M. L., Agosin, E., & Pérez-Correa, J. R., (2013). Effect of pressurized hot water extraction on antioxidants from grape pomace before and after enological fermentation. *Journal of Agricultural and Food Chemistry, 61*(28), 6929–6936.

Vermerris, W., & Nicholson, R. L., (2006). *Phenolic Compound Biochemistry*. Springer, Dordrecht.

Villalba-Villalba, A. G., Ramírez-Suárez, J. C., Valenzuela-Soto, E. M., Sánchez, G. G., Ruiz, G. C., & Pacheco-Aguilar, R., (2013). Trypsin from viscera of vermiculated sailfin catfish, *Pterygoplichthys disjunctivus*, Weber, 1991: Its purification and characterization. *Food Chemistry, 141*(2), 940–945.

Villamil, O., Váquiro, H., & Solanilla, J. F., (2017). Fish viscera protein hydrolysates: Production, potential applications and functional and bioactive properties. *Food Chemistry, 224*, 160–171.

Villanueva-Suárez, M. J., Pérez-Cózar, M. L., & Redondo-Cuenca, A., (2013). Sequential extraction of polysaccharides from enzymatically hydrolyzed okara byproduct: Physicochemical properties and *in vitro* fermentability. *Food Chemistry*, *141*(2), 1114–1119.

Vinayaka, K., Nandini, K., Rakshitha, M., Ramya, M., Shruthi, J., Shruthi, V., Prashith, K. T., & Raghavendra, H., (2010). Proximate composition, antibacterial and anthelmintic activity of *Capsicum frutescens* (L.) var. longa (Solanaceae) leaves. *Pharmacognosy Journal*, *2*(12), 486–491.

Visioli, F., Caruso, D., Plasmati, E., Patelli, R., Mulinacci, N., Romani, A., Galli, G., & Galli, C., (2001). Hydroxytyrosol, as a component of olive mill wastewater, is dose-dependently absorbed and increases the antioxidant capacity of rat plasma. *Free Radical Research*, *34*(3), 301–305.

Wang, J., Sun, B., Cao, Y., Tian, Y., & Li, X., (2008). Optimization of ultrasound-assisted extraction of phenolic compounds from wheat bran. *Food Chemistry*, *106*(2), 804–810.

Wang, L., Lu, F., Liu, Y., Wu, Y., & Wu, Z., (2018). Photocatalytic degradation of organic dyes and antimicrobial activity of silver nanoparticles fast synthesized by flavonoids fraction of *Psidium guajava* L. leaves. *Journal of Molecular Liquids*. *263,* 187–192.

Wang, L., Wang, X., Yuan, X., & Zhao, B., (2011a). Simultaneous analysis of diosgenin and sarsasapogenin in *Asparagus officinalis* byproduct by thin-layer chromatography. *Phytochemical Analysis*, *22*(1), 14–17.

Wang, X., Li, C., Liu, Y., Li, H., & Di, D., (2011b). Efficient method for screening and identification of radical scavengers in the leaves of *Olea europaea* L. *Biomedical Chromatography*, *25*(3), 373–380.

Wijngaard, H., & Brunton, N., (2009). The optimization of extraction of antioxidants from apple pomace by pressurized liquids. *Journal of Agricultural and Food Chemistry*, *57*(22), 10625–10631.

Young, T. J., Johnston, K. P., Pace, G. W., & Mishra, A. K., (2004). Phospholipid-stabilized nanoparticles of cyclosporine A by rapid expansion from supercritical to aqueous solution. *Aaps Pharmscitech.*, *5*(1), 70.

Zhang, C., Su, H., Baeyens, J., & Tan, T., (2014). Reviewing the anaerobic digestion of food waste for biogas production. *Renewable and Sustainable Energy Reviews*, *38*, 383–392.

Zhang, X. F., Han, Y. Y., Bao, G. H., Ling, T. J., Zhang, L., Gao, L. P., & Xia, T., (2012). A new saponin from tea seed pomace (*Camellia oleifera* Abel) and its protective effect on PC12 cells. *Molecules*, *17*(10), 11721–11728.

Zhang, Y., Pan, Z., Venkitasamy, C., Ma, H., & Li, Y., (2015). Umami taste amino acids produced by hydrolyzing extracted protein from tomato seed meal. *LWT-Food Science and Technology*, *62*(2), 1154–1161.

Zhao, L., Cao, X., Wang, Q., Yang, F., & Xu, S., (2013). Mineral constituents profile of biochar derived from diversified waste biomasses: implications for agricultural applications. *Journal of Environmental Quality*, *42*(2), 545–552.

Zhao, R., Li, Q., Long, L., Li, J., Yang, R., & Gao, D., (2007). Antidiabetic activity of flavone from *Ipomoea batatas* leaf in non-insulin dependent diabetic rats. *International Journal of Food Science & Technology*, *42*(1), 80–85.

Index

B

Q

R

X

Y

Z